LONDON MATHEMATICAL SOCIETY LECTURE NOTE SERIES

Managing Editor: Professor I.M.James,
Mathematical Institute, 24-29 St Giles, Oxford

1. General cohomology theory and K-theory, P.HILTON
4. Algebraic topology: a student's guide, J.F.ADAMS
5. Commutative algebra, J.T.KNIGHT
8. Integration and harmonic analysis on compact groups, R.E.EDWARDS
9. Elliptic functions and elliptic curves, P.DU VAL
10. Numerical ranges II, F.F.BONSALL & J.DUNCAN
11. New developments in topology, G.SEGAL (ed.)
12. Symposium on complex analysis, Canterbury, 1973, J.CLUNIE & W.K.HAYMAN (eds.)
13. Combinatorics: Proceedings of the British combinatorial conference 1973, T.P.McDONOUGH & V.C.MAVRON (eds.)
14. Analytic theory of abelian varieties, H.P.F.SWINNERTON-DYER
15. An introduction to topological groups, P.J.HIGGINS
16. Topics in finite groups, T.M.GAGEN
17. Differentiable germs and catastrophes, Th.BROCKER & L.LANDER
18. A geometric approach to homology theory, S.BUONCRISTIANO, C.P.ROURKE & B.J.SANDERSON
20. Sheaf theory, B.R.TENNISON
21. Automatic continuity of linear operators, A.M.SINCLAIR
23. Parallelisms of complete designs, P.J.CAMERON
24. The topology of Stiefel manifolds, I.M.JAMES
25. Lie groups and compact groups, J.F.PRICE
26. Transformation groups: Proceedings of the conference in the University of Newcastle upon Tyne, August 1976, C.KOSNIOWSKI
27. Skew field constructions, P.M.COHN
28. Brownian motion, Hardy spaces and bounded mean oscillation, K.E.PETERSEN
29. Pontryagin duality and the structure of locally compact abelian groups, S.A.MORRIS
30. Interaction models, N.L.BIGGS
31. Continuous crossed products and type III von Neumann algebras, A.VAN DAELE
32. Uniform algebras and Jensen measures, T.W.GAMELIN
33. Permutation groups and combinatorial structures, N.L.BIGGS & A.T.WHITE
34. Representation theory of Lie groups, M.F.ATIYAH et al.
35. Trace ideals and their applications, B.SIMON
36. Homological group theory, C.T.C.WALL (ed.)
37. Partially ordered rings and semi-algebraic geometry, G.W.BRUMFIEL
38. Surveys in combinatorics, B.BOLLOBAS (ed.)
39. Affine sets and affine groups, D.G.NORTHCOTT
40. Introduction to $H_p$ spaces, P.J.KOOSIS
41. Theory and applications of Hopf bifurcation, B.D.HASSARD, N.D.KAZARINOFF & Y-H.WAN
42. Topics in the theory of group presentations, D.L.JOHNSON
43. Graphs, codes and designs, P.J.CAMERON & J.H.VAN LINT
44. Z/2-homotopy theory, M.C.CRABB
45. Recursion theory: its generalisations and applications, F.R.DRAKE & S.S.WAINER (eds.)
46. p-adic analysis: a short course on recent work, N.KOBLITZ
47. Coding the Universe, A. BELLER, R. JENSEN & P. WELCH
48. Low-dimensional topology, R. BROWN & T.L. THICKSTUN (eds.)
49. Finite geometries and designs, P. CAMERON, J.W.P. HIRSCHFELD & D.R. HUGHES (eds.)
50. Commutator Calculus and groups of homotopy classes, H.J. BAUES
51. Synthetic differential geometry, A. KOCK
52. Combinatorics, H.N.V. TEMPERLEY (ed.)
53. Singularity theory, V.I. ARNOLD
54. Markov processes and related problems of analysis, E.B. DYNKIN
55. Ordered permutation groups, A.M.W. GLASS
56. Journees arithmetiques 1980, J.V. ARMITAGE (ed.)
57. Techniques of geometric topology, R.A. FENN
58. Singularities of differentiable functions, J. MARTINET
59. Applicable differential geometry, F.A.E. PIRANI and M. CRAMPIN
60. Integrable systems, S.P. NOVIKOV et al.

London Mathematical Society Lecture Note Series. 60

# Integrable Systems

## Selected Papers

S.P.NOVIKOV V.B.MATVEER

I.M.GELFAND I.M.KRICHEVER

L.A.DIKII A.M.VINOGRADOV

B.V.YUSIN B.A.KUPERSHMIDT

B.A.DUBROVIN I.S.KRASILSHCHIK

CAMBRIDGE UNIVERSITY PRESS

CAMBRIDGE

LONDON NEW YORK NEW ROCHELLE

MELBOURNE SYDNEY

Published by the Press Syndicate of the University of Cambridge
The Pitt Building, Trumpington Street, Cambridge CB2 1RP
32 East 57th Street, New York, NY 10022, USA
296 Beaconsfield Parade, Middle Park, Melbourne 3206, Australia



First published 1981

Printed in Great Britain at The Pitman Press, Bath

Library of Congress catalogue card number: 81 6093

British Library cataloguing in publication data

Integrable Systems.-(London Mathematical Society
Lecture Note Series 60 ISSN 0076-0552.)
1. Differential equations
I. Novikov, S.P.     II. Series
515.3'5     QA372

ISBN 0-521-28527-5

CONTENTS

# INTRODUCTION

George Wilson

The last 15 years have seen the creation of a new branch of mathematics: the theory of 'integrable' non-linear partial differential equations. The prototype example is the Korteweg-de Vries (KdV) equation

$$u_t = 6uu_x - u_{xxx}, \quad u = u(x, t). \tag{1}$$

The theory was at first developed mainly by mathematical physicists, but recently it has been attracting the attention of an increasing number of mathematicians too. The involvement of mathematicians has been particularly strong in the Soviet Union; the articles reprinted in this volume contain fundamental contributions from some of the leading Soviet workers in the field. As befits pioneering works, the articles do not always make easy reading; in this introduction I shall try to smooth the reader's path by setting out the central facts about the KdV equation. For brevity I shall refer to the articles by the authors' initials [GD], [DMN], [K]. (I shall not discuss the other two articles, by Vinogradov–Krasilshchik and Vinogradov–Kupershmidt, which are of a different nature, nor the two short notes by Yusin and Krichever.) The notation for references is as follows: [GD2], for example, means reference 2 in the bibliography to the article of Gel'fand and Dikii; a reference without letters, such as [2], refers to the bibliography at the end of the introduction.

The main results in the theory of the KdV equation can reasonably be divided into three classes.

A. Formal (algebraic) properties of the equation (local conservation laws, Hamiltonian formalism).

B. Integration of the equation on certain infinite-dimensional function spaces (periodic, or rapidly decreasing at infinity).

C. Integration of the equation on certain finite-dimensional function spaces (the spaces of solutions of the 'higher stationary equations').

We begin with (A). Although the basic facts here are of an algebraic nature, their significance will be understood most easily if we first state their consequences in an analytic context. So let $\mathscr{F}$ denote one of the function spaces in

(B) above; more precisely, $\mathscr{F}$ is either (a) the space of $C^\infty$ periodic functions (of one variable $x$) with some fixed period; or (b) some space of functions defined on the whole real line and 'rapidly decreasing' at $\pm\infty$, say the Schwartz space. In either case, the KdV equation has a unique solution $u(x, t)$ with prescribed initial value $u(x, 0) \in \mathscr{F}$: the equation can thus be pictured geometrically as defining a flow on $\mathscr{F}$. For either choice of $\mathscr{F}$ we have the following.

A1. The KdV equation has an infinite number of conserved quantities (first integrals), that is, functionals $I_q[u]$ that are constant on the trajectories of the KdV flow.

A2. In a sense to be clarified below, the KdV flow is the *Hamiltonian* flow corresponding to one of these integrals as Hamiltonian.

A3. The integrals $I_q$ are all in involution in the sense of Hamiltonian mechanics, that is, the Poisson brackets $\{I_r, I_s\}$ all vanish.

Before explaining the meaning of these assertions, let me mention one other fact, perhaps the most basic one of all: the connexion of the KdV equation with the Schrödinger operator

$$L = -\partial^2 + u(x, t). \tag{2}$$

(We write $\partial \equiv \partial/\partial x$.) This is the deus ex machina from which everything else follows. The connexion first appeared in the remarkable paper [GD11], where it was observed that for $u$ rapidly decreasing at infinity, the spectrum (or at any rate the discrete part of the spectrum) of the operator $L$ does not change with time if $u$ satisfies the KdV equation. Lax [GD12] then reasoned that this suggests that the operators $L(t)$ are all conjugate to a fixed one, say $L(t) = U(t)\, L(0)\, U(t)^{-1}$, or equivalently, $\partial_t(U^{-1}LU) = 0$. This last equation can be rewritten in the form

$$L_t = [P, L], \tag{3}$$

where $P = U_t U^{-1}$ (as usual, the square brackets denote the commutator $PL - LP$). Finally, Lax saw that the operator $P$ in (3) can be taken to be a differential operator: if we set

$$P = -4\partial^3 + 6u\partial + 3u_x$$

it is easy to check that $[P, L]$ is (identically in $u$) an operator of order zero, and the zero-order term is the right hand side of equation (1). Equations (1) and (3) are therefore equivalent: (3) is now called the Lax representation of the KdV equation (1).

This story provides a good illustration of how a purely algebraic assertion (the equivalence of equations (1) and (3)) can have important consequences for the KdV flows on concrete function spaces (the time independence of the spectrum). Let me now elucidate the statements A1–A3 above. First, the conserved quantities. The underlying algebraic fact here is that there is an infinite sequence of identities

$$\partial_t H_q = \partial J_q \tag{4}$$

that follow formally from (1). For example, we have

$$\partial_t(u^2) = 2uu_t = 2u(6uu_x - u_{xxx}) = \partial(4u^3 + u_x^2 - 2uu_{xx}).$$

(The first few $H_q$ are listed in [GD], p.34; they are denoted there by $R_q$.) The 'conserved densities' $H_q$ and 'fluxes' $J_q$ are differential polynomials in $u$ (that is, polynomials in $u$ and its $x$-derivatives $u^{(i)}$). In either of our analytic contexts, we get the conserved quantities $I_q$ in (A1) by integrating the $H_q$: $I_q[u] = \int H_q[u]\,dx$ (that explains the term 'conserved density'). The integrals are taken over a period, or over the whole line, depending on the choice of function space $\mathscr{F}$. In either case, it follows from (4) that if $u$ satisfies the KdV equation, then

$$\partial_t I_q[u] = \int \partial_t H_q dx = \int \frac{\partial J_q}{\partial x}\,dx = 0,$$

so that $I_q$ is indeed constant.

Notice that if we change $H_q$ by adding on a term $\partial f$ ($f$ a differential polynomial), then the corresponding conserved quantity $\int H_q dx$ will be unchanged. Thus the conserved densities $H_q$ should be regarded as well-defined only up to addition of a term $\partial f$. That explains why the lists of the $H_q$ in [GD], p.34 and [DMN], p.61 do not agree exactly (even up to multiplication by constants).

Now the Hamiltonian formalism ((A2) above). The basic algebraic fact is that the KdV equation can be written in the form

$$u_t = \partial \frac{\delta H}{\delta u}, \tag{5}$$

where $H = u^3 + \frac{1}{2}u_x^2$ is one of the conserved densities. Here $\delta H/\delta u$ is the formal variational derivative (see [GD], p.16, formula 10). In an analytic context we should write $\delta I/\delta u$ instead of $\delta H/\delta u$ (where as above $I = \int Hdx$): this would be the usual Euler–Lagrange operator of the calculus of variations, applied to the functional $u \to \int H[u]\,dx$. Now, in that context it is standard that $\delta I/\delta u$ is analogous to the exterior derivative (or gradient) of a function; thus we think of (5) as analogous to Hamilton's equations

$$\partial_t \begin{pmatrix} q \\ p \end{pmatrix} = \begin{pmatrix} 0 & 1 \\ -1 & 0 \end{pmatrix} \begin{pmatrix} \partial H/\partial q \\ \partial H/\partial p \end{pmatrix} \tag{6}$$

on a finite dimensional phase space. (Vector notation: $q = (q_1, \ldots, q_n)$, etc.) The analogy of the skew differential operator $\partial$ in (5) with the skew matrix $\begin{pmatrix} 0 & 1 \\ -1 & 0 \end{pmatrix}$ in (6) is justified by a property of the operator $\partial$ ((iii) below) whose formulation involves the Poisson bracket referred to in (A3) above. I shall explain that first in the formal context. Let $A$ be the algebra of all differential polynomials in $u$, with derivation $\partial = d/dx$ as in [GD], p.15, formula (4). For each $f \in A$, we define the 'vector field' $\partial_f : A \to A$ to be the unique derivation of $A$ that commutes with $\partial$ and satisfies

$$\partial_f u = \partial \frac{\delta f}{\delta u}. \tag{7}$$

By the theorem in [GD], p.17, $\partial_f$ depends only on the class of $f$ in $A/\partial A$. We now define the Poisson bracket $(A/\partial A) \times (A/\partial A) \to A/\partial A$ by

$$\{f, g\} = \frac{\delta g}{\delta u} \, \partial \, \frac{\delta f}{\delta u} \quad (\text{mod } \partial A).$$

By integration by parts, it is easy to check that we have

(i) $\{f, g\} = -\{g, f\}$

(ii) $\{f, g\} = \partial_f g = -\partial_g f$,

as in ordinary Hamiltonian mechanics. The crucial property of the bracket, however, which is less easy to check, is the next one:

(iii) $\partial_{\{f,g\}} = [\partial_f, \partial_g]$.

(A closely related property is that the bracket satisfies the Jacobi identity.)

The intuitive meaning of all this becomes clear when we translate to one of the analytic contexts. (The reader should be warned, however, that there is some difficulty in doing this rigorously: see [33]). Thus now each $f \in A/\partial A$ defines a function $I_f[u] = \int f[u]\,dx$ on $\mathcal{F}$; the $\partial_f$ in (7) is thought of as a vector field on the infinite dimensional manifold $\mathcal{F}$, and the Poisson bracket becomes

$$\{I_f, I_g\} = \int \frac{\delta I_g}{\delta u} \, \partial \, \frac{\delta I_f}{\delta u} \, dx$$

(integration corresponds to working modulo $\partial A$ in the formal set-up). Thus we are in the usual situation of Hamiltonian mechanics, with the operator $\partial$ defining the 'symplectic structure' on the manifold $\mathcal{F}$. More accurately, $\partial$ defines what in finite dimensions is often called a 'Poisson structure' [23] or 'Hamiltonian structure' [38] on $\mathcal{F}$; that is, $\partial$ should be thought of as a skew form on the cotangent bundle (not the tangent bundle) of $\mathcal{F}$.

I shall not say any more about the formal properties of the KdV equation: the reader will find more details, as well as proofs of most of what I have said so far, in the article [GD]. Let me now discuss the articles [DMN] and [K], which are mainly devoted to the matters mentioned in (C) above. First we have to introduce the 'higher' KdV equations. These are the Hamiltonian flows obtained by taking the various conserved quantities $I_q$ as Hamiltonians: that is, they are the equations of the form

$$u_t = \partial \frac{\delta H_q}{\delta u} \tag{8}$$

as $H_q$ runs over the conserved densities of the KdV equation. Since the $H_q$ are all in involution, the Hamiltonian formalism (more precisely, property (iii) above) shows that the corresponding vector fields $\partial_t$, and hence the corresponding 'higher KdV' flows, all commute with each other. In particular, the set of fixed points of any one of the higher KdV flows is invariant under all the others, amongst which is the flow of the KdV equation itself. Now, the fixed point set is given by setting the right hand side of (8) equal to zero; that gives an ordinary differential equation of the form

$$u^{(2q+1)} + f(u, u_x, \ldots, u^{(2q-1)}) = 0,$$

where $f$ is a polynomial. The solutions of this form a finite dimensional space $V^{2q+1}$. We have the following.

C1. The space $V^{2q+1}$ is foliated by $2q$-dimensional symplectic manifolds $M^{2q}$ on each of which the KdV flow is a completely integrable Hamiltonian system in the classical sense (see, for example, section 6 in the article of Vinogradov and Kupershmidt).

C2. More interesting still, the KdV equation with an initial value $u(x, 0) \in V^{2q+1}$ can be integrated explicitly (generally in terms of the Riemann $\theta$-function of a hyperelliptic Riemann surface).

I shall discuss (C1) only very briefly (for details see Ch.3 of [GD] and the papers [DMN64] and [9]: summaries of [DMN64] can be found in the concluding remarks to [DMN] and in the appendix 1 to [K]). Let me just explain what the manifolds $M^{2q}$ are: they are the spaces of solutions of the equations

$$\frac{\delta H_q}{\delta u} = c \tag{9}$$

for various constants $c$; clearly, as $c$ varies we get exactly the solutions of the stationary equation (8). Equation (9) can be written in the strictly Lagrangian form

$$\frac{\delta}{\delta u}(H_q - cu) = 0;$$

the symplectic structure on the space of solutions $M^{2q}$ is now obtained by a procedure that is standard in the calculus of variations, and is explained in a formal context in [GD], ch.1.

The explicit integration of these restricted KdV flows involves some algebraic geometry: the paper [11] tries to persuade us that this arises inevitably if we try to implement Liouville's procedure for integrating an integrable Hamiltonian system. However, the papers [DMN] and [K] use a more direct approach based on the Lax representation (3) of the KdV equation. We need the corresponding description of the higher KdV equations. Let $L$ as before be the Schrödinger operator (2). It is not hard to see that for each integer $q$ there is a unique differential operator of the form

$$P_q = \partial^{2q+1} + \text{(lower order terms)}$$

whose coefficients are differential polynomials in $u$ with zero constant term, such that $[P_q, L]$ is an operator of order zero. The equations

$$L_t = [P_q, L]$$

are the higher KdV equations. (A proof that this construction of the higher KdV equations agrees with our earlier, Hamiltonian, one is given in [GD], p.42.) The higher stationary equations thus take the form $[P_q, L] = 0$; more generally, we can consider equations of the form

$$[Q, L] = 0, \tag{10}$$

where $Q$ is some linear combination of the $P_q$. We want to find solutions $u(x, t)$ of the KdV equation which for each fixed value of $t$ satisfy the ordinary differential equation (10). It is possible to discuss this directly, as is done in [K], but the ideas will perhaps be more transparent if we first study equation (10) by itself: it turns out that once we have a good description of the space of solutions of (10), it is a simple matter to describe also the KdV flow on this space. The main point is that the solutions of (10) can be constructed from certain algebro-geometric data; however, to understand the motivation for the construction it is best first to see how to go in the other direction. Suppose then that we have a solution $u(x)$ (possibly complex valued) of (10); it gives us a pair of commuting differential operators $Q$, $L$. The corresponding algebro-geometric data consists essentially of the joint eigenvalues (spectrum) and eigenfunctions of these operators. In more detail, let $V_\lambda$ denote the $\lambda$-eigenspace of $L$; thus $V_\lambda$ is a two-dimensional complex vector space. Since $Q$ commutes with $L$, it preserves each space $V_\lambda$ : let $f(\lambda, \mu)$ denote the characteristic polynomial of $Q$ acting on $V_\lambda$ :

$$f(\lambda, \mu) = \det (Q \mid V_\lambda - \mu.Id).$$

The joint spectrum $S$ of $L$ and $Q$ is clearly given by the equation $f(\lambda, \mu) = 0$. We have the following facts.

(i) $f$ is a polynomial in two variables, so that the spectrum $S \subset \mathbf{C}^2$ is an affine algebraic curve.

(ii) Near infinity, $S$ looks like the curve $\mu^2 = -\lambda^{2q+1}$, (where $2q+1$ is the order of $Q$); thus $S$ has just one point at infinity. Let $\bar{S}$ denote the compact curve obtained by adjoining the non-singular point $\infty$ to $S$.

(iii) Near the point $\infty \in \bar{S}$, we can take as local parameter $k^{-1}$, where $k^2 = -\lambda$.

(iv) At each non-singular point $z = (\lambda, \mu)$ of $S$, the corresponding joint eigenspace of $L$ and $Q$ is one-dimensional.

For simplicity (as in [K]), let us suppose that $S$ is non-singular; then $\bar{S}$ is a compact Riemann surface. For each $x \in \mathbf{C}$, let $D(x) \subset S$ denote the divisor of points $z \in S$ such that the corresponding joint eigenfunction vanishes at $x$ (we assume that $u(x)$, and hence the coefficients of the operators $L$ and $Q$, are defined, and have no singularities, at least for $x$ in some neighbourhood of the origin in $\mathbf{C}$). Let $\psi(z, x)$ $(z \in S, x \in \mathbf{C})$ denote the joint eigenfunction, normalized so that $\psi(z, 0) = 1$. Naturally, this introduces a pole if $z \in D(0)$.

(v) The divisors $D(x)$ have degree $g$ = genus of $\bar{S}$, and are non-special. (Non-special means that there is no non-constant meromorphic function on $\bar{S}$ whose divisor of poles is $D(x)$ (or less). This property is a little harder to see than the others.)

(vi) The function $\psi(z, x)$ has the properties

(a) Away from $\infty$, $\psi$ is analytic except for poles when $z \in D = D(0)$ (independent of $x$).

(b) Near $\infty$, $\psi \sim \exp(kx)$, where $k$ is as in (iii).

The crucial point now is that this construction can be run backwards: suppose we are given a compact Riemann surface $\bar{S}$ of genus $g$, a point $\infty \in \bar{S}$, a non-special positive divisor $D$ of degree $g$ (not involving the point $\infty$) and a local parameter $k^{-1}$ near $\infty$. Then there is a unique function $\psi(z, x)$, $z \in \bar{S}$, $x \in$ (open neighbourhood of 0 in $\mathbf{C}$), with the properties (a) and (b) in (vi). Furthermore, we can reconstruct (uniquely) the differential operators of which $\psi$ is the joint eigenfunction: let $\lambda$ be any meromorphic function on $\bar{S}$ whose only singularity is a pole at $\infty$. Then there is a unique ordinary differential operator $L_\lambda$ in $\partial/\partial x$ such that

$$L_\lambda \psi(z, x) = \lambda(z)\, \psi(z, x) \text{ for all } z \in \bar{S} - \{\infty\}.$$

It is clear that the operators $L_\lambda$ for various $\lambda$ all commute with each other. The order of $L_\lambda$ is equal to the order of the pole of $\lambda$ at $\infty$. Thus if we want (as we do) one of the operators $L_\lambda$ to have order 2, we must choose $\bar{S}$ so that there is a function $\lambda$ on $\bar{S}$ whose only singularity is a double pole at $\infty$: that is, $\bar{S}$ must be hyperelliptic (and $\infty$ must be a 'Weierstrass point').

This construction is well explained in [K], so let me just comment on one point that Krichever passes over, the existence and uniqueness of the function $\psi$. That is most easily seen as follows. Suppose we replace (b) in (vi) by (b′): near $\infty$, $\psi \sim$ (const). $\exp(kx)$. Then conditions (a) and (b′) say that, for fixed

$x$, $\psi$ is a section of the holomorphic line bundle $\{D\} \otimes \{e^{kx}\}$ on $\bar{S}$. Here $\{D\}$ is the usual bundle associated with the divisor $D$, and $\{e^{kx}\}$ denotes the bundle defined by the single transition function $e^{kx}$ in a punctured neighbourhood of $\infty$. Now, if $x$ is near 0, this bundle is near to $\{D\}$, and hence inherits the non-special character of $\{D\}$; that is, its space of sections is one-dimensional. Hence fixing the value of the constant in (b′) gives us a unique function $\psi$. This argument also shows that as $x$ varies the corresponding bundle (which can also be described as the bundle defined by the divisor $D(x)$) moves along a straight line in the Jacobian of $S$ (which is a $g$-dimensional complex torus). That follows at once from the formula $e^{kx}e^{ky} = e^{k(x+y)}$, since the group operation in the Jacobian (tensor product of line bundles) corresponds to multiplication of transition functions.

So far we have discussed only the stationary equation (10); but it is now easy to describe the KdV flow on the space of solutions of (10). (Indeed, it is equally easy to describe the solutions of equations of 'Kadomtsev–Petviashvili' type, involving an extra variable $y$: this is taken as the starting point in [K].) Suppose we have a solution $u(x, t)$ of the KdV equation which for each $t$ also satisfies (10): thus we have everything we have said so far for each value of $t$. Equation (3) implies that the joint spectrum $S$ of $L$ and $Q$ is independent of $t$; and if we normalize the joint eigenfunction $\psi(z, x, t)$ by $\psi(z, 0, 0) = 1$, $\psi_t = P\psi$ (where $P$ is the operator in the Lax representation (3) of the KdV equation), then it is not hard to see that its asymptotics at $\infty$ will be given by $\psi \sim \exp(kx - 4k^3 t)$. (The $k^3$ reflects the fact that the operator $P$ in (3) has order 3.) By the same argument as before it follows that as $t$ changes the bundle defined by the divisor $D(x, t)$ of zeros of $\psi$ moves along a straight line in the Jacobian (of course a different one from before). I refer to [K] for the details of the inverse construction, and also for the explanation of how to translate this geometrical description of the flow into an explicit formula for $u(x, t)$: the point is that the function $\psi$, being characterized by its poles and asymptotics at $\infty$, can be written down explicitly in terms of the $\theta$-function of $\bar{S}$.

It remains to say something about the 'finite-zone' periodic solutions of the KdV equation, which play a large role in the article [DMN]. Suppose $u(x)$ is a (now real) $C^\infty$ periodic function with period $T$; let $\hat{T}$ denote the operator of translation over a period: $\hat{T}\varphi(x) = \varphi(x + T)$, and as usual let $L = -\partial^2 + u$. Clearly $L$ and $\hat{T}$ commute, so we can again consider their joint spectrum $S$, the space of 'Floquet multipliers'. This is again a Riemann surface, but in general of infinite genus: its branch points lie over the simple periodic and anti-periodic spectrum of $L$, that is, over the points $\lambda$ such that $\hat{T}$ acting on the $\lambda$-eigenspace of $L$ is not diagonalizable. By definition, the finite-zone potentials $u(x)$ are those for which there is only a finite number of such points, so that the Riemann surface $S$ has finite genus. The KdV flow on the space of such functions $u(x)$ can be described exactly as before, as a straight line flow on the Jacobian of $\bar{S} = S \cup \{\infty\}$. In fact a periodic function $u(x)$ is finite-zone precisely when it is a solution of one of the higher stationary equations (10) (see [DMN], p.78 and p.86); the Riemann surface $S$ can be described as the joint spectrum either of $L$ and $\hat{T}$, or of $L$ and $Q$. Thus the finite-zone periodic theory is really the intersection of two theories: (i) the algebro-geometric theory of Krichever that I described above (ii) the general periodic theory.

In this volume we are not really concerned with the integration of the KdV equation on the infinite dimensional spaces mentioned in (B) at the beginning of the introduction; so let me just say a very few words about it. I already mentioned above the main idea for the general periodic problem: it proceeds just like the finite-zone case, but one has to work with Riemann surfaces of infinite genus, which naturally causes some problems. Details can be found in [28]. The solution of the initial value problem in the class of functions rapidly decreasing at infinity is provided by the famous 'inverse scattering method': again the idea is that as $t$ changes the spectrum of $L$ stays constant and we study the change in the eigenfunctions: but now 'spectrum' means the $L^2$ spectrum, and the evolution of the eigenfunctions is followed by means of the 'scattering data'. An introduction to this theory is given in [DMN], p.71–73, and a rigorous account can be found in [7].

We saw above that the finite-zone periodic theory could be viewed as the intersection of the algebro-geometric and general periodic theories. It is natural to ask what is the corresponding intersection with the rapidly-decreasing-at-infinity theory, that is, what are the solutions $u(x, t)$ of the KdV equation that are rapidly decreasing at infinity, and satisfy one of the higher stationary equations (10) for each fixed value of $t$. It turns out that these are precisely the '$n$-soliton' solutions, consisting of $n$ travelling waves that appear to pass through each other and emerge unscathed from the non-linear interaction. From the algebro-geometric point of view, these solutions arise in the case when the joint spectrum of $L$ and $Q$ is a (singular) rational curve with $n$ ordinary double points. Unfortunately, I do not know of anywhere in the literature where that is explained in much detail.

Finally, a few remarks about the references that follow. These by no means represent a serious attempt to bring the bibliography up to date, but are simply a selection based on my own prejudices; furthermore, the list is confined to papers that seem to me to be a direct continuation of lines of thought to be found in the articles reprinted here. Many of the papers are concerned with the problem of generalizing the theory to the case of Lax equations $L_t = [P, L]$ in which $L$ is an operator of higher order (and possibly with matrix coefficients). Some of the physically interesting equations obtained in this way are listed in [DMN], ch.1. Indeed, the article [K] is written at this level of generality: the generalization of the algebro-geometric theory is straightforward. The formal side of the theory presents more difficulty; in particular, the Hamiltonian formalism is more complicated than in the case of the KdV equation. An introduction to this is given in [K], appendix 1; details can be found in the papers

[1, 10, 12, 13, 21, 22, 25, 40, 41, 42, K32, K33]. The papers [16, 18, 19, 26, 31, 39] discuss the 'higher rank' algebro-geometric solutions of Lax equations; these are the ones that arise when the joint eigenspaces of the commuting operators $Q$, $L$, are not one-dimensional. (That can happen only when the orders of $Q$ and $L$ are not relatively prime, which is why the KdV equation has no such solutions.) The preliminary account of this theory given in [K], section 2.3 contains some errors, which are corrected in [16]. The papers [1, 14, 22, 34, 35, 36, 37, 38] forge a connexion between this subject and Lie theory; let me just mention the result of Adler [1] and Lebedev–Manin [22] that the Hamiltonian structure of the KdV equation (and also of the more general Lax equations) can be interpreted as the standard Kirillov–Kostant structure on the dual of a certain Lie algebra of 'formal integral operators'. The papers [3, 6, 40] are those referred to as preprints in [K], pages 161 and 163; the work of Krichever mentioned in [K], p.164 is published in [15, 17]. To end, let me recommend the two survey articles [25, 27].

## References


[1] M. Adler, On a trace functional for formal pseudo-differential operators and the symplectic structure of the Korteweg-de Vries equations. Inventiones Math. **50** (1979), 219–248.

[2] M. Adler and J. Moser, On a class of polynomials connected with the Korteweg-de Vries equation. Commun. Math. Phys. **61** (1978), 1–30.

[3] H. Airault, H. P. McKean and J. Moser, Rational and elliptic solutions of the Korteweg-de Vries equation and a related many-body problem. Commun. Pure Appl. Math. **30** (1977), 95–148.

[4] S. I. Al'ber, Investigation of equations of Korteweg-de Vries type by the method of recurrence relations. J. London Math. Soc. **19** (1979), 467–480 (Russian) (sic).

[5] I. V. Cherednik, Differential equations for the Baker–Akhiezer functions of algebraic curves. Funct. Anal. Appl. **12** no.3 (1978), 45–54 (Russian), 195–203 (English).

[6] D. V. Chudnovsky and G. V. Chudnovsky, Pole expansions of non-linear partial differential equations. Il nuovo Cimento **40B** (1977), 339–353.

[7] P. Deift and E. Trubowitz, Inverse scattering on the line. Commun. Pure Appl. Math. **32** (1979), 121–251.

[8] I. Ya. Dorfman, On the formal calculus of variations in the algebra of smooth cylindrical functions. Funct. Anal. Appl. **12** no. 2 (1978), 32–39 (Russian), 101–107 (English).

[9] I. M. Gel'fand and L. A. Dikii. Lie algebra structure in the formal calculus of variations. Funct. Anal. Appl. **10** no. 1 (1976), 18–25 (Russian), 16–22 (English).

[10] I. M. Gel'fand and L. A. Dikii, The calculus of jets and non-linear Hamiltonian systems. Funct. Anal. Appl. **12** no. 2 (1978), 8–23 (Russian), 81–94 (English).

[11] I. M. Gel'fand and L. A. Dikii, Integrable non-linear equations and Liouville's theorem. Funct. Anal. Appl. **13** no. 1 (1979), 8–20 (Russian), 6–15 (English).

[12] I. M. Gel'fand and I. Ya Dorfman, Hamiltonian operators and algebraic structures connected with them. Funct. Anal. Appl. **13** no.4 (1979), 13–30 (Russian), 248–262 (English).

[13] I. M. Gel'fand and I. Ya. Dorfman, The Schouten bracket and Hamiltonian operators. Funct. Anal. Appl. **14** no 3 (1980), 71–74 (Russian).

[14] B. Kostant, The solution to a generalized Toda lattice and representation theory. Advances in Math. **34** (1979), 195–338.

[15] I. M. Krichever, On rational solutions of the Kadomtsev–Petviashvili equation and integrable systems of $N$ particles. Funct. Anal. Appl. **12** no. 1 (1978), 76–78 (Russian), 59–61 (English).

[16] I. M. Krichever, Commutative rings of ordinary differential operators. Funct. Anal. Appl. **12** no. 3 (1978), 20–31 (Russian), 175–185 (English).

[17] I. M. Krichever, Elliptic solutions of the Kadomtsev–Petviashvili equation and integrable systems of particles. Funct. Anal. Appl. **14** no. 4 (1980), 45–54 (Russian).

[18] I. M. Krichever and S. P. Novikov, Holomorphic bundles over Riemann surfaces and the Kadomtsev–Petviashvili (KP) equation. Funct. Anal. Appl. **12** no. 4 (1978), 41–52 (Russian), 276–286 (English).

[19] I. M. Krichever and S. P. Novikov, Holomorphic bundles and non-linear equations. Finite zone solutions of rank 2. Dokl. Akad. Nauk SSSR **247** (1979), 33–37; Soviet Math. Doklady **20** (1979), 650–654.

[20] M. D. Kruskal, Non-linear wave equations. Lecture notes in Physics **38** (1975), 310–354 (Springer).

[21] B. A. Kupershmidt and G. Wilson, Modifying Lax equations and the second Hamiltonian structure. Inventiones Math. **62** (1981), 403–436.

[22] D. R. Lebedev and Yu. I Manin, The Gel'fand–Dikii Hamiltonian operator and the co-adjoint representation of the Volterra group. Funct. Anal. Appl. **13** no. 4 (1979), 40–46 (Russian), 268–273 (English).

[23] A. Lichnerowicz, Les variétés de Poisson et leurs algèbres de Lie associées. J. Differential Geometry **12** (1977), 253–300.

[24] F. Magri, A simple model of the integrable Hamiltonian equation. J. Math. Phys. **19** (1978), 1156–1162.

[25] Yu. I. Manin, Algebraic aspects of non-linear differential equations. J. Sov. Math. **11** (1979), 1–122.

[26] Yu. I. Manin, Matrix solitons and bundles over curves with singularities. Funct. Anal. Appl. **12** no. 4 (1978), 53–67 (Russian), 286–295 (English).

[27] H. P. McKean, Integrable systems and algebraic curves. Lecture notes in Math. **755** (1979), 83–200 (Springer).

[28] H. P. McKean and E. Trubowitz, Hill's operator and hyperelliptic function theory in the presence of infinitely many branch points. Commun. Pure Appl. Math. **29** (1976), 143–226.

[29] R. M. Miura, C. S. Gardner and M. D. Kruskal, Korteweg-de Vries equation and generalizations. II. Existence of conservation laws and constants of motion. J. Math. Phys. **9** (1968), 1204–1209.

[30] P. van Moerbeke and D. Mumford, The spectrum of difference operators and algebraic curves. Acta Mathematica **143** (1979), 93–154.

[31] D. Mumford, An algebro-geometric construction of commuting operators and of solutions to the Toda lattice equation, Korteweg-de Vries equation and related non-linear equations. Proceedings of Kyoto conference (January 1978).

[32] P. J. Olver, On the Hamiltonian structure of evolution equations. Math. Proc. Camb. Phil. Soc. **88** (1980), 71–88.

[33] T. Ratiu, On the smoothness of the time $t$-map of the KdV equation and the bifurcation of the eigenvalues of Hill's operator. Lecture notes in Math. **755** (1979), 248–294 (Springer).
[34] T. Ratiu, The motion of the free $n$-dimensional rigid body. Indiana University Math. J. **29** (1980), 609–629.
[35] A. G. Reiman, M. A. Semenov-Tian-Shansky and I. E. Frenkel', Graded Lie algebras and completely integrable dynamical systems. Dokl. Akad. Nauk SSSR **247** (1979), 802–805; Soviet Math. Doklady **20** (1979), 811–814.
[36] A. G. Reiman and M. A. Semenov-Tian-Shansky, Reduction of Hamiltonian systems, affine Lie algebras and Lax equations. Inventiones Math. **54** (1979), 81–100.
[37] A. G. Reiman and M. A. Semenov-Tian-Shansky, Current algebras and non-linear partial differential equations. Dokl. Akad. Nauk SSSR **251** (1980), 1310–1314.
[38] W. W. Symes, Systems of Toda type, inverse spectral problems and representation theory. Inventiones Math. **59** (1980), 13–51.
[39] J.-L. Verdier, Équations différentielles algébriques. Sém. Bourbaki 1977–78, exposé 512; Lecture notes in Math. **710**, 101–122 (Springer).
[40] A. P. Veselov, On the Hamiltonian formalism of the Novikov–Krichever equations of commutativity of two operators. Funct. Anal. Appl. **13** no. 1 (1979), 1–7 (Russian), 1–6 (English).
[41] G. Wilson, Commuting flows and conservation laws for Lax equations. Math. Proc. Camb. Phil. Soc. **86** (1979), 131–143.
[42] G. Wilson, Hamiltonian and algebro-geometric integrals of stationary equations of KdV type. Math. Proc. Camb. Phil. Soc. **87** (1980), 295–305.
[43] V. E. Zakharov and A. B. Shabat, A scheme for integrating the non-linear equations of mathematical physics by the inverse scattering method. II. Funct. Anal. Appl. **13** no. 3 (1979), 13–22 (Russian), 166–174 (English).

# ASYMPTOTIC BEHAVIOUR OF THE RESOLVENT OF STURM-LIOUVILLE EQUATIONS AND THE ALGEBRA OF THE KORTEWEG-DE VRIES EQUATIONS

I. M. Gel'fand and L. A. Dikii

This paper is concerned with a group of problems associated with recent results on non-linear equations of Korteweg–de Vries type. The second chapter is basically of a survey character and investigates connections between these equations and trace formulae. In the first chapter we develop a new algebraic formalism of the calculus of variations.

## Contents



## Introduction

Work done in the last 20 years [1]–[3] (generalizing [4]) has provided an asymptotic expansion of the kernel of the resolvent of the equation

$$(1) \qquad -\frac{d^2\varphi}{dx^2}+[u(x)+\zeta]\,\varphi=0$$

in powers of $\zeta^{-1}$. The motivation for this problem came from the desire to give a meaning to traces of positive powers of differential operators (see also [6]–[8]). But the impetus for the study of the asymptotic behaviour of the resolvent came from different directions, and this study has been pursued by many authors (see [5], [9] and others) and continues to the present day [10]. Recently these problems have acquired a completely new significance owing to a close connection with the theory, dating from the mid-60's, of the integrability of the non-linear Korteweg–de Vries equation

[11]–[13]. The connection between the Korteweg–de Vries equation and the theory of traces was pointed out in [13]. Novikov and Dubrovin in their papers [14] and [15] made substantial new progress in the theory of the Korteweg–de Vries equation. Although their papers are formally concerned only with the case of periodic boundary conditions for (1), they contain many important algebraic relations of a local character that are independent of the boundary conditions. This aspect of these papers to some extent stimulated us to write this survey. Finally, we refer to the very recent paper by Marchenko [16], which also deals with the periodic problem; our account has some points of contact with this paper, which we shall note below, and we also mention the latest paper by Lax [17].

Perhaps the most surprising facet, which we did not find mentioned in any of the older papers and which became clear to us only after Novikov's work, is the fact that the coefficients of the asymptotic expansion of the kernel of the resolvent, which we had obtained, can be taken as Lagrange functions, after which we obtain a fully integrable Hamiltonian system. This fact forms the core of the present paper. We try to explain its algebraic nature, independent of the boundary conditions for the Sturm–Liouville equations and of the corresponding spectral methods. In this context we find it advantageous to develop a special algebra of polynomials in a function $u(x)$ and its derivatives $u'$, $u''$, ..., which includes elements of a formal calculus of variations and of formal Hamiltonian mechanics in the ring of such polynomials. Such an algebra arose in connection with what is now called formal geometry ([18], [19]). We hope to return to this problem in connection with work on the Gauss–Bonnet theorems for characteristic classes and on eigenvalues of partial differential operators (see also [19]).[1]

## CHAPTER 1

### The algebra of polynomials in $u$, $u'$, $u''$ ... and the formal calculus of variations

**1. Polynomials in** $u$, $u'$, $u''$ ... We consider polynomials in several symbols $u$, $u'$, $u''$, ..., $u^{(k)}$, ... with coefficients in a field of characteristic zero. These polynomials form an algebra over this field, which we denote by $A$. By $A_N$ we denote the set of all homogeneous polynomials of degree $N$. Then

(1) $$A = \sum A_N.$$

[1] NOTE IN PROOF. After this paper was sent to the printers, we became acquainted with two extremely interesting and useful preprints: P. D. Lax, Periodic solutions of the K–de V equation; H. P. McKean and P. van Moerbeck, The spectrum of Hill's equation. In his paper Lax gives explicit formulae similar to ours. We also point out the very interesting papers of M. Kac.

We define the differential order of the monomial $a_{k_1 \ldots k_r} u^{(k_1)} \ldots u^{(k_r)}$ to be $k = k_1 + \ldots + k_r$ (where $u^{(0)}$ is defined to be $u$). For brevity $[u]$ is used to denote the set of arguments $u, u', u'', \ldots$ of the polynomials.

In our algebra it makes sense to talk of linear differential operators, or vector fields,

$$\xi = \sum_{l=0}^{\infty} \xi_l [u] \frac{\partial}{\partial u^{(l)}}, \tag{2}$$

with coefficients $\xi_l \in A$. The set of such vector fields forms a module $TA$ over $A$. Among the vector fields we distinguish a canonical one, which we denote by $d/dx$; it is defined as follows:

$$\frac{d}{dx} = \sum_{l=0}^{\infty} u^{(l+1)} \frac{\partial}{\partial u^{(l)}}. \tag{3}$$

From this definition it follows that

$$\frac{d}{dx} u^{(l)} = u^{(l+1)}, \qquad u^{(l)} = \frac{d}{dx} \cdots \frac{d}{dx} u = \left(\frac{d}{dx}\right)^l u. \tag{4}$$

The commutator of two vector fields $\xi$ and $\eta$ is

$$[\xi, \eta] = \sum_{l=0}^{\infty} \sum_{k=0}^{\infty} \left[\xi_k \left(\frac{\partial}{\partial u^{(k)}} \eta_l\right) - \eta_k \left(\frac{\partial}{\partial u^{(k)}} \xi_l\right)\right] \frac{\partial}{\partial u^{(l)}}.$$

Therefore, the commutator of an arbitrary field with the canonical field $\dfrac{d}{dx}$ is

$$\left[\xi, \frac{d}{dx}\right] = \sum_{l=0}^{\infty} \left(\xi_{l+1} - \frac{d}{dx}(\xi_l)\right) \frac{\partial}{\partial u^{(l)}}. \tag{5}$$

Fields that commute with the canonical field form a submodule. These fields are characterized by the property $\xi_l\,[u] = (d/dx)^l\, \xi_0[u]$, where $\xi_0[u]$ can be any element of $A$. Between $A$ and the module of vector fields commuting with $d/dx$ there is an isomorphism (of $A$-modules). But, in addition, the vector fields form a Lie algebra, and this enables us to transfer this structure to $A$. The commutator of two elements of $A$ is given by

$$[f[u], g[u]] = \sum_{k=0}^{\infty} \left[\left(\left(\frac{d}{dx}\right)^k f\right) \frac{\partial}{\partial u^{(k)}} g - \left(\left(\frac{d}{dx}\right)^k g\right) \frac{\partial}{\partial u^{(k)}} f\right].$$

We define a differential linear form as a formal finite sum $\omega = \Sigma \omega_l [u]\; du^{(l)}$, where $\omega_l \in A$ and $du^{(l)}$ are certain new symbols. The value of the form on the vector field $\xi$ is given by $\omega(\xi) = \Sigma \omega_l \xi_l$. Differential forms of higher orders are defined similarly. Exterior products, the differential of a form, and the convolution $i\,(\xi)\omega$ of a vector field with a form are defined in the usual way. Then all the usual formulae are satisfied, for example, the Lie formula for a differential

$$(6)\qquad d\omega(\xi,\eta,\zeta,\ldots)=\xi\omega(\eta,\zeta,\ldots)-\eta\omega(\xi,\zeta,\ldots)+\ldots$$
$$\ldots-\omega([\xi,\eta],\zeta,\ldots)+\omega([\xi,\zeta],\eta,\ldots)-\ldots$$

The fact that the symbol $du^{(l)}$ is equal to the differential of $u^{(l)}$ is easily verified.

The set of all $r$-forms

$$\omega=\sum_{k_1\ldots k_r}\omega_{k_1\ldots k_r}[u]\,du^{(k_1)}\wedge\ldots\wedge du^{(k_r)},$$

with coefficients in $A_N$, is denoted by $A_N^{(r)}$. We set $A^{(r)}=\Sigma A_N^{(r)}$. We define the differential order of the form

$$cu^{(l_1)}\ldots u^{(l_t)}\,du^{(k_1)}\wedge\ldots\wedge du^{(k_r)}$$

to be $k=l_1+\ldots+l_t+k_1+\ldots+k_r$. Further, we define the action of $d/dx$ on forms by:

$$(7)\qquad \frac{d}{dx}\sum\omega_{k_1k_2\ldots k_r}du^{(k_1)}\wedge du^{(k_2)}\wedge\ldots\wedge du^{(k_r)}=$$
$$=\sum\left(\frac{d}{dx}\omega_{k_1k_2\ldots k_r}\right)du^{(k_1)}\wedge du^{(k_2)}\ldots\wedge du^{(k_r)}+$$
$$+\sum\omega_{k_1k_2\ldots k_r}du^{(k_1+1)}\wedge du^{(k_2)}\ldots\wedge du^{(k_r)}+\ldots$$
$$\ldots+\sum\omega_{k_1\ldots k_r}du^{(k_1)}\wedge du^{(k_2)}\wedge\ldots\wedge du^{(k_r+1)},$$

that is, all the $u^{(k)}$ are differentiated both where they occur in coefficients and also where they stand under the differential sign. The operator $d/dx$ commutes with $d$: $\left[\frac{d}{dx},d\right]=0$.

We define the action of the operator $\partial/\partial u^{(i)}$ on forms as

$$(8)\qquad \frac{\partial}{\partial u^{(i)}}\sum\omega_{k_1k_2\ldots k_r}du^{(k_1)}\wedge du^{(k_2)}\wedge\ldots\wedge du^{(k_r)}=$$
$$=\sum\frac{\partial\omega_{k_1k_2\ldots k_r}}{\partial u^{(i)}}du^{(k_1)}\wedge du^{(k_2)}\wedge\ldots\wedge du^{(k_r)}.$$

WARNING. Observe that $\frac{d}{dx}\,\omega\neq\Sigma u^{(l+1)}\left(\frac{\partial}{\partial u^{(l)}}\,\omega\right)$. Indeed, the right-hand side is only the first term in (7). The operator $d/dx$ is defined by (3) only in its action on functions, that is, 0-forms. We mention the commutation relations

$$(9)\qquad \left[\frac{\partial}{\partial u^{(i)}},\frac{d}{dx}\right]=\begin{cases}\dfrac{\partial}{\partial u^{(i-1)}}, & i\neq 0,\\ 0, & i=0,\end{cases}\qquad \left[\frac{\partial}{\partial u^{(i)}},d\right]=0.$$

The following operator is important: it is the variational derivative

$$(10)\qquad \frac{\delta}{\delta u}=\sum_{k=0}^{\infty}(-1)^k\left(\frac{d}{dx}\right)^k\frac{\partial}{\partial u^{(k)}}.$$

Now $\delta/\delta u$ commutes with $d$, since $d/dx$ and $\partial/\partial u^{(k)}$ do so. The operator

$d/dx$ maps

$$A_N^{(r)} \to A_N^{(r)}.$$

The operator $\delta/\delta u$ maps

$$A_N^{(r)} \to A_{N-1}^{(r)},$$

and $d$ maps:

$$A_N^{(r)} \to A_{N-1}^{(r+1)}.$$

THEOREM. *The sequence of mappings*

$$0 \to A_N^{(r)} \xrightarrow{d/dx} A_N^{(r)} \xrightarrow{\delta/\delta x} A_{N-1}^{(r)}, \quad N \neq 0$$

*is exact.*

In other words, a form $\omega$ can be written as $\frac{d}{dx}\,\omega_1$ if and only if $\frac{\delta}{\delta u}\,\omega = 0$ (the exactness of the sequence at the first term is trivial).

In one direction the theorem is very easy to prove:

$$\frac{\delta}{\delta u}\frac{d}{dx} = \sum_{k=0}^{\infty}(-1)^k\left(\frac{d}{dx}\right)^k\frac{\partial}{\partial u^{(k)}}\frac{d}{dx} =$$

$$= \sum_{k=0}^{\infty}(-1)^k\left(\frac{d}{dx}\right)^{k+1}\frac{\partial}{\partial u^{(k)}} + \sum_{k=1}^{\infty}(-1)^k\left(\frac{d}{dx}\right)^k\frac{\partial}{\partial u^{(k-1)}} = 0.$$

To prove the converse, we shall develop in the next section an auxiliary apparatus (of independent interest) that will enable us not only to prove the existence of the form, but also to construct it explicitly. For another algorithm, which is very useful in practice, see the footnote on p. 87.

NOTE ON THE DEFINITIONS. The operators $d/dx$ and $\partial/\partial u^{(i)}$ in (7) and (8) are just the Lie derivatives of $\omega$ with respect to $d/dx$ and $\partial/\partial u^{(i)}$ and are usually denoted by $L_{d/dx}$ and $L_{\partial/\partial u^{(i)}}$.

The Lie derivative with respect to the vector field $\xi$ is defined as $L_\xi = i(\xi)\,d + di\,(\xi)$. From this it follows that all Lie derivatives commute with the differential: $dL_\xi = L_\xi d = di\,(\xi)\,d$.

We claim that $L_{d/dx}\,\omega$ is, in fact, equal to the right-hand side of (7). To show this we first note that $L_{d/dx}du^{(i)} = dL_{d/dx}u^{(i)} = du^{(i+1)}$. The equation $L_\xi\omega_1 \wedge \omega_2 = (L_\xi\omega_1) \wedge \omega_2 + \omega_1 \wedge (L_\xi\omega_2)$ completes the proof. Similarly we find that $L_{\partial/\partial u^{(i)}}\,du^{(j)} = dL_{\partial/\partial u^{(i)}}\,u^{(j)} = d\delta_{ij} = 0$. Hence $L_{\partial/\partial^{(i)}}$ is equal to the right-hand side of (8).

Our reason for simply writing $d/dx$ and $\partial/\partial u^{(i)}$ rather than $L_{d/dx}$ and $L_{\partial/\partial u^{(i)}}$ is that a Lagrangian function $L$ will play a fundamental part in what follows, and formulae containing both $L_{d/dx}$ and $L$ are ugly in appearance.[1]

[1] In his excellent book on mechanics, Godbillon [21] avoids this difficulty because the Lagrangian appears only at the very end of the last chapter.

**2. The symbolic notation for polynomials in** $u$, $u'$, $u''$, ... Let $\widetilde{A}_N^{(r)}$ be the set of polynomials $\widetilde{F}(\xi; z)$ (where $\xi = (\xi_1, \xi_2, \ldots, \xi_N)$ and $z = (z_1, z_2, \ldots, z_r)$) that are symmetric in $\xi$ and antisymmetric in $z$. With each such polynomial we associate a form in $A_N^{(r)}$ by the following rule: for a monomial we set

$$a\xi_1^{k_1} \ldots \xi_N^{k_N} z_1^{l_1} \ldots z_r^{l_r} \to au^{(k_1)} \ldots u^{(k_N)} du^{(l_1)} \wedge \ldots \wedge du^{(l_r)},$$

to a sum of monomials there corresponds the sum of the forms. It is easy to see that every form is associated with one and only one polynomial in $\xi$ and $z$, which arises as follows. With the monomial $au^{(k_1)} \ldots u^{(k_N)} du^{(l_1)} \wedge \ldots \wedge du^{(l_r)}$ we associate $S_{N,r} a \xi_1^{k_1} \ldots \xi_N^{k_N} z_1^{l_1} \ldots z_r^{l_r}$, where $S_{N,r}$ is the operator of symmetrization with respect to $\xi$ and of anti-symmetrization with respect to the $z$, that is, $S_{N,r} = \dfrac{1}{N!r!} \sum_{\sigma_N, \sigma_r} (-1)^\nu \sigma_N \sigma_r$, where $\sigma_N$ ranges over all permutations of $1, \ldots, N$, and $\sigma_r$ over the permutations of $1, \ldots, r$; $\nu = 0$ if $\sigma_r$ is even and $\nu = 1$ otherwise. The action of the operators $\sigma_N$ and $\sigma_r$ on a function $F(\xi_1, \ldots, \xi_N; z_1, \ldots, z_r)$ is clear. For example,

$$uu'u''\, du \wedge du'' \leftrightarrow \frac{1}{12} (\xi_2\xi_3^2 + \xi_3\xi_2^2 + \xi_3\xi_1^2 + \xi_1\xi_3^2 + \xi_1\xi_2^2 + \xi_2\xi_1^2)(z_2^2 - z_1^2).$$

We denote the set of these symbolic polynomials by $\widetilde{A}_N^{(r)}$. The operators $\dfrac{d}{dx}, \dfrac{\delta}{\delta u}, d, \dfrac{\partial}{\partial u^{(i)}}$ on forms go over into certain operators $\dfrac{\widetilde{d}}{dx}, \dfrac{\widetilde{\delta}}{\delta u}, \widetilde{d}, \dfrac{\widetilde{\partial}}{\partial u^{(i)}}$ on symbolic polynomials. We find that

$$(11)\quad \left(\frac{\widetilde{d}}{dx} \widetilde{F}\right)(\xi_1, \ldots, \xi_N; z_1, \ldots, z_r) = \\ = (\xi_1 + \ldots + \xi_N + z_1 + \ldots + z_r)\, \widetilde{F}(\xi_1, \ldots, \xi_N; z_1, \ldots, z_r).$$

This property is obvious.

$$(12)\quad \left(\frac{\widetilde{\partial}}{\partial u} \widetilde{F}\right)(\xi_1, \ldots, \xi_{N-1}; z_1, \ldots, z_r) = N\widetilde{F}(\xi_1, \ldots, \xi_{N-1}, 0; z_1, \ldots, z_r).$$

PROOF. We note that

$$\frac{\partial}{\partial u} u^{(k_1)} \ldots u^{(k_N)} = \sum_{i=1}^{N} u^{(k_1)} \ldots \frac{\partial u^{(k_i)}}{\partial u} \ldots = \sum_{i=1}^{N} \delta_{k_i 0} u^{(k_1)} \ldots \widehat{u^{(k_i)}} \ldots;$$

here $\delta_{k_i 0}$ is the Kronecker symbol, and the symbol ˆ indicates the omission of the corresponding term. Hence

$$\frac{\widetilde{\partial}}{\partial u} S_{N,r}\xi_1^{k_1} \ldots \xi_N^{k_N} z_1^{l_1} \ldots z_r^{l_r} = S_{N,r} \sum_{i=1}^{N} \delta_{k_i 0}\xi_1^{k_1} \ldots \hat{\xi}_i^{k_i} \ldots \xi_N^{k_N} z_1^{l_1} \ldots z_r^{l_r} =$$

$$= \sum_{i=1}^{N} \widetilde{F}(\xi_1, \ldots, \xi_{i-1}, 0, \xi_{i+1}, \ldots, \xi_N; z_1, \ldots, z_r) =$$

$$= N\widetilde{F}(\xi_1, \ldots, \xi_{N-1}, 0; z_1, \ldots, z_r).$$

Similarly we find

$$(13) \quad \left(\frac{\widetilde{\partial}}{\partial u^{(k)}} \widetilde{F}\right)(\xi_1, \ldots, \xi_{N-1}, z_1, \ldots, z_r) = = N \cdot \frac{1}{k!} \frac{\partial^k}{\partial \xi_N^k} \widetilde{F}(\xi_1, \ldots, \xi_{N-1}, 0; z_1, \ldots, z_r),$$

$$(14) \quad \left(\frac{\widetilde{\delta}}{\delta u} \widetilde{F}\right)(\xi_1, \ldots, \xi_{N-1}; z_1, \ldots, z_r) =$$

$$= \sum_{k=0}^{\infty} (-1)^k (\xi_1 + \ldots + \xi_{N-1} + z_1, + \ldots + z_r)^k \times$$

$$\times N \cdot \frac{1}{k!} \frac{\partial^k}{\partial \xi_N^k} \widetilde{F}(\xi_1, \ldots, \xi_{N-1}, 0; z_1, \ldots, z_r) =$$

$$= N\widetilde{F}(\xi_1, \ldots, \xi_{N-1}, -\xi_1 - \ldots - \xi_{N-1} - z_1 - \ldots - z_r; z_1, \ldots, z_r).$$

If $\widetilde{F}_1(\xi_1, \ldots, \xi_{N_1}; z_1, \ldots, z_{r_1})$ and $\widetilde{F}_2(\xi_1, \ldots, \xi_{N_2}; z_1, \ldots, z_{r_2})$ correspond to the forms $\omega_1$ and $\omega_2$, then the form corresponding to $\omega_1 \wedge \omega_2$ is

$$(15) \quad S_{N_1+N_2, r_1+r_2}[\widetilde{F}_1(\xi_1, \ldots, \xi_{N_1}; z_1, \ldots, z_{r_1}) \times \times \widetilde{F}_2(\xi_{N_1+1}, \ldots, \xi_{N_1+N_2}; z_{r_1+1}, \ldots, z_{r_1+r_2})],$$

$$(16) \quad (\widetilde{d}\widetilde{F})(\xi_1, \ldots, \xi_{N-1}; z_1, \ldots, z_{r+1}) = = S_{N-1, r+1}\widetilde{F}(\xi_1, \ldots, \xi_{N-1}, z_{r+1}; z_1, \ldots, z_r).$$

The properties (15) and (16) are obvious.

COMPLETION OF THE PROOF OF THE THEOREM OF THE PRECEDING SECTION. By means of the symbolic calculus which we have developed, we can now prove the theorem in both directions. From (11) and (14) it is immediately clear that $\frac{\widetilde{\delta}}{\delta u} \frac{\widetilde{d}}{dx} = 0$. For by substituting $\xi_N = -\xi_1 - \ldots - \xi_{N-1} - z_1 - \ldots - z_r$ in the expression $\xi_1 + \ldots + \xi_N + z_1 + \ldots + z_r$ we obtain zero. Suppose now that the $\frac{\delta}{\delta u}\omega = 0$. Then for the corresponding function $\widetilde{F}$ we have $\frac{\widetilde{\delta}}{\delta u}\widetilde{F} = 0$. This means that the polynomial $\widetilde{F}$ vanishes on substituting $\xi_N = -\xi_1 - \ldots - \xi_{N-1} - z_1 - \ldots - z_r$. Hence, $\widetilde{F}$ is divisible by

$\xi_N - (-\xi_1 - \ldots - \xi_{N-1} - z_1 - \ldots - z_r)$, and the quotient is again a polynomial symmetric in $\xi$ and antisymmetric in $z$, say $\widetilde{F} = \frac{\widetilde{d}}{dx}\widetilde{G}$. If $\omega_1$ is the form corresponding to $\widetilde{G}$, then $\omega = \frac{d}{dx}\omega_1$, as required.

If $\omega = \frac{d}{dx}\omega_1$, then we also write $\omega_1 = \left(\frac{d}{dx}\right)^{-1}\omega$.

NOTE. $(d/dx)^{-1}$ is a well-defined operator: there are no constants of integration. We consider $d/dx$ separately acting on each $A_N^{(r)}$, where there are no constants.

**3. The Lagrangian, the Hamiltonian, 1- and 2-forms.** We fix an element $L[u] \in A$ and call it the Lagrangian. We denote by $M_L$ the ideal in $A$ generated by the elements $\frac{\delta}{\delta u}L$, $\frac{d}{dx}\frac{\delta}{\delta u}L$, $\left(\frac{d}{dx}\right)^2\frac{\delta}{\delta u}L, \ldots$, and by $A_L$ the quotient algebra $A/M_L$.

THEOREM 1. *In $A$ we have*

$$\frac{\delta}{\delta u}u'\frac{\delta}{\delta u}=0 \tag{17}$$

*(but not in $A^{(r)}$, $r \neq 0$).*

We prove the theorem by the symbolic calculus of §2. Let $F \in A_N$. Then $u'\frac{\delta}{\delta u}F$ corresponds to the function

$$\begin{aligned}\xi_N\widetilde{F}(\xi_1, \ldots, \xi_{N-1}, -\xi_1 - \ldots - \xi_{N-1}) &+\\ + \xi_1\widetilde{F}(\xi_2, \ldots, \xi_N, -\xi_2 - \ldots - \xi_N) &+\\ + \xi_2\widetilde{F}(\xi_3, \ldots, \xi_N, \xi_1, -\xi_3 - \ldots - \xi_N - \xi_1) &+ \ldots,\end{aligned}$$

and $\frac{\delta}{\delta u}u'\frac{\delta}{\delta u}F$ to

$$\begin{aligned}(-\xi_1 - \xi_2 - \ldots - \xi_{N-1})\widetilde{F}(\xi_1, \ldots, \xi_{N-1}, -\xi_1 - \ldots - \xi_{N-1}) &+\\ + \xi_1\widetilde{F}(\xi_2, \ldots, -\xi_1 - \ldots - \xi_{N-1}, \xi_1) &+\\ + \xi_2\widetilde{F}(\xi_3, \ldots, -\xi_1 - \ldots - \xi_{N-1}, \xi_1, \xi_2) &+ \ldots = 0,\end{aligned}$$

as required.

COROLLARY. *The operator*

$$H = -\left(\frac{d}{dx}\right)^{-1}u'\frac{\delta}{\delta u}. \tag{18}$$

*is well-defined.*

We call (18) the Hamiltonian operator and $H[u] = HL[u]$ the Hamiltonian corresponding to the Lagrangian $L[u]$. We consider the operator $d - du\cdot\frac{\delta}{\delta u}$ from $A$ to $A^{(1)}$:

$$\left(d - du\cdot\frac{\delta}{\delta u}\right)F[u] = dF - \left(\frac{\delta}{\delta u}F\right)du.$$

Then the following theorem holds.

THEOREM 2.

$$\frac{\delta}{\delta u}\left(d-du\frac{\delta}{\delta u}\right)=0. \tag{19}$$

PROOF. Let $F\in A_N$. Then

$$\frac{\tilde{\delta}}{\delta u}\widetilde{dF}=\frac{\tilde{\delta}}{\delta u}N\tilde{F}(\xi_1,\xi_2,\ldots,\xi_{N-1},z)=$$

$$=N(N-1)\tilde{F}(\xi_1,\xi_2,\ldots,-\xi_1-\xi_2-\ldots-\xi_{N-2}-z,z),$$

$$\frac{\tilde{\delta}}{\delta u}\tilde{F}=N\tilde{F}(\xi_1,\xi_2,\ldots,\xi_{N-1},-\xi_1-\ldots-\xi_{N-1}),$$

$$\widetilde{\frac{\delta}{\delta u}F\,du}=N\tilde{F}(\xi_1,\xi_2,\ldots,\xi_{N-1},-\xi_1-\ldots-\xi_{N-1}),$$

$$\frac{\tilde{\delta}}{\delta u}\widetilde{\frac{\delta}{\delta u}F\,du}=N(N-1)\tilde{F}(\xi_1,\xi_2,\ldots,\xi_{N-2},-\xi_1-\ldots-\xi_{N-2}-z,z),$$

that is, $\frac{\tilde{\delta}}{\delta u}\left(\widetilde{\frac{\delta}{\delta u}F\,du-dF}\right)=0$. This proves the theorem.

COROLLARY. *The operator*

$$\Omega^{(1)}=\left(\frac{d}{dx}\right)^{-1}\left(d-du\frac{\delta}{\delta u}\right) \tag{20}$$

*is well defined.*

We call it the 1-form operator and

$$\Omega^{(2)}=d\left(\frac{d}{dx}\right)^{-1}\left(d-du\frac{\delta}{\delta u}\right)=\left(\frac{d}{dx}\right)^{-1}du\wedge d\frac{\delta}{\delta u} \tag{21}$$

the 2-form operator. We call the results of applying these operators to a given Lagrangian the 1- and 2-forms of this Lagrangian,

$$\Omega^{(2)}=\left(\frac{d}{dx}\right)^{-1}du\wedge d\left(\frac{\delta}{\delta u}L\right).$$

The form $\Omega^{(2)}$ is closed, being the differential of the form $\Omega^{(1)}$.

NOTE. The form $du\wedge d\left(\frac{\delta}{\delta u}L\right)$ is highly degenerate. The operator $\left(\frac{d}{dx}\right)^{-1}$ carries it into the $\Omega^{(2)}$– form, which, as we see below, is non-degenerate, at least in the most important cases.

**4. Momenta.** We introduce the momentum operators in $A$

$$p_i=\sum_{k=0}^{\infty}(-1)^k\left(\frac{d}{dx}\right)^k\frac{\partial}{\partial u^{(k+i+1)}}\qquad(i=0,1,2,\ldots)$$

(when $i=-1$, the right-hand side is just $\frac{\delta}{\delta u}$). Clearly,[1]

---

1) Another relation $\mathbf{p}_0\frac{\delta}{\delta u}=0$ can easily be verified by means of the symbolic calculus of § 2.

$\frac{d}{dx} p_i = -p_{i-1} + \frac{\partial}{\partial u^{(i)}}$.

We define the moments $p_i = p_i L$ as the result of the actions of the $\boldsymbol{p}_i$ on the Lagrangian.

THEOREM. $H = \sum_{i=0}^{\infty} p_i u^{(i+1)} - 1$ $\Omega^{(i)} = \sum_{i}^{\infty} p_i du^{(i)}$.

PROOF.

$$\frac{d}{dx}\left(\sum_{i=0}^{\infty} p_i u^{(i+1)} - 1\right) = \sum_{i=1}^{\infty} u^{(i+1)} \sum_{k=0}^{\infty} (-1)^k \left(\frac{d}{dx}\right)^{k+1} \frac{\partial}{\partial u^{(k+i+1)}} +$$

$$+ \sum_{i=0}^{\infty} u^{(i+2)} \sum_{k=0}^{\infty} (-1)^k \left(\frac{d}{dx}\right)^k \frac{\partial}{\partial u^{(k+i+1)}} - \sum_{i=0}^{\infty} u^{(i+1)} \frac{\partial}{\partial u^{(i)}} =$$

$$= -u' \sum_{k=0}^{\infty} (-1)^{k+1} \left(\frac{d}{dx}\right)^{k+1} \frac{\partial}{\partial u^{(k+1)}} + \sum_{i=1}^{\infty} u^{(i+1)} \frac{\partial}{\partial u^{(i)}} - \sum_{i=0}^{\infty} u^{(i+1)} \frac{\partial}{\partial u^{(i)}} =$$

$$= -u' \frac{\delta}{\delta u} = \frac{d}{dx} \boldsymbol{H}.$$

The equality of these derivatives implies that of the operators themselves. The second part of the theorem is proved in exactly the same way.

COROLLARY. $\Omega^{(2)} = \sum_{i=0}^{\infty} dp_i \wedge du^{(i)}$.

Let $\xi$ and $\eta$ be two vector fields commuting with $\frac{d}{dx}$. According to (5) we have $\xi_i = \left(\frac{d}{dx}\right)^i \xi_0$, $\eta_i = \left(\frac{d}{dx}\right)^i \eta_0$. We calculate

$$\frac{d}{dx} \Omega^{(2)}(\xi, \eta) = \left(\frac{d}{dx} \Omega^{(2)}\right)(\xi, \eta) = du \wedge d\left(\frac{\delta}{\delta u} L\right)(\xi, \eta) =$$

$$= \xi_0 \eta\left(\frac{\delta}{\delta u} L\right) - \eta_0 \xi\left(\frac{\delta}{\delta u} L\right);$$

and obtain the useful identity

$$\xi_0 \sum_{i=0}^{\infty} \left[\left(\frac{d}{dx}\right)^i \eta_0\right] \frac{\partial}{\partial u^{(i)}} \frac{\delta}{\delta u} L - \eta_0 \sum_{i=0}^{\infty} \left[\left(\frac{d}{dx}\right)^i \xi_0\right] \frac{\partial}{\partial u^{(i)}} \frac{\delta}{\delta u} L = \frac{d}{dx} [\Omega^{(2)}(\xi, \eta)]. \tag{22}$$

We say that two Lagrangians are equivalent if they differ by a function of form $\frac{d}{dx} F$. For equivalent Lagrangians the elements $\frac{\delta}{\delta u} L$ coincide and so do also the ideal $M_L$, the manifold $A_L$, the Hamiltonian operators, the 2-forms and so on. From the point of view of our present theory we may always replace a Lagrangian by an equivalent one. Among all the Lagrangians equivalent to a given one we can choose one for which the leading derivative occurs raised to the smallest power. It is easy to see that the term of this Lagrangian containing the leading derivative to the highest power must have the form $a(u^{(k_1)})^{l_1} (u^{k_2})^{l_2}, \ldots, k_1 > k_2 > \ldots$, where

$l_1 \geqslant 2$, that is, the leading derivative must occur to a power higher than the first (otherwise it would be possible to lower it by adding an expression of the form $\frac{d}{dx}$ $F$).[1] A Lagrangian equivalent to the given one with the leading derivative of lowest order is called reduced. The terms containing the leading derivative to the highest power are uniquely determined in the reduced Hamiltonian.

**5. Vector fields and forms in $A_L$.** We now define the tangent space $TA_L$ to the manifold $A_L$. We consider the set of vector fields $\xi \in TA$ for which $\xi\, F \in M_L$ whenever $F \in M_L$. This is a linear subspace of $TA$, which we denote by $TA_L^*$. We distinguish in this space the set of vectors $\xi$ such that $\xi\, F \in M_L$ for every $F \in A$ (in the coordinate representation of such vector fields, all the $\xi_i \in M_L$). These fields $\xi$ form a subspace $TA_L^{(0)}$. We set $TA_L = TA_L^*/TA_L^0$ and call it the tangent space to $A_L$.

To each $\xi \in TA_L$ there corresponds the sequence of coordinates $\{\xi_i\}$, $\xi_i \in A_L$.

For an arbitrary Lagrangian $L$ the vector field $d/dx$ belongs to $TA_L^*$; we use $d/dx$ also to denote the corresponding element of $TA_L$.

Let $\omega = A^{(r)}$ be a given form. We restrict it to the vector fields $\xi \in TA_L^*$. Then the composition of the mappings $TA_L^* \times TA_L^* \times \ldots \times TA_L^* \to A \to A_L$ generates a mapping, which is a form on $TA_L^*$ with values in $A_L$. It is easy to see that this form vanishes if any one of its arguments belongs to $TA_L^0$. Hence it can be regarded as a form on $TA_L = TA_L^*/TA_L^0$. Thus, any form $\omega \in A^{(r)}$ induces a form in $A_L^{(r)}$, the space of forms on $TA_L$ with values in $A_L$. Thus, there is a mapping $A^{(r)} \to A_L^{(r)}$ and $\sum_r A^{(r)} \to \sum_r A_L^{(r)}$. It is easy to see that the kernel of this mapping is invariant under the action of $d$. Indeed, this is clear for 0-forms, that is, functions. If $F \in M_L$, then $dF(\xi) = \xi\, F \in M_L$ for $\xi \in TA_L^*$, that is, $dF$ regarded as a form in $A_L^{(1)}$ vanishes. For forms of higher dimension we must use the formula (6) for the differential, together with the fact that the commutator of two fields in $TA_L^*$ again belongs to $TA_L^*$. From this we conclude that the operator $d$ can be carried over to $A_L^{(r)}$.

---

[1] Thus we have the following algorithm for reduction of the Hamiltonian. We consider the terms with the leading derivative. If this derivative occurs to the first power, then the term is transformed according to the models

$$(u')^2 u''' = -2u'(u'')^2 + \frac{d}{dx}[(u')^2 u''] \quad \text{or} \quad u'u''u''' = -\frac{1}{2}(u'')^3 + \frac{d}{dx}\left[\frac{1}{2}u'(u'')^2\right].$$

This process is continued until the leading derivative occurs to a power higher than the first. In the special case when $L = \frac{d}{dx} F$, the leading derivative cannot occur to a power higher than the first. In this case the process of reducing the Lagrangian continues until it becomes zero, that is, the reduced Lagrangian in this case is zero. We note that incidentally we have also obtained another algorithm for determining the primitive $\left(\frac{d}{dx}\right)^{-1} L$.

We define the action of $\xi\,\omega$ for $\omega \in A_L^{(r)}$ and $\xi \in TA_L$ in exactly the same way.

The spaces $TA_L$ and $A_L^{(r)}$ are modules over $A_L$.

NOTE. We have restricted ourselves to vector fields and forms induced by fields and forms in $A$. A more general definition of a vector field would be as a derivation in $A_L$ and of a form as a linear functional on vector fields. We have no need for this more general definition in what follows.

**6. A correspondence between functions and vector fields.** A vector field $\xi \in TA_L$ and a function $F \in A_L$ are said to be associated if $dF = -i\,(\xi)\,\Omega^{(2)}$ (the form $\Omega^{(2)}$ must, of course, be understood as an element of $A_L^{(2)}$).

LEMMA 1. *The vector field* $\frac{d}{dx}$ *and the Hamiltonian $H$ are associated.*

PROOF. Considering $H$ and $\frac{d}{dx}$ as elements of $A$ and $TA$, we establish the formula

$$dH = -i\left(\frac{d}{dx}\right)\Omega^{(2)} - \left(\frac{\delta}{\delta u}L\right)du,$$

from which the lemma follows immediately by projecting onto $A_L^{(1)}$. To prove this it is sufficient to show that the result of applying $d/dx$ to the two sides of (22) coincide. We have

$$\frac{d}{dx}dH = d\frac{d}{dx}H = -d\left(u'\frac{\delta}{\delta u}L\right) = -\left(\frac{\delta}{\delta u}L\right)du' - u'd\left(\frac{\delta}{\delta u}L\right),$$

$$\frac{d}{dx}\left[-i\left(\frac{d}{dx}\right)\Omega^{(2)} - \left(\frac{\delta}{\delta u}L\right)du\right] =$$

$$= i\left(\frac{d}{dx}\right)\left(-du\wedge d\left(\frac{\delta}{\delta u}L\right)\right) - \frac{d}{dx}\left(\left(\frac{\delta}{\delta u}L\right)du\right) =$$

$$= -u'd\left(\frac{\delta}{\delta u}L\right) + \frac{d}{dx}\left(\frac{\delta}{\delta u}L\right)du - \frac{d}{dx}\left(\frac{\delta}{\delta u}L\right)du - \left(\frac{\delta}{\delta u}L\right)du' =$$

$$= -u'd\left(\frac{\delta}{\delta u}L\right) - \left(\frac{\delta}{\delta u}L\right)du'.$$

This proves the formula, and with it the lemma.

NOTE. We do not clarify in general form the conditions for the existence of a vector field associated with a given element $F \in A_L$, nor the uniqueness of the association. We introduce below a fairly important class of Lagrangians, to which, in particular, the examples in the following chapters belong. For this class, one and only one vector field $\xi \in TA_L$ is associated with each $F \in A_L$, and this field is associated with only one element $F \in A_L$.

LEMMA 2. *If $\xi_1$ and $\xi_2$ are two vector fields in $TA_L$, then the vector field $[\xi_1, \xi_2]$ and the function $\Omega^{(2)}(\xi_1, \xi_2) \in A_L$ are associated.*

PROOF. Let $\eta$ be an arbitrary field in $TA_L$. Using the fact that the form $\Omega^{(2)}$ is closed we obtain by means of (6)

$$0 = (d\Omega^{(2)})(\xi_1, \xi_2, \eta) = \xi_1\Omega^{(2)}(\xi_2, \eta) - \xi_2\Omega^{(2)}(\xi, \eta) + \eta\Omega^{(2)}(\xi_1, \xi_2) -$$
$$-\Omega^{(2)}([\xi_1, \xi_2], \eta) + \Omega^{(2)}([\xi_1, \eta], \xi_2) - \Omega^2([\xi_2, \eta], \xi_1) =$$
$$=[-d(\Omega^{(2)}(\xi_1, \xi_2)) - i([\xi_1, \xi_2])\Omega^{(2)}](\eta),$$

that is,

$$d(\Omega^{(2)}(\xi_1, \xi_2)) = i([\xi_1, \xi_2])\Omega^{(2)},$$

which establishes the lemma.

When the vector field $\xi$ associated with an element $F \in A_L$ is unique, then the function $\Omega^{(2)}(\xi_1, \xi_2) \in A_L$, where $\xi_1$ and $\xi_2$ are fields associated with $F_1$ and $F_2$, is called the Poisson bracket of $F_1$ and $F_2$ and is denoted by $(F_1, F_2)$.

7. **Invariants.** A form $\omega \in A_L^{(r)}$ (in particular, a function) is said to be invariant if

$$\frac{d}{dx}\omega = 0. \tag{23}$$

A vector field $\xi \in TA_L$ is said to be invariant if

$$\left[\xi, \frac{d}{dx}\right] = 0. \tag{24}$$

The Hamiltonian $H$ and the 2-form $\Omega^{(2)}$ are invariant. In fact, for $H \in A$ we have $\frac{dH}{dx} = -u' \frac{\delta}{\delta u} L \in M_L$, that is, regarding $H$ as an element of $A_L$, $\frac{dH}{dx} = 0$, as required. Moreover, for all $\xi, \eta \in TA_L^*$, $\left(\frac{d}{dx}\Omega^{(2)}\right)(\xi, \eta) = -(\xi u)\left(\eta \frac{\delta}{\delta u} L\right) + (\eta u)\left(\xi \frac{\delta}{\delta u} L\right) \in M_L$. Thus, in $TA_L$ $\left(\frac{d}{dx}\Omega^{(2)}\right)(\xi, \eta) = 0$, that is, $\frac{d}{dx}\Omega^{(2)} = 0$. From (24) and from the formula (5) for the commutator it follows that for an invariant vector field $\xi_i = \frac{d}{dx}\xi_{i-1}$, or

$$\xi_i = \left(\frac{d}{dx}\right)^i \xi_0. \tag{25}$$

We consider the invariants of $A$ in greater detail. An element $F \in A$ is said to be an invariant or a first integral of the equation $\frac{\delta}{\delta u} L = 0$ if $\frac{d}{dx} F \in M_L$, that is,

$$\frac{d}{dx}F = a_0\frac{\delta}{\delta u}L + a_1\frac{d}{dx}\frac{\delta}{\delta u}L + a_2\left(\frac{d}{dx}\right)^2\frac{\delta}{\delta u}L + \ldots + a_r\left(\frac{d}{dx}\right)^r\frac{\delta}{\delta u}L. \tag{26}$$

Two first integrals $F_1$ and $F_2$ are said to be equivalent, $F_1 \sim F_2$, if $F_1 - F_2 \in M_L$. In particular, if $G \in M_L$ is a first integral, then it is equivalent to zero. Thus, an equivalence class of first integrals determines an invariant function in $A_L$.

LEMMA. *Let $F$ be a first integral. Then there exists an equivalent first integral $F_1$ for which*

$$\frac{d}{dx}F_1 = a\frac{\delta}{\delta u}L.$$

*Moreover, if* (26) *holds for* $F$, *then*

$$a = \sum_{i=0}^{r}(-1)^i\left(\frac{d}{dx}\right)^i a_i. \tag{27}$$

PROOF. We show that if the $r$ in (26) is non-zero, then we can find a first integral equivalent to $F$ for which $r$ is smaller by 1 and the sum on the right-hand side of (27) is the same as for $F$. From this the lemma follows directly. We have

$$\frac{d}{dx}F = \sum_{i=0}^{r-1} a_i\left(\frac{d}{dx}\right)^i\frac{\delta}{\delta u}L + \frac{d}{dx}\left[a_r\left(\frac{d}{dx}\right)^{r-1}\frac{\delta}{\delta u}L\right] - \left(\frac{d}{dx}a_r\right)\left(\frac{d}{dx}\right)^{r-1}\frac{\delta}{\delta u}L,$$

that is, for $F_1 = F - a_r\left(\frac{d}{dx}\right)^{r-1}\frac{\delta}{\delta u}L$,

$$\frac{d}{dx}F_1 = \sum_{i=0}^{r-2} a_i\left(\frac{d}{dx}\right)^i\frac{\delta}{\delta u}L + \left(a_{r-1} - \left(\frac{d}{dx}a_r\right)\right)\left(\frac{d}{dx}\right)^{r-1}\frac{\delta}{\delta u}L.$$

That the sum (27) is unaltered is clear. This proves the lemma.

The quantity $a[u]$ given by (27) is called the characteristic of the first integral $F$. To justify this term we must prove that the characteristic is uniquely determined by the first integral. Non-uniqueness may occur because of non-uniqueness of the representation (26). We say that two characteristics are equivalent if $a_1 - a_2 \in M_L$. Then the following theorem holds.

THEOREM 1. *If* $a_1[u]$ *is a characteristic of the first integral* $F_1[u]$ *and* $a_2[u]$ *a characteristic of* $F_2[u]$ *and if* $F_1 \sim F_2$, *then* $a_1 \sim a_2$.

We prove this theorem here under an additional hypothesis on the Lagrangian, which we call non-degeneracy. We require that the relation

$$b_0\frac{\delta}{\delta u}L + b_1\frac{d}{dx}\frac{\delta}{\delta u}L + b_2\left(\frac{d}{dx}\right)^2\frac{\delta}{\delta u}L + \ldots = 0$$

implies that all the $b_i \in M_L$. We note that the class of Lagrangians to be introduced in the next section ("normal" Lagrangians) are non-degenerate. Given two expansions of form (26) for the same $F \in A$, with coefficients $a_i'$ and $a_i''$, then the condition of non-degeneracy implies that $a_i' - a_i'' \in M_L$, hence $a' \sim a$ for the characteristics.

Now let $F_1$ and $F_2$ be two equivalent first integrals. Their characteristics differ by a characteristic of the first integral $F = F_1 - F_2$, which is equivalent to zero. But $F \in M_L$, that is, $F = \sum_{i=0}^{\infty} b_i\left(\frac{d}{dx}\right)^i\frac{\delta}{\delta u}L$. So

$$\frac{d}{dx}F=\sum_{i=0}^{\infty}\left(\frac{d}{dx}b_i\right)\left(\frac{d}{dx}\right)^i\frac{\delta}{\delta u}L+\sum_{i=0}^{\infty}b_i\left(\frac{d}{dx}\right)^{i+1}\frac{\delta}{\delta u}L=$$

$$=\left(\frac{d}{dx}b_0\right)\frac{\delta}{\delta u}L+\sum_{i=1}^{\infty}\left(\frac{d}{dx}b_i+b_{i-1}\right)\left(\frac{d}{dx}\right)^i\frac{\delta}{\delta u}L.$$

The characteristic can be evaluated without difficulty:

$$\frac{d}{dx}b_0+\sum_{i=1}^{\infty}(-1)^i\left(\frac{d}{dx}\right)^i\left(\frac{d}{dx}b_i+b_{i-1}\right)=0.$$

This proves the theorem.

The importance of the characteristic of an invariant lies in the fact that under certain conditions satisfied by the Lagrangian the following theorem holds.

THEOREM 2. *The vector field constructed from the characteristic $a[u]$ of a first integral $F[u]$ by the formula*

$$\xi=\sum_{i=0}^{\infty}\left[\left(\frac{d}{dx}\right)^i a\right]\frac{\partial}{\partial u^{(i)}}, \tag{28}$$

*is tangent and is associated with the invariant $F[u]$.*

We do not know the most general conditions under which this theorem is true. It holds for the "normal" Lagrangians to be introduced in the next section, to which, in particular, the Lagrangians in the second part of this paper belong.

Let $a[u] \in A$ be an arbitrary element. We ask when there exist a Lagrangian $L \in A$ and a first integral $F \in A$ of which $a[u]$ is a characteristic; in other words, when do there exist $F,\ L \in A$ such that $\frac{d}{dx}F = a[u]\frac{\delta}{\delta u}L$ or

$$\frac{\delta}{\delta u}a[u]\frac{\delta}{\delta u}L=0. \tag{29}$$

Everything reduces to a study of the operator $\frac{\delta}{\delta u}\circ a[u]\circ\frac{\delta}{\delta u} : A \to A$. The kernel of this operator consists of all Lagrangians $L$ for which $a[u]$ is a characteristic of a first integral, and the first integral $F$ is given by the formula

$$F=\left(\frac{d}{dx}\right)^{-1}a[u]\frac{\delta}{\delta u}L. \tag{30}$$

**8. Normal Lagrangians.** We say that a Lagrangian is normal if

$$\frac{\partial}{\partial u^{(2n)}}\left(\frac{\delta}{\delta u}L\right)=1, \tag{31}$$

where $u^{(2n)}$ is the leading derivative occurring in $\frac{\delta}{\delta u}L$. This means that

$\frac{\delta}{\delta u} L = u^{(2n)} + f(u, u', \ldots, u^{(2n-1)}$, where $f$ is a polynomial. By repeated differentiation of this relation we deduce that all the derivatives of $u$ beginning with $u^{(2n)}$ can be expressed as polynomials in $u, u' \ldots u^{(2n-1)}$ and in $\frac{\delta}{\delta u} L, \frac{d}{dx} \frac{\delta}{\delta u} L, \left(\frac{d}{dx}\right)^2 L, \ldots$

Passing to $A_L$ we can reformulate this as follows: all the leading derivatives $u^{(2n)}, u^{(2n+1)}, \ldots$, as elements of $A_L$, are polynomials in the lower derivatives. Hence, every polynomial in $u, u', u'', \ldots$ can be expressed as a polynomial in $u, u', \ldots, u^{(2n-1)}$. We establish that normal Lagrangians are non-degenerate, in the sense of the preceding section. Let

$$\sum_{k=0}^{\infty} a_k [u] \left(\frac{d}{dx}\right)^k \frac{\delta}{\delta u} L = 0,$$

where $a_k [u]$ are elements of $A$. We rewrite each $a_k$ as a polynomial in the variables $u, u', \ldots, u^{(2n-1)}, \frac{\delta}{\delta u} L, \frac{d}{dx} \frac{\delta}{\delta u} L, \left(\frac{d}{dx}\right)^2 \frac{\delta}{\delta u} L, \ldots$. Since these variables are independent, the terms of the form $\varphi(u, \ldots, u^{(2n-1)}) \left(\frac{d}{dx}\right)^k \frac{\delta}{\delta u} L$ in the sum must vanish separately. Hence the coefficients $a_k$ cannot contain terms independent of $\frac{\delta}{\delta u} L, \frac{d}{dx} \frac{\delta}{\delta u} L, \ldots$, that is, must belong to $M_L$, as required.

The condition (26) is equivalent to the fact that the Lagrangian $L$, in an abbreviated form, has $u^{(n)}$ as a leading derivative, and $u^{(n)}$ occurs as a separate term $\frac{1}{2} (-1)^n [u^{(n)}]^2$, while the remaining terms contain $u^{(n)}$ in powers no higher than the first. From the formulae for the Hamiltonian and for the form $\Omega^{(2)}$

$$H = \sum_{i=0}^{n-1} p_i u^{(i+1)} - L, \qquad \Omega^{(2)} = \sum_{i=0}^{n-1} dp_i \wedge du^{(i)}$$

it follows that $H$ depends only on $u, u', \ldots, u^{(2n-1)}$, and

$$\Omega^{(2)} = \sum_{k,\, l=0}^{2n-1} \Omega^{(2)}_{k,\, l}\, du^{(k)} \wedge du^{(l)},$$

where the matrix $\Omega^{(2)}_{k,l}$ has the form

$$\begin{pmatrix} * & \cdot & \cdot\cdot & * & 1 \\ \cdot & \cdot\cdot & \cdot\cdot & -1 & \cdots \\ \cdot & \cdot\cdot & \cdot\cdot & \cdot & \cdots \\ * & 1 & \cdot\cdot & \cdot & \\ -1 & & & & 0 \end{pmatrix}.$$

Only elements above the diagonal can be non-zero, and they depend on $u, u', \ldots, u^{(2n-1)}$.

What does $TA_L$ represent in the case of a normal Lagrangian? We claim that the first $2n$ coordinates of a vector field $\xi \in TA_L^*$ can be arbitrary elements of $A_L$, while the remaining ones are uniquely determined by the former. Indeed, for a vector field $\xi \in TA$ to belong to $TA_L^*$ it is necessary and sufficient that $\xi \left(\frac{d}{dx}\right)^i \frac{\delta}{\delta u} L \in M_L$ $(i = 0, 1, 2, \ldots)$, that is,

$$\xi_0 \frac{\partial}{\partial u}\left(\frac{\delta}{\delta u} L\right) + \ldots + \xi_{2n-1} \frac{\partial}{\partial u^{(2n-1)}}\left(\frac{\delta}{\delta u} L\right) + \xi_{2n} \in M_L,$$

$$\xi_0 \frac{\partial}{\partial u}\left(\frac{d}{dx}\frac{\delta}{\delta u} L\right) + \ldots \qquad\qquad + \xi_{2n+1} \in M_L,$$

$$\ldots\ldots\ldots\ldots\ldots\ldots\ldots\ldots\ldots\ldots$$

or, to within terms belonging to $M_L$ $\xi_{2n}, \xi_{2n+1}, \ldots$, are linear combinations of the first $2n$ coordinates. As elements of $A_L$, the coordinates $\xi_{2n}, \xi_{2n+1} \ldots$ are linear combinations of the first $2n$ coordinates, which are completely arbitrary. Thus, $TA_L$ is a $2n$-dimensional module over $A_L$.

LEMMA 1. *If the Lagrangian is normal, then with each $F \in A_L$ there is associated one and only one vector field $\xi \in TA_L$. Every vector field is associated with at most one $F \in A_L$ (to within a constant.)*

PROOF. Let $F \in A_L$ be an arbitrary element of $A_L$. It can be regarded as a polynomial in $u, u', \ldots, u^{(2n-1)}$. Then the equality $-i(\xi)\, \Omega^{(2)} = dF$ means that

$$\sum_{j=0}^{2n-1} \Omega_{ji}^{(0)} \xi_j = -\frac{\partial F}{\partial u^{(i)}} \qquad (i = 0, \ldots, 2n-1).$$

These equations can be solved successively if we recall the form of the matrix $\Omega_{ji}^{(2)}$. After the first $2n$ coordinates have been found (as polynomials in $u, u', \ldots, u^{(2n-1)}$), the remaining ones are uniquely determined, as explained above.

There remains the last statement of the lemma. Suppose that one field $\xi \in TA_L$ is associated with two functions $F_1, F_2 \in A_L$. Then the difference $F = F_1 - F_2$ satisfies $dF = 0$. If we take for $F \in A$ the representative of the residue class $F \in A_L$ that depends on $u, u', \ldots, u^{(2n-1)}$, then from $\partial F/\partial u^{(i)} \in M_L$ we obtain $\partial F/\partial u^{(i)} = 0$, that is, $F = \text{const}$, as required.

We consider now invariant functions and vector fields for normal Lagrangians.

First we prove that if $F \in A_L$ is an invariant, then its associated vector field $\xi \in TA_L$ is invariant. For by Lemma 6 in §6, $\left[\xi, \frac{d}{dx}\right]$ is the vector field corresponding to $\Omega^{(2)}\left(\xi, \frac{d}{dx}\right) = [i(\xi)\, \Omega^{(2)}\left(\frac{d}{dx}\right) = -\frac{d}{dx} F = 0$. Hence

$\left[\xi, \frac{d}{dx}\right] = 0$, as required.

Now we show that Theorem 2 of the preceding section holds for normal Lagrangians.

Let $F \in A$ be a first integral. We may regard it as depending only on $u, u', \ldots u^{(2n-1)}$. Then $\frac{d}{dx} F$ contains derivatives up to $u^{(2n)}$, and the leading derivative occurs linearly. In addition, $\frac{d}{dx} F \in M_L$. Hence $\frac{d}{dx} F = a[u] \frac{\delta}{\delta u} L$, where $a$ depends on $u, u', \ldots, u^{(2n-1)}$. Clearly, $a = \partial F/\partial u^{(2n-1)}$. Now from $-i(\xi)\,\Omega^{(2)} = dF$, recalling the form of $\Omega^{(2)}_{ji}$, we obtain what is required:

$$-\sum_{j=0}^{2n-1} \Omega^{(2)}_{j,\,2n-1}\xi_j = \frac{\partial F}{\partial u^{(2n-1)}}, \quad -\xi_0 = \frac{\partial F}{\partial u^{(2n-1)}} = a, \quad \xi_i = \left(\frac{d}{dx}\right)^i \xi_0 = -\left(\frac{d}{dx}\right)^i a.$$

# CHAPTER 2

## The asymptotic expansion of the resolvent of a Sturm–Liouville equation

**1. The asymptotic series for the resolvent.** We consider the second order linear differential equation

$$-\varphi'' + [u(x) + \zeta]\varphi = 0. \tag{1}$$

A kernel of the resolvent is defined to be a function $R(x, y; \zeta)$ that

a) is continuous in the pair of variables $x, y$;

b) is symmetric: $R(x, y; \zeta) = R(y, x; \zeta)$;

c) satisfies (1) as a function of either of the variables $x, y$ when the other is held fixed and $x \neq y$.

d) is such that for fixed $y$, the derivative $R'_x$ has a jump discontinuity of height 1 at $x = y$. By symmetry, this is equivalent to

$$\lim_{x \to y} (R'_x - R'_y) = 1; \tag{2}$$

e) $R(x, y; \zeta)$ converges to zero as $\zeta \to +\infty$ through positive values faster than any power of $\zeta$.

The notion of resolvent defined here includes as a particular case the Green's function for any fixed self-adjoint boundary conditions (and even for a broad class of conditions depending on $\zeta$).

It is clear that a linear combination of two resolvents such that the sum of the coefficients is 1 is again a resolvent. The difference of two resolvents is a smooth solution of (1) in each of the variables; from this we can deduce that this difference decreases exponentially as $\zeta \to \infty$ not only off the diagonal, but also on it. We denote the restriction of $R(x, y; \zeta)$ to the diagonal by $R(x; \zeta)$.

We now construct an asymptotic expansion

$$(3) \qquad R(x;\zeta) = \sum_{l=0}^{\infty} \frac{R_l[u]}{\zeta^{l+\frac{1}{2}}}$$

as $\zeta \to \infty$. The coefficients $R_l[u]$ are polynomials in $u$ and its derivatives $u, u', \ldots$ Two different resolvents have one and the same asymptotic series, since they differ by an exponentially small quantity. In what follows, other asymptotic expansions occur, not just for $R(x;\zeta)$. We always understand the equality of two functions to mean equality of their asymptotic expansions as $\zeta \to +\infty$ (that is, as equality of formal power series in powers of $\zeta^{-1/2}$).

The method of obtaining the asymptotic expansion of $R(x;\zeta)$, which we now describe briefly, was developed more fully in [1]. It is based on the expansion of the operator $\left(-\frac{d^2}{dx^2} + u + \zeta\right)^{-1}$ in powers of $\left(-\frac{d^2}{dx^2} + \zeta\right)^{-1}$:

$$\left(-\frac{d^2}{dx^2}+u+\zeta\right)^{-1} = \left(-\frac{d^2}{dx^2}+\zeta\right)^{-1}\left[1+u\left(-\frac{d^2}{dx^2}+\zeta\right)^{-1}\right]^{-1} =$$

$$= \left(-\frac{d^2}{dx^2}+\zeta\right)^{-1} - \left(-\frac{d^2}{dx^2}+\zeta\right)^{-1} u \left(-\frac{d^2}{dx^2}+\zeta\right)^{-1} + \ldots$$

For the kernel of the resolvent we obtain accordingly the asymptotic series

$$R = R_0 - R_0 \circ u \circ R_0 + R_0 \circ u \circ R_0 \circ u \circ R_0 - \ldots,$$

where $R_0$ is the kernel of the resolvent of (1) for $u = 0$, while $R_0 \circ u \circ R_0 \ldots$ is the superposition

$$\int \ldots \int R_0(x, x_1; \zeta)\, u(x_1)\, R_0(x_1, x_2; \zeta)\, u(x_2) \ldots dx_1 \ldots dx_N.$$

The limits of integration are arbitrary, except that the point $x$ must be within the interval of integration, since altering the limits only adds an exponential term. The equation is to be understood as one between formal (asymptotic) power series; it is meaningful, since for each fixed power of $\zeta$ only finitely many non-zero terms contribute to the right-hand side. As a representative of the resolvent $R_0(x, y; \zeta)$ we can take $e^{-\sqrt{\zeta}|x-y|}/2\sqrt{\zeta}$.

For the general term of the series we obtain the expression

$$\frac{(-1)^N}{2^{N+1}\zeta^{\frac{N+1}{2}}} \int \ldots \int e^{-\sqrt{\zeta}\,(|x-x_1|+|x_1-x_2|+\ldots+|x_N-x|)} u(x_1) \ldots u(x_N)\, dx_1 \ldots dx_N.$$

To find an asymptotic expansion of this expression, we represent $u(x_i)$ as a series

$$u(x_i) = \sum \frac{(x_i - x)^{k_i}}{k_i!} u^{(k_i)}(x).$$

We obtain the asymptotic form as

$$\sum_{N=0}^{\infty}\sum_{(k)}(-1)^N \frac{M_{k_1\ldots k_N}u^{(k_1)}\ldots u^{(k_N)}}{2^{N+1}(\sqrt{\zeta})^{k_1+\ldots+k_N+2N+1}},$$

where

$$M_{k_1\ldots k_N}=\frac{1}{k!\ldots k_N!}\times \tag{4}$$

$$\times\int_{-\infty}^{\infty}\ldots\int_{-\infty}^{\infty}e^{-\{|\eta_1|+|\eta_1-\eta_2|+\ldots+|\eta_N|\}}\eta_1^{k_1}\ldots\eta_N^{k_N}\,d\eta_1\ldots d\eta_N.$$

The coefficients are non-zero only if $k_1+\ldots+k_N$ is even, that is, the expansion is in odd powers of $\sqrt{\zeta}$. So we have obtained an expansion of the form (3), with

$$R_l[u]=\sum_{N=l}^{\infty}\sum_{k_1+\ldots+k_N=2l-2N}M_{k_1\ldots k_N}u^{(k_1)}\ldots u^{(k_N)}. \tag{5}$$

This is, in fact, a polynomial in $u, u'\ldots$, and is homogeneous in the grading $k_1+\ldots+k_N+2N$, that is, in the sum of twice the degree $N$ in the variables $u, u', \ldots$ and of the differential order $k_1+k_2+\ldots+k_N$. The resulting expression for the coefficients in the form of integrals is complicated. These formulae acquire a more lucid and practically useful form if the symbolic calculus of polynomials in $u, u', u''\ldots$ developed in Chapter I, §2 is used. This is done in Appendix I. In Appendix II we describe another method, leading to recurrence formulae for $R_l[u]$, which was developed in [2], [3]. Although these recurrence formulae are more complicated than those occurring in the subsequent sections, they have the advantage that the method used to obtain them is suitable not only for ordinary differential equations, but also in more general cases (see [9]).

**2. A third order linear equation for the resolvent and the Riccati equation.** We prove that $R(x;\zeta)$ satisfies the equation

$$-2RR''+(R')^2+4(u+\zeta)R^2=1. \tag{6}$$

(By our convention, this equation is to be interpreted as meaning that the left- and right-hand sides differ by a term that is exponentially small as $\zeta\to\infty$). The dashes always denote $\frac{d}{dx}$. We find

$$R'(x,\zeta)=\left[\frac{\partial}{\partial x}R(x,y;\zeta)+\frac{\partial}{\partial y}R(x,y;\zeta)\right]_{y=x-0},$$

$$R''(x,\zeta)=\left[\frac{\partial^2}{\partial x^2}R(x,y;\zeta)+2\frac{\partial^2}{\partial x\,\partial y}R(x,y;\zeta)+\frac{\partial^2}{\partial y^2}R(x;y;\zeta)\right]_{y=x-0}=$$
$$=[(u(x)+\zeta)R(x,y;\zeta)+(u(y)+\zeta)R(x,y;\zeta)+2R_{xy}(x,y;\zeta)]_{y=x-0}.$$

We now observe that $RR_{xy}=R_xR_y+c(\zeta)$, where $c(\zeta)$ does not depend on $x$, $y$ and tends exponentially to zero as $\zeta\to\infty$. For both $R_x$ and $R$ satisfy (1) with respect to $y$. Consequently, the Wronskian $R(R_x)_y-R_y(R_x)$ does not depend on $y$. By symmetry, it does not depend on $x$ either. Off the diagonal $R(x, y; \zeta)$ is exponentially small, hence so is $c(\zeta)$. Furthermore, we have

$$-2RR''+(R')^2+4(u(x)+\zeta)R^2=$$
$$=-2R[(u(x)+\zeta)R(x,y;\zeta)+(u(y)+\zeta)R(x,y;\zeta)+2R_{xy}(x,y;\zeta)]_{y=x-0}+$$
$$+[R_x(x,y;\zeta)+R_y(x,y;\zeta)]^2_{y=x-0}+4(u(x)+\zeta)R^2(x,x;\zeta)=$$
$$=[-4R_xR_y+R_x^2+2R_xR_y+R_y^2]_{y=x-0}-4c(\zeta)=(R_x-R_y)^2_{y=x-0}-4c(\zeta)=$$
$$=1-4c(\zeta),$$

as was claimed.

As a consequence of (6) we see that $R(x; \zeta)$ satisfies the third order linear equation

$$-R'''+4(u(x)+\zeta)R'+2u'(x)R=0. \tag{7}$$

This can be verified by differentiating (6) and dividing by $R$. (But (7) could have obtained more simply. If $\varphi_1, \varphi_2$ is a fundamental system of solutions of (1), then $R(x: \zeta)$ is always a linear combination $\varphi_1^2, \varphi_2^2, \varphi_1\varphi_2$. Each of these expressions satisfies (7), as we can see by direct verification. Now (6) contains rather more information than (7) (the constant of integration depends on the normalization of $R$). Equation (7) is also used in [16].[1]

The equations (6) and (7) yield a recurrence method for determining the expansion coefficients $R_l$. Substituting (3) in (6) we obtain

$$R_0=\frac{1}{2},\quad R_1=-\frac{1}{4}u,\quad R_{l+1}=2\sum_{k=0}^{l-1}R_kR''_{l-k}-\sum_{k=1}^{l-1}R'_kR'_{l-k}-$$
$$-4u\sum_{k=0}^{l}R_kR_{l-k}-4\sum_{k=1}^{l}R_kR_{l-k+1}\qquad (l=1,2,\ldots). \tag{8}$$

But (7) gives simpler formulae than these. Substituting (3) in (7) we obtain

$$R'_{l+1}=\frac{1}{4}R'''_l-uR'_l-\frac{1}{2}u'R_l, \tag{9}$$

that is, $R'_{l+1}=\frac{1}{4}R''_1-u\,R_l+\frac{1}{2}\left(\frac{d}{dx}\right)^{-1}u'R_l$. The simplicity of this relation is due to the linearity of (7). However, here we do not obtain $R_{l+1}$ but only its derivative. In Chapter 1 we have described an algorithm

---

1) Apparently (7) plays an important part in Sturm–Liouville theory. It arises, for example, in the following situation (see [20]). Consider the equation $y''+q(x)y=0$. Under the change of variables $x=F(\xi)$, $y=\sqrt{F'(\xi)}\,Y(\xi)$ it goes over into $Y''+Q(\xi)Y=0$, $Q=(F')^2q-\frac{3}{4}(F'')^2(F')^2\frac{1}{2}F'''/F$. Under an infinitesimal deformation of the variables $Y(\xi)=y/\sqrt{1+\ \varepsilon f'}$ where $\varepsilon$ is a small parameter we have $Q=q+(4qf'+2q'f+f''')\,2\varepsilon\ +0(\varepsilon^2)$. If $f$ satisfies $f'''+2q'f+4qf'=0$, which is the same as (7), then $Q=q+0(\varepsilon)^2$, that is, the equation is invariant to within terms of smaller than the highest order. From this it follows that there is a one-parameter group of transformations $x=F_t(\xi)$ (with a corresponding transformation of $y$) under which the equation is invariant. This group is given by the equation $d\xi/dt=R(\xi)$ with the condition $\xi|_{t=0}=x$. $R$ is one solution of (7). There are three such independent groups corresponding to the three independent solutions of (7).

for calculating $\left(\frac{d}{dx}\right)^{-1}$.

We exhibit the first few coefficients $R_l$:

$$(10)\quad \begin{cases} R_0 = \frac{1}{2}, \\ R_1 = -\frac{1}{4}u, \\ R_2 = \frac{1}{16}(3u^2 - u''), \\ R_3 = -\frac{1}{64}(10u^3 - 10uu'' - 5(u')^2 + u^{\mathrm{IV}}), \\ R_4 = \frac{1}{256}(35u^4 - 70u(u')^2 - 70u^2u'' + 21(u'')^2 + \\ \qquad\qquad + 28u'u''' + 14uu^{\mathrm{IV}} - u^{\mathrm{VI}}), \\ R_5 = -\frac{1}{1024}(126u^5 - 630u^2(u')^2 + 504uu'u''' + 462(u')^2u'' + \\ \qquad + 378u(u'')^2 - 54u'u^{\mathrm{V}} - 114u''u^{\mathrm{IV}} - 69(u''')^2 - \\ \qquad\qquad - 420u^3u'' + 126u^2u^{\mathrm{IV}} - 18uu^{\mathrm{VI}} + u^{\mathrm{VIII}}. \end{cases}$$

We now write (6) in another form and show how it is connected with the Riccati equation. We introduce two functions:

$$(11)\qquad \chi_R = \frac{1}{2iR}, \qquad \chi_I = -\frac{\chi_R'}{2\chi_R}.$$

Then (6) transforms into the system

$$(12)\qquad \begin{cases} \chi_R' + 2\chi_R\chi_I = 0, \\ \chi_I' + \chi_R^2 - \chi_I^2 + u + \zeta = 0, \end{cases}$$

which can be combined into a single equation by setting $\chi = \chi_R \; = + i\varkappa_I$, so we obtain

$$(13)\qquad i\chi' + \chi^2 + u + \zeta = 0,$$

that is, a Riccati equation. If we expand $\chi$ by means of (13) as an asymptotic series in powers of $\sqrt{\zeta}$, then, as is easily seen from (12), all the odd powers refer to $\chi_R$ and the even ones to $\chi_I$. For the coefficients of the expansion

$$(14)\qquad \chi = \sum_{l=-1}^{\infty} \frac{\chi_l}{(\sqrt{\zeta})^l} = \sum_{l=-1}^{\infty} \frac{\chi_l^{(R)}}{\zeta^{l+\frac{1}{2}}} + i\sum_{l=0}^{\infty} \frac{\chi_l^{(I)}}{\xi^l}$$

we obtain the recurrence formulae

$$(15)\qquad \begin{cases} \chi_{-1} = i, \quad \chi_0 = 0, \quad \chi_1 = iu/2, \\ \chi_l = -\frac{1}{2}\chi_{l-1}' + \frac{i}{2}\sum_{r=1}^{l-2} \chi_r\chi_{l-1-r}. \end{cases}$$

**3. Variational relations between the coefficients of the asymptotic form.**

FIRST RELATION.

$$\frac{\delta}{\delta u} R(x;\zeta) = \frac{\partial}{\partial \zeta} R(x;\zeta), \tag{16}$$

or in terms of the coefficients $R_l$;

$$\frac{\delta}{\delta u} R_l[u] = -\left(l-\frac{1}{2}\right) R_{l-1}[u]. \tag{16'}$$

A purely algebraic proof of this relation, based on the formulae obtained previously for the coefficients $R_l[u]$, is given in Appendix 1. Here we give a very simple proof, using the spectral representation of the resolvent.

Let $R(x, y; \zeta)$ be a resolvent for some fixed self-adjoint boundary conditions at the ends of an arbitrary interval containing $x$. The coefficients $R_l[u]$, as we know, do not depend on the specific form of the resolvent. If $\lambda_n$ are the eigenvalues of the boundary problem and $\varphi_n$ are the eigenfunctions, then

$$R(x;\zeta) = \sum \frac{\varphi_n^2(x)}{\lambda_n+\zeta}.$$

On variation we obtain

$$\frac{\delta}{\delta u}\int R(x;\zeta)\,dx = \frac{\delta}{\delta u}\sum \frac{1}{\lambda_n+\zeta} = -\sum \frac{\delta\lambda_n}{\delta u}\cdot\frac{1}{(\lambda_n+\zeta)^2}.$$

By perturbation theory,

$$\Delta\lambda_n = \int \varphi_n^2(x)\,\delta u(x)\,dx + O(\delta u^2),$$

from which, by definition of the variational derivative $\frac{\delta\lambda_n}{\delta u} = \varphi_n^2(x)$, we obtain

$$\frac{\delta}{\delta u}\int R(x;\zeta)\,dx = -\sum \varphi_n^2(x)\cdot\frac{1}{(\lambda_n+\zeta)^2} = \frac{\partial}{\partial\zeta}\sum\frac{\varphi_n^2(x)}{(\lambda_n+\zeta)} = \frac{\partial R(x,\zeta)}{\partial\zeta}.$$

In accordance with the accepted usage we replace $\frac{\delta}{\delta u}\int \ldots dx$ by $\frac{\delta}{\delta u} R$, which proves the relation.

SECOND RELATION.

$$\frac{\partial}{\partial u} R_l[u] = -\left(l-\frac{1}{2}\right) R_{l-1}[u]. \tag{17}$$

This relation can be established without difficulty by induction, using the recurrence formulae (8).[1] Symbolically it can be written as

$$\frac{\partial}{\partial u} R(x;\zeta) = \frac{\partial}{\partial\zeta} R(x;\zeta). \tag{17'}$$

(We say symbolically because this equation is meaningful only for asymptotic series; otherwise it is not clear how to define $\partial/\partial u$-differentiation

---

[1] (16′) and (17) show that $\frac{\delta}{\delta u} R_l = \frac{\partial}{\partial u} R_l$ or $p_0 R_l = 0$, where $p_0$ is the first momentum operator (Chapter 1.) The note on page 85 makes it clear that this is a consequence of the fact that $R_l$ is the variational derivative.

with respect to $u$ for fixed $u'$, $u''$, ... ). Intuitively, the equation is obvious; it is a consequence of the fact that $R(x, \zeta)$ depends on $u$ and $\zeta$ not separately but only through their sum.

EXAMPLE. $R_2 = \frac{1}{16}(3u^2 - u'')$, $R_3 = -\frac{1}{64}(10u^3 - 10uu'' - 5(u')^2 + u^{\text{IV}})$. Here the relation can be verified without difficulty.

THIRD RELATION

$$\frac{\delta}{\delta u}\chi(x;\zeta) = \frac{\partial}{\partial \zeta}\chi(x;\zeta) + \frac{d}{dx}(\ ), \tag{18}$$

where $\frac{d}{dx}(\ )$ denotes the derivative of a function that is expanded as an asymptotic series whose coefficients are polynomials in $u$, $u''$ ... In terms of the coefficients (18) means that

$$\frac{\delta}{\delta u}\chi_l[u] = -\frac{l-2}{2}\chi_{l-2}[u] + \frac{d}{dx}(\ ). \tag{18'}$$

PROOF. First we show that

$$\frac{\partial}{\partial u}\chi_l[u] = -\frac{l-2}{2}\chi_{l-2}[u], \tag{19}$$

or, symbolically,

$$\frac{\partial}{\partial u}\chi(x;\zeta) = \frac{\partial}{\partial \zeta}\chi(x;\zeta). \tag{19'}$$

(19) is proved by induction, using the recurrence formula (15). The intuitive meaning is the same as that of (17) and (17').

It remains for us to note that the partial derivative $\partial\chi_l/\partial u$ coincides with the variational derivative to within terms of the form $\frac{d}{dx}(\ )$.

The relation obtained for $\chi_l$ is weaker than the relation for $R_l$ by the presence of an additional undetermined term of the type of a derivative.

FOURTH RELATION ([15]).

$$\frac{\delta}{\delta u}\chi_R(x;\zeta) = -\frac{1}{2\chi_R(x;\zeta)}. \tag{20}$$

We first prove a weaker identity having an additional term of the type of a derivative.

The function $\chi_R$ satisfies the equation obtained by eliminating $\chi_I$ from the system (12):

$$\frac{\chi_R''}{2\chi_R} - \frac{3}{4}\frac{(\chi_R')^2}{\chi_R^2} + \chi_R^2 + u + \zeta = 0. \tag{21}$$

We differentiate this equation with respect to $\zeta$ and, for simplicity, write $\chi_R = f$, $\partial\chi_R/\partial\zeta = g$:

$$\frac{g''}{2f} - \frac{f''g}{2f^2} - \frac{3}{2}\frac{f'g'}{f^2} + \frac{3}{2}\frac{f'^2g}{f^3} + 2fg + 1 = 0,$$

$$g = -\frac{1}{2f} - \frac{g''f^2 - ff''g - 3ff'g' + 3f'^2g}{4f^4} = -\frac{1}{2f} - \frac{1}{4}\left(\frac{g'f - gf'}{f^3}\right)',$$

that is,

$$\frac{\partial \chi_R}{\partial \zeta} = -\frac{1}{2\chi_R} + \frac{d}{dx}(\ ).$$

Now we take the variational derivative, using the connection (11) between $\chi_R$ and $R$ and the first variational relation:

$$\frac{\partial}{\partial \zeta}\frac{\delta}{\delta u}\chi_R = -\frac{\delta}{\delta u}\frac{1}{2\chi_R} = -i\frac{\delta}{\delta u}R = -i\frac{\partial R}{\partial \zeta} = -\frac{1}{2}\frac{\partial}{\partial \zeta}\frac{1}{\chi_R},$$

that is,

(22)
$$\frac{\partial}{\partial \zeta}\left(\frac{\delta}{\delta u}\chi_R + \frac{1}{2\chi_R}\right) = 0,$$

from which (21) follows, since the expression in brackets tends to zero as $\zeta \to \infty$.

The relation (21) can be rewritten as

(23)
$$\frac{\delta}{\delta u}\chi_R = -iR$$

or, in terms of the coefficients,

(24)
$$\frac{\delta}{\delta u}\chi_l^{(R)} = -iR_l.$$

Comparing this with (16) and (17), we see that the $R_l$ are obtained as the variational derivatives of $\chi_l^{(R)}$ and $R_{l+1}$. Compared with $\chi(x; \zeta)$, the function $R(x; \zeta)$ has the advantage that it is self-reproducing under the action of the operator of variational derivative. As we shall see below, what is important in fact, are the variational derivatives of the coefficients $R_l$ (or $\chi_l$), that is, the coefficients $R_l$.

**4. The basic property of the coefficients $R_l$.** The main application of the coefficients $R_l$ to the Korteweg–de Vries equation rests on the following result.

THEOREM. *The quantities $R_k R_l'$ are derivatives, that is, there exist polynomials $P_{k,l}[u]$ in $u$, $u'$ ... such that*

(25)
$$R_k R_l' = \frac{d}{dx} P_{k,l}.$$

PROOF. By means of the recurrence relation (9) it can easily be verified that

$$R_k R_{l+1}' - R_{k+1}' R_l = \left(\frac{1}{4} R_k R_l'' + \frac{1}{4} R_k'' R_l - R_k' R_l' - u R_k R_l\right)'.$$

Moreover, it is clear that $R_k R_l' + R_k' R_l = (R_k R_l)'$. Therefore, the quantity $R_k R_l'$ differs from $R_{k+1} R_{l-1}'$ by the sign and by a term of the type of a derivative. The chain of these equations can be continued until the second index becomes zero. But $R_0' = 0$.

COROLLARY. $\dfrac{\delta}{\delta u}(R_k R_l') = 0.$

Since the quantities $P_{k,l}$ play an important role in what follows, some formulae for them will be useful. We can write down a generating function

$$P(x;\zeta_1,\zeta_2)=\sum_{k,\,l=0}^{\infty}\frac{P_{k,\,l}[u]}{\zeta_1^{k+\frac{1}{2}}\zeta_2^{l+\frac{1}{2}}}. \tag{26}$$

In terms of this generating function, (25) indicates that

$$R(x;\zeta_1)\,R'(x;\zeta_2)=\frac{d}{dx}P(x;\zeta_1,\zeta_2). \tag{27}$$

We now use the third order equation (7):

$$-R'''(x;\ \zeta_1)+4(u+\zeta_1)R'(x;\ \zeta_1)+2u'R(x;\ \zeta_1)=0,$$
$$-R'''(x;\ \zeta_2)+4(u+\zeta_2)R'(x;\ \zeta_2)+2u'R(x;\ \zeta_2)=0.$$

Multiplying the first equation by $R(x;\zeta_2)$, the second by $R(x;\zeta_1)$, and adding, we obtain

$$\begin{aligned}(28)\qquad & 2(\zeta_1-\zeta_2)[P(x;\ \zeta_1,\zeta_2)-P(x;\ \zeta_2,\zeta_1)]=\\ &=-R''(x;\ \zeta_1)R(x;\ \zeta_2)-R(x;\ \zeta_1)R''(x;\ \zeta_2)+R'(x;\ \zeta_1)R'(x;\ \zeta_2)+\\ &\quad+4uR(x;\ \zeta_1)R(x;\ \zeta_2)+2(\zeta_1+\zeta_2)R(x;\ \zeta_1)R(x;\ \zeta_2)+c(\zeta_1,\zeta_2).\end{aligned}$$

Here $c$ is a quantity independent of $x$. It is completely determined by the condition that the right-hand side is divisible by $\zeta_1-\zeta_2$. To evaluate it, we must replace $R$ on the right-hand side by the first term of the expansion $1/(2\sqrt{\zeta})$ and require that the result should vanish. We compute without trouble that $c(\zeta_1,\zeta_2)=-(\zeta_1+\zeta_2)/2\sqrt{\zeta_1\zeta_2}$. Now (28) defines the antisymmetric part of the generating function $P(x;\zeta_1,\zeta_2)$. The symmetric part can be obtained more simply:

$$P(x;\ \zeta_1,\ \zeta_2)+P(x;\ \zeta_1,\ \zeta_2)=R(x;\ \zeta_1)R(x;\ \zeta_2). \tag{29}$$

# CHAPTER 3

## The Korteweg–de Vries equation

**1. The Novikov equations (the stationary Korteweg-de Vries equation).** We define a Novikov equation (or stationary Korteweg-de Vries equation) as an ordinary differential equation of the form

$$\sum_{k=0}^{n+1}c_kR_k[u]=0 \tag{1}$$

for the function $u$, where the $c_k$ are arbitrary fixed constants and $c_{n+1}=1$. The order of this equation is $2n$. By (17) in Chapter 2 this equation is equivalent to

$$\frac{\delta}{\delta u}\sum_{k=0}^{n+2}d_kR_k[u]=0,\qquad d_k=-c_{k-1}\Big/\left(k-\frac{1}{2}\right), \tag{2}$$

which is the Euler–Lagrange variational equation for the Lagrangian $L(u,u',\ldots,u^{(2n+2)})=\sum_{k=0}^{n+2}d_kR_k$. Hence we can apply here the theory of

Chapter 1. First we construct a Hamiltonian. This satisfies $\frac{dH}{dx} = -u' \frac{\delta}{\delta u} L$. This is the general formula. But in our case, when the Lagrangian is a linear combination of the $R_k$, we can go further and write down explicitly not only the derivative of the Hamiltonian, but also the Hamiltonian itself. We observe that the correspondance between Lagrangians and Hamiltonians is linear. Therefore, it is sufficient to consider the case when the Lagrangian is equal to a single $R_k$. We can write down a generating function for the Hamiltonians $H_k$ corresponding to the Lagrangians $R_k$:

$$H(x;\zeta) = \sum_{k=0}^{\infty} \frac{H_k[u]}{\zeta^{k+\frac{1}{2}}}.$$

Then

$$\frac{dH}{dx} = -u' \frac{\delta}{\delta u} R = -u' \frac{\partial R}{\partial \zeta}. \tag{3}$$

Recalling the third order equation (7) of Chapter 2, we obtain

$H' = -\frac{\partial}{\partial \zeta}\left[\frac{1}{2} R''' - 2(uR)' - 2\zeta R'\right]$ ; hence

$$H = \frac{\partial}{\partial \zeta}\left(\frac{1}{2} R'' - 2uR - 2\zeta R\right),$$

and this means that

$$H_k = (2k-1)\left(R_k - \frac{1}{4} R''_{k-1} + uR_{k-1}\right). \tag{4}$$

This formula expresses $H_k$ in terms of $R_k$. From (3) we obtain

38

$$H'_k = \left(k - \frac{1}{2}\right) u' R_{k-1}, \tag{5}$$

which enables us to express $R_k$ in terms of $H_k$.

2. **First integrals of the Novikov equations.** The system (1) or (2) has $n$ independent first integrals. For multiplying (1) by $R'_l$ and using (25) of Chapter 2 we have

$$\frac{d}{dx} \sum_{k=0}^{n+1} c_k P_{k,l}[u] = 0,$$

that is, the quantities

$$I_l = \sum_{k=0}^{n+1} c_k P_{k,l}[u] \qquad (l = 1, 2, \ldots, n) \tag{4}$$

are invariant.

The independence of these integrals is established as follows: the $l$-th integral contains the term $\pm[u^{(n+l-1)}]^2 \cdot 2^{-2(n+l+1)}$. The remaining terms of this integral and also all the preceding integrals can contain $u^{(n+l-1)}$ only in products by lower derivatives. If we now put all the variables equal to zero except for $u^{(n+l-1)}$, we see that $I_l$ is independent of all the preceding integrals. The independence of all the integrals in the set follows from this. We note, that in (6) we can take $l$ greater than $n$ and still obtain first integrals, but they are no longer independent of the preceding ones.

The vector fields $\xi_{I_l} = \xi_l$ corresponding to the integrals $I_l$ according to Theorem 3 in Ch.1 §7, can be found as follows. We must find a characteristic of each invariant, that is, an element $a$ such that $\frac{dI_l}{dx} = a\,\frac{\delta}{\delta u}\,L$. In our case $a = R_l'$. Then the vector field is $\xi_l = -\sum_i R_l^{(i+1)}\,\frac{\partial}{\partial u^{(i)}}$. This equation must hold in $A_L$, that is, for solutions of the equation. Thus, the vector field corresponding to the invariant $I_l$ is

$$\xi_l = -\sum_{i=0}^{\infty} R_l^{(i+1)} \frac{\partial}{\partial u^{(i)}}, \tag{7}$$

where the derivatives higher than $u^{(2n-1)}$ in the coefficients must be eliminated by means of the equation. It is remarkable that the vector field $\xi_1$ in this form does not depend on the equation, only the method of eliminating the higher derivatives depends on the equation.

The commutator of two such vector fields is equal to

$$[\xi_l, \xi_m] = \sum_i \sum_k \left( R_l^{(k+1)} \frac{\partial R_m^{(i+1)}}{\partial u^{(k)}} - R_m^{(k+1)} \frac{\partial R_l^{(i+1)}}{\partial u^{(k)}} \right) \frac{\partial}{\partial u^{(i)}}. \tag{8}$$

In [14] Novikov states that integrals that are polynomials in $u, u', u'', \ldots$, must be in involution, that is, their Poisson brackets vanish, or the corresponding vector fields commute. This proposition reduces to the fact that the right-hand sides of (8) vanish, except for the leading derivatives with the help of any of the equations for all $n$. Hence these right-hand sides must be identically equal to zero, with the exception of the leading derivatives:

$\sum_k \left( R_l^{(k+1)} \frac{\partial R_m^{(i+1)}}{\partial u^{(k)}} - R_m^{(k+1)} \frac{\partial R_l^{(i+1)}}{\partial u^{(k)}} \right) = 0$ $(i = 0, 1, \ldots)$. This is equivalent to the single identity

$$\sum_{k=0}^{\infty} \left( R_l^{(k+1)} \frac{\partial R_m}{\partial u^{(k)}} - R_m^{(k+1)} \frac{\partial R_l}{\partial u^{(k)}} \right) = 0. \tag{9}$$

A general proof of this identity will be published separately. It is trivial to verify for $l = 1$, when $I_l = H$. For $R_1 = -\frac{1}{4}\,u$ and

$$\sum_{k=0}^{\infty} \left( u^{(k+1)} \frac{\partial R_m}{\partial u^{(k)}} - R_m^{(k+1)} \frac{\partial u}{\partial u^{(k)}} \right) = R_m' - R_m' = 0.$$

It can also easily be verified for $l = 2$ and arbitrary $m$.

In conclusion we give a table of invariants for the first few Novikov equations.

1) The Lagrangian $L = R_3$. The equation is $3u^2 - u'' = 0$. The invariant is $I_1 = H = u^3 - \frac{1}{2}\,(u')^2$.

2) The Lagrangian $R_4$. The equation is

$10u^3 - 10uu'' - 5(u')^2 + u^{IV} = 0$. The invariants are:

$$I_1 = H = \frac{5}{2}u^4 - 5u(u')^2 + u'''u' - \frac{1}{2}(u'')^2$$

and

$$I_2 = 12u^5 - 10u^3u'' - 15u^2(u')^2 + 6uu'u''' - (u')^2u'' + 2u(u'')^2 - \frac{1}{2}(u''')^2.$$

3) The Lagrangian $R_5$. The equation is

$$35u^4 - 70uu'^2 - 70u^2u'' + 21(u'')^2 + 28u'u''' + 14uu^{IV} - u^{VI} = 0.$$

The invariants are

$$I_1 = H = 7u^5 - 35u^2(u')^2 + 14(u')^2u'' + 14uu'u''' - 7u(u'')^2 - u'u^V + u''u^{IV} - \frac{1}{2}(u''')^2.$$

$$I_2 = 35u^6 - 35u^4u'' - 140u^3(u')^2 + 84u^2u'u''' - \\ - 7u^2(u'')^2 + 70u(u')^2u'' - \frac{35}{2}(u')^4 - 6uu'u^V + 6(u')^2u^{IV} + \\ + 6uu''u^{IV} - 10u(u''')^2 - 18u'u''u''' + 6(u'')^3 + u'''u^V - \frac{1}{2}(u^{IV})^2.$$

$$I_3 = 150u^7 - 350u^5u'' - 525u^4(u')^2 + 280u^3u'u''' + \\ + 35u^4u^{IV} + 210u^3(u'')^2 + 700u^2(u')^2u'' - 30u^2u'u^V - \\ - 10u(u')^2u^{IV} - 40u^2u''u^{IV} + 10(u')^3u''' - 180uu'u''u''' - \\ - 50u^2(u''')^2 - 205(u')^2(u'')^2 - 10u(u'')^3 + 20u'u''u^V + \\ + 10uu'''u^V + (u'')^2u^{IV} - 2u'u'''u^{IV} + 2u(u^{IV})^2 - \frac{1}{2}(u^V)^2.$$

**3. The Korteweg–de Vries equation.** We assume that the function $u$ depends additionally on a parameter $t$. As before, we use dashes to denote the derivatives with respect to $x$. (Generalized) Korteweg–de Vries equations are of the form

$$\frac{\partial u}{\partial t} = \frac{\partial}{\partial x}\sum_{k=0}^{n+1} c_k R_k[u]. \tag{10}$$

Thus, Novikov equations are Korteweg–de Vries equations for stationary functions $u$ (independent of $t$). The original (non-generalized) Korteweg–de Vries equation is the particular case of (10)

$$\frac{\partial u}{\partial t} = \frac{\partial}{\partial x}R_2[u].$$

In contrast to the preceding, we must now choose a class of functions $u$. In fact, the integral $\int \ldots dx$ must have a meaning and an integral of a function of the type of a derivative must vanish. For example, the functions may be damped at infinity, or periodic, or almost periodic, and so on.

THEOREM. *The quantities*

$$I_l = \int R_l[u]\,dx \tag{11}$$

*are invariants of the solutions of* (10).

PROOF.

$$\frac{dI_l}{dt}=\int \frac{\partial}{\partial t} R_l\,[u]\,dx=\int \sum_{k=0}^{\infty} \frac{\partial R_l}{\partial u^{(k)}} u_t^{(k)}\,dx=\int \Big(\sum_{k=0}^{\infty} (-1)^k \Big(\frac{d}{dx}\Big)^k \frac{\partial R_l}{\partial u^{(k)}}\Big)\cdot u_t\,dt=$$

$$=\int \frac{\delta}{\delta u} R_l\,[u]\cdot \sum_{k=0}^{n+1} c_k R_k'\,[u]\,dx=-\Big(l-\frac{1}{2}\Big)\sum_{k=0}^{n+1} c_k \int R_{l-1}\,[u]\,R_k'\,[u]\,dx=0,$$

since $R_{l-1}R_k'$ is an expression of the type of a derivative.[1]

Since the paper of Lax [12] it has been usual to base the theory of the Korteweg–de Vries equation on the construction of so-called $(\mathscr{L}, A)$-pairs. We have avoided this above, but for completeness we now describe this method. Let $\mathscr{L}$ be the operator

$$\mathscr{L}=-\frac{d^2}{dx^2}+u\,(x).$$

Let $A$ be a differential operator of arbitrary order whose coefficients are polynomials in $u$, $u'$, ..., whose commutator with $\mathscr{L}$ is the operator of multiplication by some function $\widetilde{A}[u]$. Then we call $u_t = \widetilde{A}[u]$ a generalized Korteweg–de Vries equation. It is equivalent to the operator equation $\mathscr{L}_t = [A, \mathscr{L}]$. Hence it follows that if $u$ is a solution of $u_t = \widetilde{A}[u]$, then under a change of the parameter $t$ the operator $\mathscr{L}$ is transformed into a similar operator $\mathscr{L}(t) = T^{-1}(t)\,\mathscr{L}(0)\,T(t)$, and its spectrum is invariant. Therefore, the quantities (11) must be invariants of the equation, because they are the coefficients of the asymptotic expansion of the trace of the resolvent and therefore expressible in terms of the spectrum.

How can such an operator $A$ be constructed? One of the possible ways is given in [15]. These operators appear in the expansion of $\zeta$,

$$\mathscr{A}=\Big(\frac{1}{2}R\frac{d}{dx}-\frac{1}{4}R'\Big)(\mathscr{L}+\zeta)^{-1}=\sum_{k=0}^{\infty}\frac{A_k}{\zeta^{k+\frac{1}{2}}}$$

in powers of $\sqrt{\zeta}$. The fact that the resulting operators have the required property follows from the formula $[\mathscr{L}, \mathscr{A}] = R'(x;\zeta)$, which is most

---

1) In an infinite-dimensional space we can introduce a symplectic structure by means of the 2-form (see [13]):

$$\omega\,(\delta u, \delta v)=\int_{-\infty}^{\infty}\int_{-\infty}^{x} [\delta u\,(x)\,\delta v\,(x_1)-\delta u\,(x_1)\,\delta v\,(x)]\,dx_1\,dx.$$

Then for any functional $H = \int H\,[u]\,dx$ as Hamiltonian, we can construct the Hamiltonian equation $u_t = \frac{\partial}{\partial x}\frac{\delta}{\delta u}H$. The equation (10) corresponds to the Hamiltonian $H = \sum_{k=0}^{n+1} \frac{c_k}{-(k+\frac{1}{2})} R_{k+1}$. The fact that $R_l R_k'$ is an expression of the type of a derivative is in this context equivalent to the fact that the Poisson brackets of the functionals (11) vanish. Therefore, the quantities $I_l$ are integrals in involution of the Korteweg–de Vries equation.

easily verified using the third order equation (7) in Chapter 2:

$$[\mathcal{L}, \mathcal{A}] = \frac{1}{2}\left[\mathcal{L}, R\frac{d}{dx} - \frac{1}{2}R'\right](\mathcal{L}+\zeta)^{-1} = \frac{1}{4}\left(R''' - 4R'\frac{d^2}{dx^2} - 2Ru'\right)(\mathcal{L}+\zeta)^{-1} =$$

$$= \frac{1}{4}\left(4uR' - 4R'\frac{d^2}{dx^2}\right)(\mathcal{L}+\zeta)^{-1} = R'(\mathcal{L}+\zeta)(\mathcal{L}+\zeta)^{-1} = R' = \frac{d}{dx}\sum_{l=0}^{\infty}\frac{R_l}{\zeta^{l+\frac{1}{2}}}.$$

So we arrive at the equation $\frac{\partial}{\partial t}u = \frac{\partial}{\partial x}R_l$, that is, the Korteweg–de Vries equation.

## APPENDIX 1

### A generating function for the polynomials $R_l[u]$

The expressions (4) in Chapter 2 for the coefficients of the polynomials $R_l[u]$ are complicated and difficult to apply in practice. We now write the polynomials $R_l[u]$ in symbolic form and find for them a generating function of a completely transparent form.

We denote by $R_l^{(N)}[u]$ the homogeneous part of $R_l[u]$ of degree $N$ in the variables $u, u', u'' \dots$ In agreement with the symbolic notation of Chapter 1, §2, the function $R_l^N[u]$ is associated with a polynomial $\widetilde{R}^{(N)}(\xi_1, \dots, \xi_N)$. Specifically,

$$(1)\qquad \widetilde{R}_l^{(N)}(\xi_1, \xi_2, \dots, \xi_N) = \frac{(-1)^N}{2^{N+1}} S_N\left(\sum_{k_1+\dots+k_N=2l-2N} M_{k_1\dots k_N}\xi_1^{k_1}\dots\xi_N^{k_N}\right).$$

We consider the series

$$(2)\qquad \widetilde{R}^{(N)}(\xi_1, \xi_2, \dots, \xi_N; \zeta) = \sum_{l=N}^{\infty}\frac{R_l^{(N)}(\xi_1, \xi_2, \dots, \xi_N)}{\zeta^{l+\frac{1}{2}}}.$$

whose coefficients in the expansion in powers of $\sqrt{\zeta}$ are given by (1). We transform the expressions (4) for $M_{k_1\dots k_N}$ in Chapter 2 using the Fourier transform

$$\frac{1}{2\sqrt{\zeta}}e^{-\sqrt{\zeta}|\eta|} = \frac{1}{2\pi}\int\frac{e^{i\alpha\eta}}{\alpha^2+\zeta}d\alpha.$$

Then we have

$$(3)\qquad \widetilde{R}^{(N)}(\xi_1, \xi_2, \dots, \xi_N; \zeta) = \frac{(-1)^N}{(2\pi)^{N+1}}S_N\left(\int\dots\int\frac{e^{i\alpha_1\eta_1+i\alpha_2(\eta_1-\eta_2)+\dots+i\alpha_{N+1}\eta_N}}{(\alpha_1^2+\zeta)\dots(\alpha_{N+1}^2+\zeta)}\times\right.$$
$$\times e^{\xi_1\eta_1+\dots+\xi_N\eta_N}d\alpha_1\dots d\alpha_{N+1}d\eta_1\dots d\eta_N.$$

We now evaluate these integrals. Integrating first with respect to $\eta_1, \ldots, \eta_N$, and then with respect to $\alpha_1, \ldots, \alpha_{N+1}$, we obtain

$$\tilde{R}^{(N)}(\xi_1, \ldots, \xi_N; \zeta) = \frac{(-1)^N}{2\pi} \times$$

$$\times S_N\left(\int \frac{\delta(i\xi_1+\alpha_1-\alpha_2)\,\delta(i\xi_2+\alpha_2-\alpha_3)\ldots\delta(i\xi_N+\alpha_N-\alpha_{N+1})}{(\alpha_1^2+\zeta)\ldots(\alpha_{N+1}^2+\zeta)}\,d\alpha_1\ldots d\alpha_{N+1}\right) = \frac{(-1)^N}{2\pi}\times$$

$$\times S_N\Big(\int \frac{d\alpha_{N+1}}{[-(\xi_1+\ldots+\xi_N+i\alpha_{N+1})^2+\zeta][-(\xi_2+\ldots+\xi_N+i\alpha_{N+1})^2+\zeta]\ldots[-(\xi_N+i\alpha_{N+1})^2+\zeta]\cdot[-(i\alpha_{N+1})^2+\zeta]}\Big).$$

Finally

(4) $$\tilde{R}^{(N)}(\xi_1, \ldots, \xi_N; \zeta) = \frac{(-1)^N}{2\sqrt{\zeta}} \times$$

$$\times s_N\Big(\sum_{r=1}^{N+1} \frac{1}{[\zeta-(\sqrt{\zeta}-\xi_1-\ldots-\xi_{r-1})^2]\ldots[\zeta-(\sqrt{\zeta}-\xi_{r-1})]^2[\zeta-(\sqrt{\zeta}+\xi_r)^2]\ldots[\zeta-(\sqrt{\zeta}+\xi_r+\ldots+\xi_N)^2]}\Big).$$

The first term is absent for $r = 1$, and the second for $r = N + 1$. We now give another form of this expression, as an expansion in elementary fractions:

(5) $$\tilde{R}^{(N)}(\xi_1, \xi_2, \ldots, \xi_N; \zeta) =$$

$$= \frac{1}{2\sqrt{\zeta}} \sum_{r=1}^{N+1} \sum_{\substack{k=1\\(r\neq k)}}^{N+1} (-1)^{N+r+k} b_r \cdot b_k \cdot \begin{cases} \dfrac{(\xi_k+\ldots+\xi_{r-1})^2}{4\zeta-(\xi_k+\ldots+\xi_{r-1})^2}, & k<r,\\[2ex] \dfrac{(\xi_r+\ldots+\xi_{k-1})^2}{4\zeta-(\xi_r+\ldots+\xi_{k-1})^2}, & k>r,\end{cases}$$

$$b_r = \prod_{j=1}^{r-1}(\xi_j+\ldots+\xi_{r-1})^{-1}\cdot\prod_{j=r}^{N}(\xi_r+\ldots+\xi_j)^{-1}.$$

In the expression for $b_r$ the first term is absent for $r = 1$ and the second for $r = N + 1$.

We show how the generating function $\tilde{R}^{(N)}(\xi_1, \ldots, \xi_N)$ is used to prove the basic variational relation (16) or (17) of Chapter 2. It is sufficient to prove this relation separately for each homogeneous part of $R_l[u]$, that is, for $R_l^{(N)}[u]$. We claim that

$$\frac{\tilde{\delta}}{\delta u}\tilde{R}_l^{(N)} = -\left(l-\frac{1}{2}\right)\tilde{R}_{l-1}^{(N-1)} \quad\text{or}\quad \frac{\tilde{\delta}}{\delta u}\tilde{R}^{(N)} = -\frac{\partial}{\partial\zeta}\tilde{R}^{(N-1)}.$$

To prove this we choose the generating function in the form (3), without integrating. Then

$$\frac{\tilde{\delta}}{\delta u}\tilde{R}^{(N)}(\xi_1, \ldots, \xi_{N-1}; \zeta) = \frac{(-1)^N N}{(2\pi)^{N+1}}\cdot\frac{1}{N!}\times$$

$$\times\sum_{(k)}\int\ldots\int \frac{e^{-i\alpha_1\eta_1+i\alpha_2(\eta_1-\eta_2)+\ldots+i\alpha_{N+1}\eta_N}}{(\alpha_1^2+\zeta)\ldots(\alpha_{N+1}^2+\zeta)}\,e^{\xi_1\eta_{k_1}+\xi_2\eta_{k_2}+\ldots+\xi_{N-1}\eta_{k_{N-1}}+(-\xi_1-\ldots-\xi_{N-1})\eta_{k_N}}\times$$

$$\times d\alpha_1\ldots d\alpha_{N+1}d\eta_1\ldots d\eta_N.$$

Here $(k) = (k_1, k_2, \ldots, k_N)$ comprises all permutations of the indices $1, \ldots, N$. We divide all the terms of the sum into three types, for $k_N = 1$, $k_N = N$ and $k_N \neq 1, N$. In the case $k_N = 1$ we set $\eta_k - \eta_2 = \tilde{\eta}_{ki}$ $(i = 1, \ldots, N-1)$. Then we have

$$\frac{(-1)^N}{(2\pi)^{N+1}} \cdot \frac{1}{(N-1)!} \times$$

$$\times \sum_{(k_1, \ldots, k_N)} \int \cdots \int \frac{e^{-i\alpha_1\eta_1 + i\alpha_2(-\tilde{\eta}_2) + i\alpha_3(\tilde{\eta}_2 - \tilde{\eta}_3) + \ldots + i\alpha_{N+1}\tilde{\eta}_N + i\alpha_{N+1}\eta_1}}{(\alpha_1^2 + \zeta) \ldots (\alpha_{N+1}^2 + \zeta)} \times$$

$$\times e^{\xi_1 \tilde{\eta}_{k_1} + \ldots + \xi_{N-1} \tilde{\eta}_{k_{N-1}}} d\alpha_1 \ldots d\alpha_{N+1} \, d\eta_1 \, d\tilde{\eta}_2 \ldots d\tilde{\eta}_N.$$

Integrating with respect to $\eta_1$, $\alpha_1$ and relabelling $\alpha_2, \ldots, \alpha_{N+1} \to \alpha_1, \ldots, \alpha_N$; $\tilde{\eta}_2, \ldots, \tilde{\eta}_N \to \eta_1, \ldots, \eta_{N-1}$, we obtain

$$-\frac{(-1)^{N-1}}{(2\pi)^N} \cdot \frac{1}{(N-1)!} \sum_{(k_1, \ldots, k_{N-1})} \int \cdots \int \frac{e^{-i\alpha_1\eta_1 + i\alpha_2(\eta_1 - \eta_2) + \ldots + i\alpha_N \eta_{N-1}}}{(\alpha_1^2 + \zeta) \ldots (\alpha_{N-1}^2 + \zeta)(\alpha_N^2 + \zeta)^2} \times$$

$$\times e^{\xi_1 \eta_{k_1} + \ldots + \xi_{N-1} \eta_{k_{N-1}}} d\alpha_1 \ldots d\alpha_N \, d\eta_1 \ldots d\eta_{N-1}.$$

This is equal to the term in the derivative $\frac{\partial}{\partial \zeta} \tilde{R}^{(N-1)}(\xi_1, \ldots, \xi_{N-1}; \zeta)$ arising from the differentiation with respect to $\zeta$ of the last bracket of the denominator. The other two cases $k_N = N$ and $k_N \neq 1, N$ are dealt with in exactly the same way. We obtain the terms in the derivative corresponding, respectively, to the first and the remaining brackets of the denominator. We omit the details, which differ little from the case above. This proves the formula.

## APPENDIX 2

### Another recurrence method for obtaining the coefficients of the asymptotic expansion of the resolvent

This method was proposed in [2], [3]. In [9] it was developed in full generality for arbitrary elliptic pseudo-differential operators on manifolds of any dimension.

The method is also based on an expansion of the kernel of the resolvent as a series in the iterations of the kernel of the simplest resolvent, for the equation $-\frac{d^2\varphi}{dx^2} + \zeta\varphi = 0$. But instead of the series used in Chapter 2 §1, now write down a series in which the factors of each term occur in a different order: on the right the iterations of the kernel $R_0$, and on the left the operators of multiplication by a function. Explicitly, we write

$$(1) \qquad R(x, y; \zeta) = \sum_{l=0}^{\infty} \sum_{m=0}^{l} (-1)^{\frac{l+m}{2}} B_{l,m}[u] D^m R_0^{\left(1+\frac{l+m}{2}\right)}(x, y; \zeta),$$

where

$$D = i\frac{d}{dx}, \qquad R_0^{(n)} = R_0 \circ R_0 \circ \ldots \circ R_0.$$

The sum is over pairs $l$, $m$ such that $l + m$ is even, the $B_{l,m}[u]$ are polynomials in $u, u', \ldots$, which are chosen so that the series (1) is asymptotic: if the series is truncated at $l = T$, then the difference $R - R_T$ between the left- and the right-hand sides is $O(\zeta^{-T/2})$. This can be done as follows. Both sides of (6) are multiplied on the left by $D^2 + u + \zeta$. On the right-hand side, using the commutation relations for the operators of differentiation and of multiplication by a function, the operators are permuted, so that they occur in the same order as in (1), that is, that on the left there is a multiplication by a function, then a differentiation, then the iterations of the kernel $R_0$. Here we must bear in mind

$$(D^2 + \zeta)R_0(x, y; \zeta) = \delta(x - y), \; (D^2 + u + \zeta)R(x, y; \zeta) = \delta(x - y),$$
$$(D^2 + \zeta)R_0^{(n)}(x, y; \zeta) = R_0^{n-1}(x, y; \zeta).$$

Then the $B_{l,m}$ are chosen so that all the terms are mutually annihilated. This leads to recurrence relations to determine the coefficients:

$$(2) \qquad \begin{cases} B_{0,0} \equiv 1, \; B_{l,m} \equiv 0 \quad \text{for} \quad m < 0 \quad \text{or} \quad l < 0, \\ B_{l,m} = -B''_{l-2,m} + uB_{l-2,m} + 2iB'_{l-1,m-1}. \end{cases}$$

Each term of (1) can be evaluated directly, by means of the Fourier transform

$$D^m (D^2 + \zeta)^{-n} \frac{1}{2\pi} \int e^{i\xi(x-y)}\, d\xi = \frac{1}{2\pi} \int (-\xi)^m (\xi^2 + \zeta)^{-n} e^{i\xi(x-y)}\, d\xi.$$

For $x = y$ these expressions are non-zero only when $m$ is even, and then they are equal to

$$\frac{1}{2\pi} \frac{\Gamma\left(\frac{m+1}{2}\right)\Gamma\left(n - \frac{m+1}{2}\right)}{\Gamma(n)} \zeta^{-n + \frac{m+1}{2}}$$

([22], 3.241). It is now straightforward to show that the series (1) has the required property of being asymptotic. We find, ultimately,

$$R(x; \zeta) = \sum_{l=0}^{\infty} \sum_{m=0}^{\infty} B_{l,m}[u] \cdot \frac{1}{2} \begin{pmatrix} \frac{l-1}{2} \\ \frac{l+m}{2} \end{pmatrix} (-1)^{\frac{l}{2}} \zeta^{-\frac{1}{2} - \frac{l}{2}}$$

Here $l$ and $m$ are even. We write the result as follows:

$$R_l[u] = \frac{1}{2} A_{2l,\, l-\frac{1}{2}}[u] \cdot (-1)^l,$$

where

$$A_{l,s}=\sum_{m=0}^{l}\binom{s}{\frac{l+m}{2}}B_{l,m}[u],$$

and $B_{l,m}$ is determined by the recurrence relation (2).[1]

The technique just developed throws a new light on the procedure for finding Lax pairs ($\mathscr{L}$, $A$) (see Ch.3 §3). To the asymptotic expansion of the resolvent (1) there corresponds the asymptotic expansion of the operators

$$(D^2+u+\zeta)^{-1}=\sum_{l=0}^{\infty}\sum_{m=0}^{l}(-1)^{\frac{l+m}{2}}B_{l,m}[u]\,D^m\,(D^2+\zeta)^{-1-\frac{l+m}{2}},$$

where the operators $D^2 + u + \zeta$ and $D^2 + \zeta$ must, of course, be taken under certain boundary conditions. The results are purely local and depend on the function $u(x)$ only in a neighbourhood of the point $x$ in question, and not on the choice of boundary conditions. It is simplest to assume zero boundary conditions at the ends of an interval containing $x$ and $u(x)$ to vanish in a neighbourhood of the end-points. The asymptotic character of the series is to be understood in the sense that if the series is truncated at the $l$-th stage, the remainder is $O(D^{-2-l})$. The latter means that if it is multiplied by $D^{2+l}$ the remainder is a bounded operator (for greater detail see [2]). Now if both sides of the equation are multiplied by $(-\zeta)^s$, and then integrated with respect to $z = -\zeta$ over a contour enclosing the whole spectrum of $D^2 + u$ and $D^2$ (but not $z = 0$; we may assume that zero is not a point of the spectrum), we obtain

$$(D^2+u)^s=\sum_{l=0}^{\infty}A_{l,s}[u]\,D^l\,(D^2)^{s-l},$$

where the $A_{l,s}[u]$ are the polynomials in $u$, $u'$, ... defined above; the series is asymptotic in the sense explained.

Now let $s = n + \frac{1}{2}$, where $n$ is an integer. We denote by $Q$ the part of the series corresponding to negative powers of $D$:

$$Q=\sum_{l=0}^{2n+1}A_{l,\,n+\frac{1}{2}}[u]\,D^{2n+1-l}.$$

The operator $(D^2 + u)^{n+\frac{1}{2}}$ commutes with $D^2 + u$, but

$$(D^2+u)^{n+\frac{1}{2}}=Q+A_{2n+2,\,n+\frac{1}{2}}D\,(D^2)^{-1}+O\,(D^{-2}).$$

[1] For comparison with Seeley's paper [9] it should be noted that there the coefficients $\gamma_{j,k}$ are connected with $B_{l,m}$ by the relations $\gamma_{k,j} = (-1)^{j-k}B_{j,\,2k-j}\,\xi^{2k-j}$. Seeley's recurrence relations are somewhat more complex: the difference occurs because these recurrence formulae arise from multiplying (1) by $D^2 + u + \zeta$ not on the left, as in our case, but on the right.

Hence

$$[Q, D^2+u] = -[A_{2n+2,\, n+\frac{1}{2}} D(D^2)^{-1}, D^2+u] + O(D^{-1})$$

or

$$[Q, D^2+u] = 2iA'_{2n+2,\, n+\frac{1}{2}} + O(D^{-1}).$$

The remaining terms must vanish, because the commutator of two differential operators can only be a differential operator. Thus, the $(2n+1)$-th order operator $Q$ has the required property – its commutator with $D^2+u$ is the operator of multiplication by a function,

$$[Q, D^2+u] = 2iA'_{2n+2,\, n+\frac{1}{2}} = 4i\frac{d}{dx}R_{n+1}[u].$$

As might be expected, the commutator is the right-hand side of the Korteweg–de Vries equation.

## References


[1] I. M. Gel'fand, On identities for eigenvalues of a differential operator of the second order, Uspekhi Mat. Nauk **11**:1 (1956), 191–198. MR **18**–129.

[2] L. A. Dikii, The zeta-function of an ordinary differential equation on a finite interval, Izv. Akad. Nauk SSSR Ser. Mat. **19** (1955), 187–200. MR **17**–619.

[3] L. A. Dikii, Trace formulae for Sturm–Liouville operators, Uspekhi Mat. Nauk **13**:3 (1958), 111–143. MR **20**–6655.

[4] I. M. Gel'fand and B. M. Levitan, On a simple identity for the eigenvalues of a differential operator of the second order, Dokl. Akad. Nauk SSSR **88** (1953), 593–596. MR **13**–558.

[5] S. Minakshisundaram, On a generalization of Epstein's zeta-function, Canadian J. Math. **1** (1949), 320–327. MR **11**–357.

[6] M. G. Krein, On a trace formula in perturbation theory, Mat. Sb. **33** (1953), 597–626. MR **15**–720.

[7] L. D. Fadeev, An expression for the trace of the difference between two singular differential operators of Sturm–Liouville type, Dokl. Akad. Nauk SSSR **115** (1957), 878–880. MR **20**–1029.

[8] V. S. Buslaev and L. D. Faddeev, On trace formulae for a singular differential Sturm–Liouville operator, Dokl. Akad. Nauk SSSR **132** (1960), 13–16.
= Soviet Math. Dokl. **1** (1960), 451–454.

[9] R. T. Seeley, The index of elliptic systems of singular integral operators, J. Math. Anal. Appl. **7** (1963), 289–309. MR **28** #2464.
= Matematika **12** (1968), 96–112.

[10] V. S. Buslaev, On the asymptotic behaviour of spectral characteristics of exterior problems for the Schrödinger operator, Izv. Akad. Nauk SSSR Ser. Mat. **39** (1975), 149–235.

[11] C. S. Gardner, J. M. Greene, M. D. Kruskal and R. M. Miura, A method for solving the Korteweg–de Vries equation, Phys. Rev. Letters **19** (1967), 1095–1097.

[12] P. Lax, Integrals of non-linear equations of evolution and solitary waves, Comm. Pure Appl. Math. **21** (1968), 467–490. MR **38**–3620.

[13] V. E. Zakharov and L. D. Faddeev, The Korteweg–de Vries equation as a completely integrable Hamiltonian system, Funktsional. Anal. i Prilozhen. **5**:4 (1971), 18–27. = Functional Anal. Appl. **5** (1971), 280–287.

[14] S. P. Novikov, The periodic problem for the Korteweg–de Vries equation. I, Funktsional. Anal. i Prilozhen. **8**:3 (1974), 54–66. = Functional Anal. Appl. **8** (1974), 236–246.

[15] B. A. Dubrovin, The periodic problem for the Korteweg–de Vries equation in a class of finite zone potentials, Funktsional. Anal. i Prilozhen. **9**:3 (1975), 41–52. = Functional Anal. Appl. **9** (1975), 215–223.

[16] V. A. Marchenko, The periodic Korteweg–de Vries problem, Mat. Sb. **95** (1974), 331–356.

[17] P. Lax, Periodic solutions of the Korteweg–de Vries equations, Lectures in Appl. Math. **15** (1974), 85–96.

[18] I. M. Gel'fand, The cohomology of infinite-dimensional Lie algebras, some questions of integral geometry, Actes Congrès Internat. Mathématiciens, Nice 1970, Vol.1, 35–111 (in English).

[19] A. M. Gabrielov, I. M. Gel'fand and M. V. Losik, Combinatorial calculation of characteristic classes, Funktsional. Anal. i Prilozhen. **9**:2 (1975), 12–28. = Functional Anal. Appl. **9** (1975), 103–115.

[20] V. F. Lazutkin and T. V. Pankratova, Normal forms and versal deformations for the Hill equation, Funktsional. Anal. i Prilozhen. **9**:4 (1975), 41– 47. = Functional Anal. Appl. **9** (1975), 306–311.

[21] C. Godbillon, Géométrie différentielle et mécanique analytique, Hermann, Paris 1969. MR **39** #3416.
Translation: *Differentsial'naya geometriya i analiticheskaya mekhanika*, Mir, Moscow 1973.

[22] I. S. Gradshtein and I. M. Ryzhik, *Tablitsy integralov, sum, ryadov i proizvedenii*, (Tables of integrals, sums, series and products), Fizmatgiz, Moscow 1963.

Received by the Editors, 26 May 1975

Translated by G. and R. Hudson

# PROOF OF A VARIATIONAL RELATION BETWEEN THE COEFFICIENTS OF THE ASYMPTOTIC EXPANSION OF THE RESOLVENT OF A STURM-LIOUVILLE EQUATION

B. V. Yusin

In [1] an important variational relation is introduced between the coefficients of the asymptotic expansion of the diagonal of the resolvent for a Sturm–Liouville equation (see (2) below). Two proofs are given there, one simple but not algebraic, the other algebraic but fairly complicated. In this note we present a simple algebraic proof. Let $R(x, y; \zeta)$ be the kernel of the resolvent of a Sturm–Liouville equation. Then we have the asymptotic expansion

$$R(x, x; \zeta) = \sum_{k=0}^{\infty} \frac{R_k[u]}{\zeta^{k+1/2}} \tag{1}$$

as $\zeta \to +\infty$ (see [1]). The coefficients $R_k[u]$ are polynomials in $u, u', u'', \ldots$, where $u$ is the coefficient of the Sturm–Liouville equation.

THEOREM.

$$\frac{\delta R_k[u]}{\delta u} = -\left(k - \frac{1}{2}\right) R_{k-1}[u], \tag{2}$$

*where* $\dfrac{\delta}{\delta u} = \sum_{i=0}^{\infty} \left(-\dfrac{d}{dx}\right)^i \dfrac{\partial}{\partial u^{(i)}}$.

We denote by $A$ the ring of polynomials in $u, u', u'', \ldots$ over a field of characteristic 0, and by $B$ the ring of formal power series in $\zeta^{-1/2}$ with coefficients in $A$. We regard $R = R(x, x; \zeta)$ as an element of $B$. It is completely determined by the equation

$$-2RR'' + (R')^2 + 4(u + \zeta) R^2 = 1 \tag{3}$$

(see [1]). The operators $\dfrac{d}{dx}, \dfrac{\partial}{\partial u^{(i)}}, \dfrac{\delta}{\delta u}$ are defined in $A$, and also in $B$ when the action is coefficientwise. Moreover, $\dfrac{\partial}{\partial \zeta}$ also acts in $B$. Let $A^{(1)}$ be the module of first differentials of $A$, as in [1], and $B^{(1)}$ the module whose elements are the formal sums

$$\sum_{k=0}^{\infty} \omega_k \zeta^{-k/2},$$

where $\omega_k \in A^{(1)}$. Now $dF \in B^{(1)}$ is defined for any $F \in B$, and the Lie derivative $\dfrac{d}{dx}$ in $B^{(1)}$, which is determined by $\dfrac{d}{dx}(P\omega) = \dfrac{d}{dx}(P)\,\omega + P\dfrac{d}{dx}\omega$ and $d\dfrac{d}{dx}d$, where $P \in B$ and $\omega \in B^{(1)}$.

LEMMA. *If $F \in B$, then $dF$ has a unique representation in the form $K\,du + \dfrac{d}{dx}\omega$, where $K \in B$ and $\omega \in B^{(1)}$. Here $K = \dfrac{\delta F}{\delta u}$.*

PROOF. $dF = \sum_{i=0}^{\infty} \dfrac{\partial F}{\partial u^{(i)}} du^{(i)}$, hence, by using the identity

$$G\,du^{(k)} = -G'\,du^{(k-1)} + \frac{d}{dx}(G\,du^{(k-1)})$$

we come to the required result. To prove the uniqueness we have to verify that $Pdu$ is a total differential. Let

$$P\,du=\frac{d}{dx}\sum_{i,\,j\geqslant 0} G_{ij}\,du^{(i)}\zeta^{-j/2}, \tag{4}$$

where $G_{ij}\in A$. We denote by $j_0$ any one of the $j$ for which there is an $i$ with $G_{ij}\neq 0$. Let $i_0$ be the largest index such that $G_{ij_0}\neq 0$. Then on the right-hand side of (4) there is the term $G_{i_0j_0}du^{(i_0+1)}\zeta^{-j_0/2}$ which cannot cancel out, but not on the left-hand side.

PROOF OF THE THEOREM. The identity (2) is equivalent to

$$\frac{\delta R}{\delta u}=\frac{\partial R}{\partial \zeta} \tag{5}$$

By the lemma, (5) is equivalent to

$$dR-\frac{\partial R}{\partial \zeta}du\in\frac{d}{dx}B^{(1)}. \tag{6}$$

To prove (6) we apply $d$ to $\frac{\partial}{\partial \zeta}$ in (3):

$$-R\,dR''-R''\,dR+R'\,dR'+4(u+\zeta)R\,dR+2R^2\,du=0, \tag{7}$$

$$-R\frac{\partial R''}{\partial \zeta}-R''\frac{\partial R}{\partial \zeta}+R'\frac{\partial R'}{\partial \zeta}+4(u+\zeta)R\frac{\partial R}{\partial \zeta}+2R^2=0. \tag{8}$$

We multiply (7) by $\frac{1}{2R^2}\frac{\partial R}{\partial \zeta}$ and apply the same process of "transferring dashes" from one factor to another, as in the proof of the lemma: we obtain

$$\left[\left(\frac{1}{2R}\frac{\partial R}{\partial \zeta}\right)''+\frac{R''-4(u+\zeta)R}{2R^2}\frac{\partial R}{\partial \zeta}+\left(\frac{R'}{2R^2}\frac{\partial R}{\partial \zeta}\right)'\right]dR-\frac{\partial R}{\partial \zeta}du\in\frac{d}{dx}B^{(1)}. \tag{9}$$

Removing the brackets we obtain

$$\left(\frac{\partial R''}{\partial \zeta}R+R''\frac{\partial R}{\partial \zeta}-R'\frac{\partial R'}{\partial \zeta}-4(u+\zeta)R\frac{\partial R}{\partial \zeta}\right)\frac{dR}{2R^2}-\frac{\partial R}{\partial \zeta}du\in\frac{d}{dx}B^{(1)}. \tag{10}$$

Using (8) we come to (6) as required, and hence to (2).

The author wishes to thank I. M. Gel'fand, L. A. Dikii, and Yu. I. Manin for their interest in this work.

## Reference

[1] I. M. Gel'fand and L. A. Dikii, The asymptotic behaviour of the resolvent of Sturm–Liouville equations, and the algebra of the Korteweg-de Vries equations, Uspekhi Mat. Nauk 30:5 (1975), 67–100.
= Russian Math. Surveys 30:5 (1975), 77–113.

Received by the Editors 20 April 1976

# NON-LINEAR EQUATIONS OF KORTEWEG–DE VRIES TYPE, FINITE-ZONE LINEAR OPERATORS, AND ABELIAN VARIETIES

B. A. Dubrovin, V. B. Matveev, and S. P. Novikov

The basic content of this survey is an exposition of a recently developed method of constructing a broad class of periodic and almost-periodic solutions of non-linear equations of mathematical physics to which (in the rapidly decreasing case) the method of the inverse scattering problem is applicable. These solutions are such that the spectrum of their associated linear differential operators has a finite-zone structure. The set of linear operators with a given finite-zone spectrum is the Jacobian variety of a Riemann surface, which is determined by the structure of the spectrum. We give an explicit solution of the corresponding non-linear equations in the language of the theory of Abelian functions.

Contents

## Introduction

In 1967 a remarkable mechanism was discovered relating some important non-linear wave equations with the spectral theory of auxiliary linear operators. This connection makes it possible, in a certain sense, to "integrate" these non-linear equations (see [18]). The first such equation, the famous Korteweg-de Vries (K-dV) equation, was reduced in [18] to the inverse scattering problem for the Schrödinger operator $L = -d^2/dx^2 + u(x)$, to solve the Cauchy problem for the K-dV equation in the class of rapidly decreasing functions $u(x)$. Subsequently, this mechanism was perfected and interpreted from different points of view by Lax [19], Zakharov and Faddeev [21], and Gardner [22]; later, other important non-linear equations were found to which a similar mechanism can also be applied. The first after the K-dV equation was the non-linear Schrödinger equation (Zakharov and Shabat [25], [26]) then the standard sine-Gordon equation, the Toda chain, the non-linear equation of the string, and a number of others (see [23]–[35]); for all these equations an analogue of the method of Gardner-Green-Kruskal-Miura makes it possible to "integrate" the Cauchy problem for rapidly decreasing functions of $x$, by means of the scattering theory for an auxiliary linear operator. In particular, this method enables us to investigate the asymptotic behaviour of solutions in time and to obtain some important particular solutions called "multisoliton", which describe the interaction of a finite number of "solitons" (isolated waves of the form $u(x-ct)$).

Up to the end of 1973 there were no papers in which this mechanism was used successfully to study the Cauchy problem for functions periodic in $x$, even for the original K-dV equation. This is not surprising: one of the reasons is that the inverse problem for the Schrödinger equation with periodic potential was, in essence, unsolved. There were no effective methods whatever of finding the potential from the spectral data. (For almost periodic potentials this problem had not even been raised.) Recently, Novikov, Dubrovin, Matveev and Its have published a number of papers (see [38]–[43], [46], [47]) in which a method is developed that permits us to find a broad set of exact solutions of the K-dV equation, periodic and almost periodic in $x$, which give a natural generalization of the multisoliton solutions. It is fairly clear that this set of solutions is dense among the periodic functions. This was noted in [42] and [43], but has not yet been established rigorously. The method of the present authors, which is expounded in this survey, required a substantial perfection of the algebraic mechanism mentioned above, and also an appeal to ideas of algebraic geometry. This method is applicable not only to the K-dV equation, but also to other non-linear equations of this type, for the investigation of the periodic problem, as was pointed out in [42] and [43]. The appropriate modifications of the method will be indicated in the survey.

At the International Congress of Mathematicians in Vancouver it became known that simultaneously with Novikov's first paper [38], the work of Lax [50] appeared, which contains as its principal result part of the main theorem of Novikov's paper [38]; Lax's proof is not constructive and differs from Novikov's method (see Ch. 2, §2, p. 84). Furthermore, Marchenko has completed his works [44] and [45] in which he develops a method of successive approximations to solutions of the K-dV equation, which is based on the spectral theory of the Schrödinger operator $L$. Some of his arguments overlap with specific technical arguments of [38].

In studying a periodic or almost-periodic problem by the method of the authors an important part is played by the class of potentials $u(x)$ for which the Schrödinger operator $L = -d^2/dx^2 + u(x)$ has only finitely many forbidden zones (lacunae or zones of instability) in the spectrum on the whole real $x$-axis.

We must bear in mind that the physical interpretation of the K-dV equation in the theory of non-linear waves (see [16]) is such that its natural Cauchy problem must be posed without definite boundary conditions (for example, periodic functions with a given period in $x$ are insufficient). We must find as wide as possible a set of solutions $u(x, t)$ belonging to various classes of functions that are bounded in $x$. The most natural such classes are, apart from the class of rapidly decreasing functions already mentioned, those of periodic functions with arbitrary periods in $x$, and also those of almost periodic functions of $x$ with an arbitrary group of periods (in any case, the dynamics in $t$ turns out to be almost-periodic a posteriori).

The formulation in which we shall solve the inverse problem for the Schrödinger operator $L$ automatically gives not only periodic, but also almost-periodic real and complex potentials and the corresponding solutions of the K-dV equation. Our approach, based on the consideration of finite-zone potentials, is closely connected both with the theory of the K-dV equation and with the algebraic geometry of Riemann surfaces and Abelian varieties (see Ch. 2). This is a constructive approach and enables us to obtain exact analytic solutions. It is noteworthy that the connection with the K-dV equation gives non-trivial new results in the theory of Abelian varieties itself, in particular, it leads to explicit formulae for the universal fibering of the Jacobian varieties of hyperelliptic Riemann surfaces (for example, even the fact that this covering space is unirational was not known hitherto).

Examples of finite-zone potentials (but in a context unrelated to algebraic geometry and the K-dV equation) have in the past aroused the interest of various mathematicians in isolation, beginning with Ince who noted in 1940 that the potential of the Lamé problem, which coincides with the doubly elliptic function of Weierstrass, has only one lacuna, but the $n(n+1)/2$ – fold Lamé potentials have $n$ lacunae (see [7]). In 1961, Akhiezer [10] began to construct examples of finite-zone potentials on the half-line $x \geqslant 0$ without knowing of Ince's work [7]. He proposed an interesting (in essence, algebraic-geometric) method of constructing the eigenfunctions without constructing the potentials themselves. The potentials which result from Akhiezer's construction, turn out under analytic continuation in $x$ to be even almost-periodic functions of $x$; true, he paid no attention to this and considered the whole problem only on the half-line and pursued the analysis to its conclusion only in the single-zone case. The development of Akhiezer's ideas, which was undertaken by Dubrovin [40] and Matveev and Its [41], plays a large part in the method of the present authors (see Ch. 2).

In 1965, Hochstadt [9], without knowing of Akhiezer's work [10], raised the inverse problem for finite-zone periodic potentials and began to solve it. He proved that there are no potentials other than elliptic functions with a single lacuna (for $n = 1$ this theorem is the converse of the result of Ince). For a number $n > 1$ of lacunae Hochstadt could only prove that a continuous $n$-lacuna potential is an infinitely smooth function. Such is the history of this problem prior to the recent papers covered in this survey.

We are particularly concerned with the properties of the one-dimensional Schrödinger (Sturm-Liouville) equation with an almost-periodic potential $u(x)$. Before the publication of [38] there were in the literature neither serious general results, nor integrable cases of this problem. In fact, the translation matrix $\hat{T}$, the law of dispersion $p(E)$, and the Bloch eigenfunctions $\psi_\pm$ are formally meaningless. Although the Bloch eigenfunction $\psi_\pm$ $(x, x_0, E)$ can here also be defined by requiring that the function

$$\chi(x,\ E) = \frac{d \ln \psi_\pm}{dx}$$

has the same group of periods as the potential (instead of the condition

$$\psi_\pm(x+T) = e^{\pm ip(E)}\psi_\pm(x)),$$

it is no longer clear whether it exists for all real and complex values of the spectral parameter $E$, $L\psi_\pm = E\psi_\pm$ (the strongest results on this problem were obtained quite recently in [59]); the usual spectrum of the operator $L$ in $\mathcal{L}_2(-\infty, \infty)$ is obtained when $\psi_\pm$ is bounded, but we need it for all complex values of $E$. We shall study the class of real and complex almost-periodic potentials $u(x)$ that possess what we call *correct analytic properties*: for these an analogue of the Bloch eigenfunction $\psi_\pm(x, x_0, E)$ exists for all complex $E$ and it has the following properties (see Ch. 2, §2):

1) $L\psi_\pm = E\psi_\pm$;
2) $\psi_\pm(x, x_0, E)\,|_{x=x_0} \equiv 1$;
3) $\psi_\pm \sim e^{\pm ik(x-x_0)}$ as $E \to \infty$, $k^2 = E$;
4) $\psi_\pm = \dfrac{d \ln \psi_\pm}{dx}$ is almost periodic with the same group of periods as the potential $u(x)$;
5) $\psi_\pm(x, x_0, E)$ is meromorphic on a Riemann surface $\Gamma$ doubly covering the $E$-plane and $\psi_\pm$ and $\psi_-$ are obtained from one another by exchanging the two sheets.

It is permitted here also to consider potentials with poles at certain points; in this case, if the potential is unbounded, we assume the function $u(x)$ to be complex analytic (meromorphic) in a certain strip $x + iy$ around the $x$-axis.

We say that the potential $u(x)$ (or the operator $L$) is *finite-zoned* (finitely lacunary) if the Riemann surface $\Gamma$ has finite genus. $\Gamma$ itself is called the *spectrum* of the operator $L$, its *genus* is the number of forbidden zones (or lacunae), and the *branch-points* are "boundaries of the zones" (or lacunae), although a direct spectral interpretation of these concepts is valid only in the case of bounded real potentials $u(x)$.

Novikov in [38] observed that it is natural to consider periodic potentials in the solution of inverse scattering problems. He proved there that every stationary periodic solution $u(x)$ of the so-called "higher order K-dV equations" is a finite-zone potential and that these stationary K-dV equations are themselves totally integrable Hamiltonian systems with $n$ degrees of freedom ($n$ being the number of zones). Hence their general solution is an almost-periodic function, possibly meromorphic. These equations depend on $n+1$ constants $c_0, \ldots, c_n$ and have $n$ commuting integrals $J_1, \ldots, J_n$; furthermore, symmetric functions of the boundaries of the forbidden zones (lacunae) can be expressed in terms of these constants, and the levels of the integrals $(c_0, \ldots, c_n, J_1, \ldots, J_n)$. From this there follows an important result: *for any given boundaries of zones (lacunae) we can find a potential by solving stationary higher K-dV equations, but, generally speaking, it is almost-periodic with group of periods* $T_1, \ldots, T_n$ *and may possibly have poles. All the right-hand sides of the K-dV equations and of the integrals* $J_\alpha$ *are polynomials, and naturally there arise also complex meromorphic almost-periodic potentials.*

Soon after the completion of the papers of Dubrovin [40] and Its and Matveev [41], Novikov and Dubrovin showed [43] that the complex level surface of the integrals $J_1, \ldots, J_n$ is an Abelian variety, explicitly embedded in projective space; all the potentials $u(x)$ have regular analytic properties and are almost-periodic (if $x$ is complex) with a group of $2n$ "real and imaginary" periods $T_1, \ldots, T_n, T_1', \ldots, T_n'$, and their time evolution by the K-dV equation and all its higher order analogues for these potentials is given by a rectilinear (complex) development of this Abelian variety (the $2n$-dimensional complex torus). This Abelian variety is called the "Jacobian variety" of $\Gamma$.

These results can also be applied directly to the theory of Abelian varieties (see Ch. 2). To see this it suffices to note that in the solution of the inverse problem of the spectral theory for the Schrödinger operator $L$ by the method of the present authors all reference to results from the theory of the Schrödinger operator with periodic potential is easily eliminated, and all the results are true for the almost-periodic case when the boundaries of the zone (branch-points of $\Gamma$) are given arbitrarily. The group of periods of $u(x)$ is determined by the spectrum (by $\Gamma$); to specify the potential $u(x)$ we need one arbitrary point on the complex torus – the Jacobian variety $\mathbf{J}(\Gamma)$. Explicit formulae are given for the potential $u(x)$. There are formulae of several types (see Ch. 2); the best of them seems to be that due to Matveev and Its (see [41] and Ch. 2, §3): it expresses the potential by a simple explicit formula in terms of the Riemann $\theta$-function. A number of useful formulae for the time dynamics according to K-dV can be found in [46]. We recall that in the usual statement of the inverse problem for a periodic potential as "scattering data" all the eigenvalues $(E_n)$ are given for which the eigenfunctions are periodic with the same period $T$ as the potential, as well as the residual spectrum of the Sturm-Liouville problem on the half-line (see, for example, [6]; uniqueness theorems were already proved by Borg [1]). In this form even finite-zoned potentials cannot be effectively distinguished.

In our formulation (for example, for the finite-zone case) the boundaries of the zones are specified arbitrarily, then all the potentials with this spectrum which have a common group of periods and form the Jacobian variety $\mathbf{J}(\Gamma)$ are found explicitly. If the branch-points (boundaries of zones) are all real and the poles of the function $\psi_\pm(x, x_0, E)$ lie at points of the "spectrum" $\Gamma$ over finite lacunae, then we obtain a family of bounded almost-periodic real potentials, forming a real torus $T^n$, on which the higher equations of K-dV type form rectilinear developments in the corresponding "angular" variables (see [46]).

If even the potential $u(x)$ obtained is real and periodic in $x$, then under continuation into the complex domain in $x$, it becomes conditionally periodic along the imaginary axis, generally speaking, with a group of $n$ imaginary periods $T'_1, \ldots, T'_n$. The only exceptions are the cases when the potential reduces to elliptic functions. Such potentials are given, for example, in Ch. 2 and in [42]. The simplest is the Lamé potential.

It is not difficult to obtain the "closure" of our formulations, as the number of zones tends to infinity. However, in this case the relevant theorems become "existence and uniqueness theorems" and cease to be analytically constructive. It has not yet been proved rigorously that any (periodic) potential can be approximated by finite-zone potentials.

In Ch. 3 we give generalizations of our theorems to some other non-linear equations. As was pointed out to the authors by A. N. Tyurin, of particular interest from the point of view of algebraic geometry is the translation of the theory to the case of first order $(n \times n)$-matrix operators with $n > 2$, in which non-hyperelliptic Riemann surfaces appear. If arbitrary Riemann surfaces can, in fact, occur in the periodic theory of Zakharov-Shabat operators (see Ch. 3, §2) as "spectra" of a finite-zone linear operator, then this leads to a proof of the celebrated conjecture that the space of moduli of Riemann surfaces of arbitrary genus is unirational. Indeed, according to the scheme of [43] (see Ch. 2, §3) we can show that the space of the universal fibering of Jacobian varieties is unirational. But then the base also (that is, the variety of moduli of the curves themselves) is also unirational. However, the question of whether all Riemann surfaces $\Gamma$ can occur in this way, is not yet settled.

Thus, in naturally occurring totally Liouville integrable Hamiltonian systems connected with the scattering theory for the auxiliary local operator $L$, the level surfaces of the commuting integrals are not simply real tori $T^n$, but under continuation into the complex domain, they are Abelian varieties $T^{2n}$ which are the Jacobian varieties $\mathbf{J}(\Gamma)$ of Riemann surfaces arising as spectra of the operator $L$. An interesting question is whether there are natural classes of totally integrable systems in which there arise similarly Abelian varieties that are not the Jacobian varieties of any Riemann surfaces.

Incidentally, as a concluding remark we point out that in the classical non-trivial integrable cases of Jacobi (geodesics on a 3-dimensional ellipsoid) and of Sonya Kovalevski (the case of the heavy gyroscope) Abelian varieties also occur as level surfaces of commuting integrals: the direct product of one-dimensional ones for Jacobi's case and non-trivial two-dimensional Abelian varieties for the Kovalevski case. (This can be guessed from the formulae given, for example, in [57], although the result has not been stated anywhere explicitly.) Here we quote from a letter written by Sonya Kovalevski in December 1886.

"He (Picard) reacted with great disbelief when I told him that functions of the form

$$y=\frac{\theta(cx+a,\ c_1x+a_1)}{\theta_1(cx+a,\ c_1x+a_1)}$$

can be useful in the integration of certain differential equations" (quoted in [57]). The analysis of the present authors shows that for 90 years after the work of Sonya Kovalevski, until the 1974 papers on the K-dV equations, Picard's scepticism was justified.

The present survey consists of three chapters and applications. The first chapter is a brief survey of the non-linear equations integrable by the method of the inverse scattering problem that were known up to the end of 1974. The second and third chapters and the Appendices contain an account of the results of the authors, Its and Krichever.

# CHAPTER 1

## EXAMPLES OF NON-LINEAR EQUATIONS ADMITTING A COMMUTATION REPRESENTATION. METHODS OF FINDING THEM

### §1. The K-dV equation and its higher order analogues

As Lax pointed out in 1968 (see [19]), the Kruskal–Gardner–Green–Miura method [18] of integrating the Cauchy problem for functions rapidly decreasing in $x$ for the K-dV equation $u_t = 6uu_x - u_{xxx}$ can be obtained, by reduction to the inverse scattering problem for the Schrödinger operator, from the operator representation of this equation

$$\frac{dL}{dt}=[A,\ L], \tag{1.1.1}$$

in which $L = -\frac{d^2}{dx^2} + u(x,\ t)$ and $A = -4\frac{d^3}{dx^3} + 3\left(u\frac{d}{dx} + \frac{d}{dx}u\right)$. The operators $L$ and $A$ act on functions on $x$, $dL/dt$ is the operator of multiplication by the function $u_t$, while $[A,\ L]$ is the operator of multiplication by $6uu_x - u_{xxx}$. It follows from (1.1.1) that the discrete spectrum $\lambda_i(L)$ does not change in time. This is true both for functions $u(x,\ t)$ rapidly decreasing in $x$ and also for periodic functions (although in the latter case this gives a set of integrals equivalent to the Kruskal–Zabusskii integrals [17]. Lax and Gardner (see [19], [22]) indicated a series of evolution equations of the form

$$u_t=Q_n\left(u,\ \frac{\partial u}{\partial x},\ \frac{\partial^2 u}{\partial x^2},\ \ldots,\ \frac{\partial^{2n+1}u}{\partial x^{2n+1}}\right), \tag{1.1.2}$$

admitting an analogous representation

$$\frac{dL}{dt}=[A_n,\ L], \tag{1.1.1'}$$

where $A_n = \frac{d^{2n+1}}{dx^{2n+1}} + \sum_{i=0}^{2n} P_i \frac{d^i}{dx^i}$, $P_i$ and $Q_n$ are polynomials in $u, u_x, u_{xxx}, \ldots$ with constant coefficients, $L$ is the Schrödinger operator $L = -\frac{d^2}{dx^2} + u(x, t)$,

$A_0 = \frac{d}{dx}$, $A_1 = A$, and

$$A_2 = 16\frac{d^5}{dx^5} - 20\left(u\frac{d^3}{dx^3} + \frac{d^3}{dx^3}u\right) + 30u\frac{d}{dx}u + 5\left(u''\frac{d}{dx} + \frac{d}{dx}u''\right).$$

All these are called "higher order K-dV equations" and have the form $Q_n = \frac{\partial}{\partial x}\frac{\delta I_n}{\delta u(x)}$, where $I_{n-1} = \int \chi_{2n+1}(u, u_x, \ldots, u^{(n-1)}dx$ and the $\chi_{2n+1}$ are polynomials to be specified below; for example,

$$\text{(1.1.3)} \quad \begin{cases} I_{-1} = \int u\,dx, \quad I_0 = \int u^2\,dx, \quad I_1 = \int\left(\frac{u_x^2}{2} + u^3\right)dx, \\ I_2 = \int\left(\frac{1}{2}u_{xx}^2 - \frac{5}{2}u^2u_{xx} + \frac{5}{2}u^4\right)dx. \end{cases}$$

As was shown first in [17], the quantities $I_n$ are conserved in time by the K-dV equation. From the representation $Q_n = \frac{\partial}{\partial x}\frac{\delta I_n}{\delta u(x)}$ it follows, as Gardner noted, that all the "higher order K-dV equations" are Hamiltonian systems with infinitely many degrees of freedom, since $\frac{\partial}{\partial x}$ is a skew-symmetric operator. Faddeev and Zakharov [21] have shown that all the K-dV equations are totally integrable Hamiltonian systems in which the scattering data of $L$ (see Ch. 2, §1) are "action-angle" canonical variables, and the eigenvalues of the discrete spectrum $\lambda_i(L)$ commute (have vanishing Poisson brackets). From this it follows that the integrals $I_n$ also commute. This last fact was proved independently by Gardner [22] by another method. An algorithm for discovering the integrals $I_n$ is as follows (these integrals were first found in 1965). Let $L\psi = E\psi$ and $i\chi(x, k) = \frac{d\ln\psi}{dx}$, where $k^2 = E$. Then the quantity $\chi$ satisfies the Riccati equation

$$\text{(1.1.4)} \qquad -i\chi' + \chi^2 + u - E = 0$$

and admits, by (1.1.4), the formal expansion as $E \to \infty$:

$$\text{(1.1.5)} \qquad \chi(x, k) \sim k + \sum_{n=1}^{\infty}\frac{\chi_n(x)}{(2k)^n}, \qquad k^2 = E.$$

All the polynomials $\chi_{2n}(x)$ are purely imaginary and are total derivatives, while the polynomials $\chi_{2n+1}(x)$ are real and depend on $u, u_x, u_{xx}, \ldots, u^{(n-1)}$.

For all $k$, the integral $I_k = \int_{-\infty}^{\infty} \chi(x, k)dx$ is time-invariant under all "higher order K-dV equations", hence, all the quantities $I_n$ are also conserved. The scattering data and also the method of integrating the periodic problem will be explained later (see Ch. 2). We know that the modified K-dV equation $u_t = 6|u|^2 u_x - u_{xxx}$ can be reduced to the K-dV equation, hence is also totally integrable.

## §2. The non-linear equation of the string and the two-dimensional K-dV equation

Closest to the K-dV equation from the present point of view are the non-linear equations of the string (see [23])

$$\frac{3}{4}\beta\frac{\partial u}{\partial y} = -\frac{\partial w}{\partial x}, \quad \beta\frac{\partial w}{\partial y} = \frac{1}{4}(6uu_x + u_{xxx}) + \lambda u_x \tag{1.2.1}$$

and the "two-dimensional K-dV equation" (see [24], [37])

$$\begin{cases} \frac{3}{4}\beta\frac{\partial u}{\partial y} = -\frac{\partial w}{\partial x}, \\ \beta\frac{\partial w}{\partial y} = \alpha\frac{\partial u}{\partial t} + \lambda u_x + \frac{1}{4}(6uu_x + u_{xxx}), \end{cases} \tag{1.2.2}$$

which occurred earlier in the study of transverse perturbations of long waves in non-linear media with dispersion. Here the operators $L$ and $A$ have the form

$$L = \frac{d^2}{dx^2} + u, \quad A = 4\frac{d^3}{dx^3} + 3\left(u\frac{d}{dx} + \frac{d}{dx}u\right) + \lambda\frac{d}{dx} + w.$$

The equation (1.2.1) is equivalent to the following:

$$\beta\frac{\partial A}{\partial y} = [L, A]. \tag{1.2.3}$$

and (1.2.2) to the relation

$$\beta\frac{\partial A}{\partial y} + \alpha\frac{\partial L}{\partial t} = [L, A]. \tag{1.2.4}$$

In fact, in (1.2.1) and (1.2.3) the operators $L$ and $A$ have exchanged their roles in comparison with the K-dV equation, and the two-dimensional K-dV equation ((1.2.2) and (1.2.4)) unites them.

## §3. The non-linear Schrödinger equation

This equation has the form

$$\alpha\frac{\partial u}{\partial t} = \frac{\partial^2 u}{\partial x^2} \pm |u|^2 u \tag{1.3.1}$$

and was after the K-dV equation the first for which Zakharov and Shabat discovered in 1970 a mechanism for the reduction to the scattering problem for the auxiliary linear operator, which is no longer a Schrödinger operator (see [25], [26]). Here the operators $L$ and $A$ have the form

$$L=I\frac{d}{dx}+\begin{pmatrix}0 & u\\ v & 0\end{pmatrix}(l_1-l_2),\qquad I=\begin{pmatrix}l_1 & 0\\ 0 & l_2\end{pmatrix},$$
$$A=\frac{d^2}{dx^2}+2\begin{pmatrix}0 & u_x\\ v_x & 0\end{pmatrix}.$$

If $u = \bar{v}$, then the equation

$$\alpha\frac{dL}{dt}=[A,\ L] \tag{1.3.2}$$

is equivalent to (1.3.1):

$$-\alpha\frac{\partial u}{\partial t}=\frac{l_1+l_2}{l_2-l_1}u_{xx}-2\frac{l_1+l_2}{l_1l_2}|u|^2u$$

for an appropriate choice of constants $l_1$ and $l_2$. The integration of (1.3.1) for rapidly decreasing functions can be found in [25] and [26].

## §4. First order matrix operators

Suppose that $L$ and $A$ are first order matrix operators in $x$. We look for them in the form

$$L=l_1\frac{d}{dx}+[l_1,\ \xi],\qquad A=l_2\frac{d}{dx}+[l_2,\ \xi], \tag{1.4.1}$$

where $[l_1, l_2] = 0$, $l_1$, $l_2$ are constant $(N \times N)$-matrices and $\xi = (\xi_{ij})$. The matrices $l_1$ and $l_2$ can be assumed to be diagonal, $l_1 = a_i\delta_{ij}$, $l_2 = b_i\delta_{ij}$. The equation

$$\alpha\frac{\partial L}{\partial t}+\beta\frac{\partial A}{\partial y}=[L,\ A]$$

can easily be brought to the form

$$(a_i-a_j)\frac{\partial\xi_{ij}}{\partial t}=(b_i-b_j)\frac{\partial\xi_{ij}}{\partial y}+\ldots \tag{1.4.2}$$

In a number of cases, by imposing on $\xi_{ij}$ constraints of the type $\xi^+ = I\xi I$, where $I^2 = 1$, we can reduce (1.4.2) to a system with fewer unknowns. Some important cases of this kind for $N = 3$ were first given in [27], with $I = \begin{pmatrix}1 & & 0\\ & 1 & \\ 0 & & 1\end{pmatrix}$ or $I = \begin{pmatrix}1 & & 0\\ & -1 & \\ 0 & & 1\end{pmatrix}$. A general method of obtaining these systems was developed in [37].

The degenerate case in which some of the components of the matrices $l_1$, $l_2$ vanish, $a_i = b_i = 0$, is very interesting. For fourth order matrices we have here systems for which the corresponding non-linear equation reduces to the standard "sine-Gordon" equation (see [30]):

$$u_{tt}-u_{xx}=\sin u. \tag{1.4.3}$$

As a consequence of the degeneracy $a_i = b_i$ mentioned above, the spectral problem for the corresponding operator $L$ turns out to be non-trivial; various difficulties are overcome in [30]. More straightforward is the variant of the "sine-Gordon" equation

$$u_{\xi\eta} = -\sin u, \tag{1.4.4}$$

in which the initial data are placed on a single characteristic. This case is simpler. It was studied previously in [29] and [28]. Here the spectral problem is not degenerate, and the operators $L$, $A$ for (1.4.4) have the form

$$L = \begin{pmatrix} 0 & -1 \\ 1 & 0 \end{pmatrix} \frac{d}{dx} + \frac{i}{2} u_{\xi} \begin{pmatrix} 0 & 1 \\ 1 & 0 \end{pmatrix}, \tag{1.4.5}$$

$$A\psi(x) = = \frac{1}{4} \int_{-\infty}^{x} \begin{pmatrix} \exp\left(-\frac{i}{2}(u(x, t) + u(x', t))\right) & 0 \\ 0 & \exp\left(\frac{i}{2}(u(x, t) + u(x', t))\right) \end{pmatrix} \psi(x')\, dx'.$$

## §5. Discrete systems
## The Toda chain and the "K-dV equation difference"

We consider a chain of particles in a straight line with coordinates $x_n$ and the interaction Hamiltonian $H = \sum_n (e^{x_n - x_{n-1}} + \dot{x}_n^2/2)$, which was first discussed in [31]. The integrals found by Henon in [32] for the Toda chain explicitly showed its integrability. In [33]–[35], the $L-A$ pair for this chain was found and the commutativity of the integrals previously found by Henon was proved. For the "rapidly decreasing" case, where $c_n \to 1, v_n \to 0$ ($c_n = e^{x_n - x_{n-1}}, v_n = \dot{x}_n$), the Toda chain was integrated by the scattering theory method. Here the equations have the form

$$\dot{v}_n = c_{n+1} - c_n, \qquad \dot{c}_n = c_n(v_n - v_{n-1}), \tag{1.5.1}$$

and the operators $L$ and $A$ are as follows:

$$\begin{cases} L_{mn} = i\sqrt{c_n}\,\delta_{n,m+1} - i\sqrt{c_m}\,\delta_{n+1,m} + v_n\delta_{mn}, \\ A_{mn} = \frac{i}{2}(\sqrt{c_n}\,\delta_{n,m+1} + \sqrt{c_m}\,\delta_{n+1,m}). \end{cases} \tag{1.5.2}$$

(1.5.1) is equivalent to the equation $dL/dt = [A, L]$.

In addition to the Toda chain (1.5.1), in [35] another "K-dV difference equation" is considered:

$$\dot{c}_n = c_n(c_{n+1} - c_{n-1}), \tag{1.5.3}$$

which can be obtained from the $L-A$ pair (1.5.2) with the same operator $L$ under the condition $v_n \equiv 0$ and the new operator $A \to \bar{A}$, where

$$\bar{A} = -\frac{1}{2}(\sqrt{c_n c_{n-1}}\,\delta_{n,m+2} - \sqrt{c_m c_{m-1}}\,\delta_{n+2,m}); \tag{1.5.4}$$

then (1.5.3) is equivalent to the equation $dL/dt = [\bar{A}, L]$, $v_n \equiv 0$. The operator $L$ itself, given by (1.5.2), is a difference analogue to a Sturm-Liouville operator and was considered previously for other purposes [12], where the inverse problem on the half-line $n \geqslant 0$ was solved for it.

## §6. The method of Zakharov and Shabat of constructing non-linear equations that have an $L-A$ pair

Let $\hat{F}$ be a linear integral operator acting on the vector function $(\psi_1, \ldots, \psi_N) = \psi$ of the variable $x$ $(-\infty < x < \infty)$:

$$\hat{F}\psi = \int_{-\infty}^{\infty} F(x, z)\,\psi(z)\,dz, \tag{1.6.1}$$

where $\psi$ and $\hat{F}$ depend in addition on two parameters $t$ and $y$. Let us assume that the operator $\hat{F}$ admits the following representation:

$$1 + \hat{F} = (1 + K_+)^{-1}(1 + K_-), \tag{1.6.2}$$

in which $K_+$ and $K_-$ are Volterra integral operators with

$$\begin{cases} K_+(x, z) = 0, & z < x, \\ K_-(x, z) = 0, & z > x. \end{cases} \tag{1.6.3}$$

From (1.6.1) and (1.6.2) there follows the (Gel'fand–Levitan) equation for the kernel $K_+$:

$$F(x, z) + K_+(x, z) + \int_x^{\infty} K_+(x, s)\,F(s, z)\,ds = 0, \tag{1.6.4}$$

and the kernel $K_-$ can be found from the formula

$$K_-(x, z) = F(x, z) + \int_x^{\infty} K_+(x, s)\,F(s, z)\,ds.$$

We consider the operator $M_0 = \alpha \dfrac{\partial}{\partial t} + \beta \dfrac{\partial}{\partial y} + L_0$, acting on $\psi(t, y, x)$, where $L_0 = \sum_n l_n\, \partial^n/\partial x^n$, and the $l_n$ are constant $(N \times N)$-matrices. By analogy with the theory of the inverse scattering problems we can establish the following fact: if the operators $M_0$ and $\hat{F}$ commute, $[M_0, \hat{F}] = 0$, then the transformed operator $M = (1 + K_+)M_0(1 + K_+)^{-1}$ is also a differential operator with variable coefficients

$$M = (1 + K_+)\,M_0\,(1 + K_+)^{-1} = \alpha\frac{\partial}{\partial t} + \beta\frac{\partial}{\partial y} + L, \tag{1.6.5}$$

where $L$ is an operator involving only differentials in $x$. In scattering theory this fact was known for the Shrödinger operator and was used for the operator $L_0 = \dfrac{d^2}{dx^2}$.

We assume now that $M_0^{(1)}$ and $M_0^{(2)}$ are two operators such that

(1.6.6)
$$\begin{cases} M_0^{(1)} = \alpha \frac{\partial}{\partial t} + L_0^{(1)}, & L_0^{(1)} = \sum l_n^{(1)} \frac{\partial^n}{\partial x^n}, \\ M_0^{(2)} = \beta \frac{\partial}{\partial y} + L_0^{(2)}, & L_0^{(2)} = \sum l_n^{(2)} \frac{\partial^n}{\partial x^n} \end{cases}$$

and $[M_0^{(1)}, M_0^{(2)}] = 0$.

If $\hat{F}$ commutes with both of these, $[\hat{F}, M_0^{(1)}] = [\hat{F}, M_0^{(2)}] = 0$, then we obtain the differential operators

(1.6.7)
$$\begin{cases} M^{(1)} = (1+K_+) M_0^{(1)} (1+K_+)^{-1} = \alpha \frac{\partial}{\partial t} + L^{(1)}, \\ M^{(2)} = (1+K_+) M_0^{(2)} (1+K_+)^{-1} = \beta \frac{\partial}{\partial y} + L^{(2)}, \\ [M^{(1)}, M^{(2)}] = 0. \end{cases}$$

The relations (1.6.7) lead to an equation for the operators $L^{(1)}$ and $L^{(2)}$:

(1.6.8)
$$\alpha \frac{\partial L^{(2)}}{\partial t} - \beta \frac{\partial L^{(1)}}{\partial y} = [L^{(2)}, L^{(1)}],$$

which is equivalent to a system of non-linear equations for the coefficients of $L^{(1)}$ and $L^{(2)}$. Here the kernel $F(x, z, t, y)$ of $\hat{F}$ satisfies a system of linear equations with constant coefficients, which follow from the identities

(1.6.9)
$$[\hat{F}, M_0^{(1)}] = [\hat{F}, M_0^{(2)}] = 0.$$

We can easily solve the Cauchy problem in time $t$ for the equations (1.6.9) in the usual way, and then, having solved the Gel'fand–Levitan equation (1.6.4) define the coefficients of the operators $L^{(1)}$ and $L^{(2)}$ at any time $t$. In principle, this procedure can lead to the integration of the Cauchy problem for (1.6.8) only for functions rapidly decreasing in $x$ as coefficients of $L^{(1)}$ and $L^{(2)}$. We note that the method of Zakharov–Shabat permits us to construct a number of new systems with an $L-A$ pair, among them some of physical interest (see [37]); this method both constructs the systems and produces a way of solving the inverse problem – the Gel'fand–Levitan equation. Some of the above examples of $L-A$-pairs were first found in this manner.

# CHAPTER 2

# THE SCHRÖDINGER OPERATOR AND THE K-dV EQUATION. FINITE-ZONE POTENTIALS

## § 1. General properties of the Schrödinger operator with a periodic and rapidly decreasing potential

We first consider the usual Schrödinger operator $L = -\frac{d^2}{dx^2} + u$ from the requisite point of view. It is convenient to fix a basis in the space of solutions of the equation $L\varphi = E\varphi$. Let $x_0$ be a given point. We specify a solution $\varphi(x, x_0, \pm k)$ by setting

$$(2.1.1)\qquad \begin{cases} \text{a)} & L\varphi(x, x_0, k) = k^2\varphi(x, x_0, k), \\ \text{b)} & \varphi(x_0, x_0, k) = 1, \\ \text{c)} & \varphi'(x, x_0, k) = ik \text{ when } x = x_0 \; (k^2 = E). \end{cases}$$

Then we have a basis $\varphi(k), \varphi(-k)$ for all $k \neq 0$; for real $k$ (or $E > 0$) we have $\varphi(-k) = \overline{\varphi(k)}$.

We obtain another basis $c(x, x_0, E)$, $s(x, x_0, E)$ as follows:

$$(2.1.2)\qquad \begin{cases} c = 1, & c' = 0, \\ s = 0, & s' = 1 \end{cases} \quad \text{when } x = x_0.$$

If the potential $u(x)$ is periodic with period $T$, then the operator of translation (or "monodromy") is defined by

$$(2.1.3)\qquad (\hat{T}\psi)(x) = \psi(x + T).$$

The translation operator becomes a matrix of the second order in the bases (2.1.1) and (2.1.2)

$$(2.1.4)\qquad \begin{cases} \hat{T}\varphi = a\varphi + b\overline{\varphi}, & \hat{T}\overline{\varphi} = \overline{b}\varphi + \overline{a}\overline{\varphi}, \\ \hat{T}c = \alpha_{11}c + \alpha_{12}s, & \hat{T}s = \alpha_{21}c + \alpha_{22}s. \end{cases}$$

From the invariance of the Wronskian it follows that $\det \hat{T} = 1$, or $|a|^2 - |b|^2 = 1$ for real $k$, and $\alpha_{11}\alpha_{22} - \alpha_{21}\alpha_{12} = 1$ for all $E$. The matrix $\hat{T}$ depends on $x_0$ and $E$ (or $k$). In the basis (2.1.2), $\hat{T}$ is an integral function of $E$. Under a change of the parameter $x_0$ the matrix $\hat{T}(x_0, k)$ changes to a similar matrix. Hence the dependence on $x_0$ is governed by the very useful equation

$$(2.1.5)\qquad \frac{d\hat{T}}{dx_0} = [Q, \hat{T}],$$

in which the matrix $Q$ is easily evaluated and in the bases (2.1.1) and (2.1.2) has the form

$$Q = ik\begin{pmatrix} 1 & 0 \\ 0 & -1 \end{pmatrix} - \frac{iu}{2k}\begin{pmatrix} 1 & -1 \\ 1 & -1 \end{pmatrix} \qquad \text{(basis (2.1.1))},$$

$$Q = \begin{pmatrix} 0 & E-u \\ -1 & 0 \end{pmatrix} \qquad \text{(basis (2.1.2))}.$$

The eigenfunctions ("Bloch" functions) are characterized by the requirements

$$(2.1.6)\qquad \hat{T}\psi_\pm = e^{\pm ip(E)}\psi_\pm,$$

where $p(E)$ is called the *quasi-momentum*. For the Schrödinger operator it is convenient to normalize them by setting

$$(2.1.7)\qquad \psi_\pm(x, x_0, E) = 1 \quad \text{for} \quad x = x_0.$$

If the potential $u(x)$ is real, then the solution zones (points of the spectrum) are determined, $p(E)$ being real and the functions $\psi_\pm$ almost periodic. The complements of the solution zones are called forbidden zones, lacunae, or zones of instability. As a rule, a typical potential has infinitely many lacunae, the lengths of which decrease rapidly as $E \to \infty$. The rate of decrease depends on the smoothness of the potential. If the potential is analytic, then the rate of decrease of the lengths of the lacunae is exponential.

The eigenvalues of the translation matrix $\hat{T}$ do not depend on the basis or on the choice of $x_0$; they determine the quasi-momentum $p(E)$ and the boundaries of the solution zones and forbidden zones. The trace of $\hat{T}$ has the form

$$\operatorname{Sp}\hat{T}=\alpha_{11}+\alpha_{22}=2a_{\mathrm{Re}} \quad \text{(basis (2.1.1))},$$
$$\operatorname{Sp}\hat{T}=a+\bar{a}=2a_{\mathrm{Re}} \quad \text{(basis (2.1.2))},$$

and the eigenvalues $\mu_\pm = e^{\pm ip(E)}$ are:

$$(2.1.8) \qquad \begin{cases} \mu_\pm(E)=a_{\mathrm{Re}}\pm i\sqrt{1-a_{\mathrm{Re}}^2}, \\ a_{\mathrm{Re}}=\cos p(E)=\frac{1}{2}\operatorname{Sp}\hat{T}. \end{cases}$$

It follows, clearly, from (2.1.8) that the periodic levels $\psi_n(x+T)=\psi_n(x)$, $E=E_n$, are determined by

$$(2.1.9) \qquad \frac{1}{2}\operatorname{Sp}\hat{T}=a_{\mathrm{Re}}=1.$$

The antiperiodic levels $\psi_m(x+T)=-\psi(x)$ are determined by

$$(2.1.9') \qquad \frac{1}{2}\operatorname{Sp}\hat{T}=a_{\mathrm{Re}}=-1.$$

The periodic and antiperiodic levels can be simple (that is, singly degenerate) or doubly degenerate. In both cases we have $\mu_\pm(E)=\pm 1$, but in the degenerate case $\hat{T}$ is diagonal:

$$(2.1.10) \quad \hat{T}=\pm\begin{pmatrix}1&0\\0&1\end{pmatrix}, \quad \text{or} \quad b(E_n,\ x_0)\equiv 0, \quad \alpha_{21}(E_n,\ x_0)\equiv 0.$$

In the non-degenerate case $\hat{T}$ is a Jordan matrix and $|b(E_n,x_0)|\neq 0$, $\alpha_{21}\neq 0$. From the condition $|a|^2-|b|^2=1$ we have for $E=E_n$

$$(2.1.10') \qquad |a_{\mathrm{Re}}|=1, \quad |a_{\mathrm{Im}}|=|b|, \quad E=E_n,$$

where $|b|\neq 0$. We already know that the boundaries of the forbidden and the solution zones are precisely the non-degenerate periodic and antiperiodic levels for which $|b|\neq 0$, $\alpha_{21}\neq 0$. Degenerate levels can be visualised as a forbidden zone contracted to a point. For example, if we increase the period to an integer multiple of itself, $T\to mT$, then we obtain

$$(2.1.11) \qquad \hat{T}\to\hat{T}^m, \quad e^{ip(E)}\to e^{imp(E)}.$$

The transformation (2.1.11) conserves the solution zones and forbidden zones, but inside the solution zones new degenerate levels appear when $mp(E)$ is a multiple of $2\pi$.

As we know, $1-\frac{1}{2}\operatorname{Sp}\hat{T}$ is an entire function of order $\frac{1}{2}$ in the variable

$E$, whose zeros, by (2.1.9), precisely determine all the periodic levels $E_n$. Therefore, the function $1 - \frac{1}{2}$ Sp $\hat{T}$ can be represented as an infinite product, which is completely determined by the zeros. Hence, by (2.1.9′) the antiperiodic levels are determined by the complete set of periodic levels (including the degenerate ones). However, we use as basic parameters only the boundaries of the zones, or the non-degenerate part of the periodic and antiperiodic levels (that is, the spectrum of the operator $L$ over the whole line).

The following simple proposition is extremely important for our situation.

LEMMA. *For any (real or complex) smooth periodic potential $u(x)$, the Bloch eigenfunction $\psi_\pm(x, x_0, E)$ defined by the conditions* (2.1.6) *and* (2.1.7) *is meromorphic on a two-sheeted Riemann surface $\Gamma$, covering the E-plane and having branch-points (for a real potential) at the ends of the zones. In general, this Riemann surface has infinite genus; however, if the number of lacunae is finite (this case is especially important for the theory of the K-dV equation), then $\Gamma$ is hyperelliptic and has finite genus equal to the number of lacunae.*

This proposition naturally leads to the following definition.

DEFINITION 1. A periodic potential $u(x)$ is said to be *finite-zoned* if the eigenfunction $\psi_\pm(x, x_0, E)$ defined by the conditions (2.1.6) and (2.1.7) is meromorphic on a hyperelliptic Riemann surface $\Gamma$ of finite genus; the branch-points of the Riemann surface are said to be the "boundaries of the zones".

For later purposes we also need the following definition.

DEFINITION 1′. An almost periodic (real or complex) potential $u(x)$ is said to be *finite-zoned* if it has for all $E$ an eigenfunction $\psi(x, E)$, meromorphic on a hyperelliptic Riemann surface $\Gamma$ of finite genus, doubly covering the $E$-plane, with (2.1.6) modified as follows: the logarithmic derivative $\frac{d \ln (x)}{dx}$ is an almost-periodic function with the same group of periods as the potential $u(x)$.

It is also required that an $E \to \infty$ the function $\psi(x, E)$ has the asymptotic form $\psi \sim \exp\{\pm ik(x - x_0)\}$, $k^2 = E$. Then we denote $\psi$ by $\psi(x, x_0, E)$, where $\psi|_{x=x_0} = 1$, as in the periodic case. The logarithmic derivative $i\chi(x, E) = \frac{d \ln \psi}{dx}$ does not depend on $x_0$ and is a quantity of great importance in the theory of the operator $L = -\frac{d^2}{dx^2} + u$ and in the theory of the K-dV equation, as we have already indicated in Ch. 1, § 1. It satisfies the Riccati equation (1.1.4). If the potential is real, and in the solution zones for $E$ we set $\chi = \chi_{Re} + i\chi_{Im}$ then from the Riccati equation we deduce the relation

$$\chi_{Im} = \frac{1}{2}(\ln \chi_{Re})'. \tag{2.1.12}$$

In the solution zones we have for the Wronskian $W$ of the Bloch functions

(2.1.13) $$W(\psi, \bar{\psi}) = W(\psi_+, \psi_-) = 2i\chi_{\mathrm{Re}}(x_0, E).$$

By (2.1.13) the function equal to

$\chi_R = \frac{1}{2i} W(\psi_+, \psi_-) = \frac{1}{2i}(\psi_+'\psi_- - \psi_+\psi_-')$ is defined for all $E$; it is as the analogue to $\chi_{\mathrm{Re}}$ for complex potentials. We always denote this function by $\chi_R$. By definition, we have from (2.1.12)

(2.1.14) $$\psi_\pm(x, x_0, E) = \sqrt{\frac{\chi_R(x_0, E)}{\chi_R(x, E)}} \exp\left\{\pm i \int_{x_0}^{x} \chi_R(x, E)\, dx\right\},$$

from which it follows, by the definition of the quasi-momentum, that

(2.1.15) $$p(E) = \int_{x_0}^{x_0+T} \chi_R(x, E)\, dx.$$

Since $\cos p(E) = \frac{1}{2}\operatorname{Sp}\hat{T}$ by (2.1.8), and since the function $1 - \frac{1}{2}\operatorname{Sp}\hat{T}$ is completely determined by the periodic spectrum, for example, as an infinite product, the integral $\int_{x_0}^{x_0+T} \chi_R(x, k)dx = \int_{x_0}^{x_0+T} \chi(x, k)dx$ and all the coefficients of its expansion in $\frac{1}{\sqrt{E}} = \frac{1}{k}$ as $E \to \infty$ (see (2.1.18) and below) are expressible in terms of the spectrum of the periodic problem (including the degenerate levels). These expressions are called "trace identities". In what follows it will become clear that all these quantities can be expressed in terms of only the boundaries of the zones.

It is convenient to express the function $\psi(x, x_0, E)$ in the basis (2.1.2):

(2.1.16) $$\psi(x, x_0, E) = c(x, x_0, E) + i\alpha(x_0, E)s(x, x_0, E).$$

Evaluating the Wronskian $W(\psi, c)$, we obtain

(2.1.16′) $$\alpha(x_0, E) = \chi(x_0, E).$$

For $p(E)$ we can obtain the important general relations when the potential varies

(2.1.17) $$\frac{dp(E)}{dE} = \int_{x_0}^{x_0+T} \frac{dx}{2\chi_R(x, E)}, \qquad \frac{\delta p(E)}{\delta u(x)} = -\frac{1}{2\chi_R(x, E)},$$

where $\delta p/\delta u$ is the variational derivative. A proof of these relations can be found in [40], [46].

As $|E| \to \infty$, $\chi(x, E)$ admits the asymptotic expansion

(2.1.18) $$\chi(x, E) \sim k + \sum_{n\geq 1} \frac{\chi_n(x)}{(2k)^n},$$

in which the expansion coefficients $\chi_n(x)$ can be found by recurrence formulae derived from the Riccati equation (1.1.4); they are polynomials in $u$, $u'$, $u''$, .... All the polynomials $\chi_{2n}(x)$ are purely imaginary and by (2.1.12) they are total derivatives. Hence their integrals over a period vanish. A list of the first few polynomials $\chi_{2n+1}(x)$ is given in Ch. 1, §1 (see (1.1.3)).

Using the representation of $\psi_{\pm}(x, x_0, E)$ in the form (2.1.16), (2.1.16′) it is not difficult to obtain an expression for $\chi(x, E)$ in terms of the coefficients of the translation matrix $\hat{T}$, since $\psi$ is an eigenvector of this matrix. In the bases (2.1.1) and (2.1.2) we obtain

$$(2.1.19)\qquad \begin{cases} \chi_R(x, E) = \dfrac{k\sqrt{1-a_{\mathrm{Re}}^2}}{a_{\mathrm{Im}}+b_{\mathrm{Im}}} & \text{(basis (2.1.1))}, \ k^2 = E, \\[2ex] \chi_R(x, E) = \dfrac{\sqrt{1-\frac{1}{4}(\alpha_{11}+\alpha_{22})^2}}{\alpha_{21}} & \text{(basis (2.1.2))}, \end{cases}$$

where in the basis (2.1.1) the formula is valid only in a permitted zone. The function $\psi_{\pm}(x, x_0, E)$ can have poles at certain points $P_1, P_2, \ldots$ on $\Gamma$, depending, in general, on $x_0$ and $x$. For a real potential, these poles lie at the points of $\Gamma$ over points of the lacunae (or their boundaries) $\gamma_1(x_0), \ldots, \gamma_m(x_0), \ldots$, independent of $x$, one each for every lacuna (a pole occurs only on one sheet of $\Gamma$ over the point $\gamma_j(x_0)$). As is clear from (2.1.14), the zeros of $\psi_{\pm}$ lie over the points $\gamma_j(x)$, independent of $x_0$. If the potential is finite-zoned, having $n$ lacunae, then there are altogether $n$ zeros $P_1(x), \ldots, P_n(x)$ of $\psi_{\pm}(x, x_0, E)$, lying over the points $\gamma_1(x), \ldots, \gamma_n(x)$ of the $E$-plane, and $n$ poles $P_1(x_0), \ldots, P_n(x_0)$ over the points $\gamma_1(x_0), \ldots, \gamma_n(x_0)$. For complex potentials the zeros and poles can be located arbitrarily on $\Gamma$.

We conclude this section by considering rapidly decreasing potentials for which the analytic properties of the eigenfunctions and the scattering matrices can be found in [5]. From our point of view, rapidly decreasing potentials are a degenerate limiting case of periodic potentials when $T \to \infty$, $u(x) \to 0$, as $|x| \to \infty$, and $u', u'', \ldots \to 0$ as $|x| \to \infty$ (more precisely: $\int_{-\infty}^{\infty} (1 + |x|)\,|u(x)|\,dx < \infty$). Suppose to begin with that the potential $u(x)$ is of compact support (that is, $u$ is a finite function). We consider the bases of solutions analogous to (2.1.1) in which we set $x_0 = \pm\infty$. We have two bases (the right and the left):

$$(2.1.20)\qquad \begin{cases} f_+(x, k) \to e^{ikx}, \quad f_-(x, k) \to e^{-ikx}, \\ \qquad (x \to +\infty), \\ g_+(x, k) \to e^{ikx}, \quad g_-(x, k) \to e^{-ikx}, \\ \qquad (x \to -\infty). \end{cases}$$

Here the translation matrix for the period $T = \infty$ is the matrix of the transition from the basis $(g_+, g_-)$ to the basis $(f_+, f_-)$. For real $k$ this matrix has the form

$$\hat{T} = \begin{pmatrix} a & b \\ \bar{b} & \bar{a} \end{pmatrix}, \quad |a|^2 - |b|^2 = 1 \tag{2.1.21}$$

and $f_+ = ag_+ + bg_-$, $\hat{T} = \hat{T}(k)$.

The scattering matrix $\hat{S}$ is the matrix expressing the basis $(f_-, g_+)$ in terms of the basis $(g_-, f_+)$; it is unitary and has the form

$$\hat{S} = \begin{pmatrix} \frac{1}{a} & -\frac{b}{a} \\ \frac{\bar{b}}{a} & \frac{1}{a} \end{pmatrix}. \tag{2.1.22}$$

The coefficients $s_{11} = 1/a$ and $-s_{12} = b/a$ are called transmission and reflection coefficients (the scattering amplitudes in the directions 0 and $\pi$).

We note that for finite potentials $\hat{T}(k)$ can be continued analytically as an entire function of $k$; if $\operatorname{Im} k > 0$, then $f_+$ decreases and $g_-$ decreases, as $x \to +\infty$. From the equation $f_+ = a(k)g_+ + b(k)g_-$ it follows that we obtain a solution that decreases in both directions ($|x| \to \infty$) when $a(k_n) = 0$ (or the amplitude $s_{11}(k) = \frac{1}{a}$ has a pole). There are finitely many discrete levels $E_n = k_n^2$; the functions $f_+(x, k)$ and $a(k)$ are analytic for $\operatorname{Im} k > 0$, and

$$a(k) \to 1 + O(1/k), \qquad |k| \to \infty, \qquad \operatorname{Im} k > 0. \tag{2.1.23}$$

The eigenfunctions of the discrete spectrum for the levels $E_n = k_n^2$, where $a(k_n) = 0$, have the form

$$\psi_n(x) = g_-(x, k_n); \tag{2.1.24}$$

the normalization factors are easily evaluated:

$$\int_{-\infty}^{\infty} |\psi_n|^2 \, dx = \frac{1}{c_n} = \frac{i \left.\frac{da}{dk}\right|_{k=i\varkappa_n}}{b(i\varkappa_n)}. \tag{2.1.25}$$

In the general case of a rapidly decreasing potential, according to [5] we have functions $a(k)$, $b(k)$, where $a(k)$ is analytic in the upper half-plane $\operatorname{Im} k > 0$, has the asymptotic form (2.1.23) and the relation $|a|^2 - |b|^2 = 1$ holds on the real axis $\operatorname{Im} k = 0$. In addition, there are finitely many zeros $k_n$, $a(k_n) = 0$ (all the $k_n$ are purely imaginary) and corresponding numbers $c_n$ given by (2.1.25). From the uniqueness theorem of Marchenko [4] it follows that $(a(k), b(k), k_n, c_n)$ is a set of "scattering data", completely determining the potential $u(x)$. Moreover, the kernel

(2.1.26) $$F(z+t)=\int_{-\infty}^{\infty} e^{-ik(z+t)}\frac{b(k)}{a(x)}\,dk+\sum_n c_n e^{-ik_n(z+t)}$$

is well-defined, and we can write down the Gel'fand–Levitan equation

(2.1.27) $$K(k,\ z)+F(x+z)+\int_{-\infty}^{x} F(s+x)K(z,\ s)\,ds=0.$$

The potential itself is obtained from the formula $u(x) = 2\,\frac{d}{dx}\,K(x,\ x)$. When $a(k)$ has the indicated analytic properties [5], the Gel'fand–Levitan equation (2.1.27) has one and only one solution. An important special case is that for which $b \equiv 0$ on the real $k$-axis. Such potentials are called "non-reflecting". In this case the equation (2.1.27) becomes algebraic, and the potential $u(x)$ is a rational function of the exponentials $e^{\varkappa_1 x}, e^{\varkappa_2 x}, \ldots, e^{\varkappa_n x}$, in which $k_n = i\varkappa_n = \sqrt{E_n}$. The function $a(k)$ then has the form

(2.1.28) $$a(k)=\prod_n \frac{k-i\varkappa_n}{k+i\varkappa_n}\,.$$

These potentials were first found by Borg [1] and Bargmann [2]. They play an important part in the theory of the K-dV equation. They are also important to us in our concern with periodic problems, because as $T \to \infty$, the finite-zone potentials degenerate into non-reflecting ones; the solution zones shrink to isolated points of the discrete spectrum inside the solution zones $b \equiv 0$ for $T = \infty$, by (2.1.10) and (2.1.11), and $\Gamma$ degenerates into a rational surface, since the pairs of branch-points coalesce.

## §2. A new commutation representation of the K-dV and "higher order K-dV" equations. An algorithm for finding finite-zone potentials and their spectra

In Ch. 1, §1 we have recalled the Lax representation (1.1.1) and (1.1.1′) of the K-dV equation and its higher order analogues in the form $\frac{dL}{dt} = [A_n,\ L]$, where $\mathring{u} = Q_n(u,\ u',\ u'',\ \ldots,\ u^{(2n+1)})$. We now reproduce the algorithm [18] for integrating the Cauchy problem for K-dV equations with rapidly decreasing functions. This algorithm is based on the following equation for the translation (or monodromy) matrix $\hat{T}(k)$, discovered in [18]:

(2.2.1) $$\dot{a}(k) = 0, \qquad \dot{b}(k) = -8ik^3 b(k).$$

For finite potentials we obtain from (2.2.1) the following equation for the discrete levels, using (2.1.24) and (2.1.25):

(2.2.1′) $\dot{\varkappa}_n = (\sqrt{-E_n})^{\bullet} = 0, \qquad \dot{c}_n = -8\varkappa_n^3 c_n,$

since $k_n = i\varkappa_n$ are the zeros of the function $a(k)$, and $c_n = \left(\frac{da}{dk}\right)_{k=k_n} \frac{1}{b(k_n)}$. Therefore, (2.2.1) and (2.2.1′) hold for any rapidly decreasing potential and completely determine the time-development of the scattering data (and hence of the potential itself). Let us derive the equations (2.2.1). We consider the eigenfunction $f_+(x, k)$ defined by (2.1.20). Clearly, $\frac{\partial}{\partial t} f_+(x, k) \to 0$, as $x \to +\infty$. Since $Lf_+ = k^2 f_+$, for the derivative $((L-k^2)f_+)^{\bullet}$ we have

$$\dot{L}f_+ + (L-k^2)\dot{f}_+ = (AL - LA)f_+ + (L-k^2)\dot{f}_+ = (L-k^2)(\dot{f}_+ - Af_+).$$

From this we obtain

(2.2.2) $\dot{f}_+ = Af_+ + \lambda(k)f_+ + \mu(k)f_-.$

By letting $x \to \infty$, we obtain

$$0 \sim (Af_+)_{x\to+\infty} + (\lambda f_+)_{x\to+\infty} + (\mu f_-)_{x\to+\infty},$$

consequently, since $A$ has the form $A = -4\frac{d^3}{dx^3} + 3\left(u\frac{d}{dx} + \frac{d}{dx}u\right)$ and $u, u' \to 0$, we finally have

(2.2.3) $(Af_+)_{x\to+\infty} \to 4ik^3 e^{ikx}, \quad \mu(k) \equiv 0, \quad \lambda(k) = -4ik^3.$

We now recall the equation $f_+ = ag_+ + bg_-$ (2.1.21) and let $x \to -\infty$. Then $g_+ \to e^{ikx}$, $g_- \to e^{-ikx}$. From this it follows that $\dot{f}_+ \to \dot{a}e^{ikx} + \dot{b}e^{-ikx}$, as $x \to -\infty$. Comparing these formulae with (2.2.2) and bearing (2.2.3) in mind, we finally obtain (2.2.1): $\dot{a} = 0$, $\dot{b} = -8ik^3 b$. Now (2.2.1′) follows from this, as indicated above, and the integration of the K-dV equation for rapidly decreasing functions is complete.

For the higher order K-dV equations we have

$$A_n = \frac{d^{2n+1}}{dx^{2n+1}} + \sum_{i=0}^{2n} P_i(u, u', \ldots)\frac{d^i}{dx^i}, \quad \text{where } P_i \equiv 0 \text{ when } u \equiv 0.$$

By analogy with the preceding derivation it follows for $\hat{T}(k)$ that

(2.2.4) $\dot{a} = 0, \qquad \dot{b} = \text{const}(ik)^{2n+1}.$

We see from (2.2.4) that all the higher order K-dV equations commute as dynamical systems in the function space of rapidly decreasing functions.

Gardner [22] has shown that all the higher order K-dV equations have the Hamiltonian form

(2.2.5) $\dot{u} = \frac{\partial}{\partial x}\frac{\delta I_n}{\delta u(x)},$

as already stated in Ch. 1, §1, where the $I_n$ are the Hamiltonians. From the commutativity of these dynamical systems it follows that the Poisson brackets of the functionals $I_n$ vanish:

(2.2.6) $$[I_n, I_m] = 0.$$

since the skew-symmetric form is determined by $\frac{\partial}{\partial x}$, (2.2.6) is equivalent to

(2.2.6′) $$\int_{-\infty}^{\infty} \frac{\delta I_n}{\delta u(x)} \frac{\partial}{\partial x} \frac{\delta I_m}{\delta u(x)} dx = 0$$

for any rapidly decreasing function $u(x)$. Since $I_n$ and $I_m$ are integrals of expressions that are polynomials in $u$, $u'$, $u''$, with constant coefficients, it now becomes absolutely clear that the identity (2.2.6) also holds for any periodic function $u(x)$. Gardner proved in [22] the commutativity relation (2.2.6) by direct calculations. Simultaneously, Faddeev and Zakharov in [21] computed all the Poisson brackets of all the "scattering data" $a(k)$, $b(k)$, $\varkappa_n c_n$ and proved that

(2.2.7) $$\begin{cases} P_k = \frac{2k}{\pi} \ln |a(k)|, & P_n = \varkappa_n^2, \\ Q_k = \arg b(k), & Q_n = \ln b_n, \\ \text{where } b_n = ic_n (da/dk)|_{k=i\varkappa_n}, \end{cases}$$

are canonical variables ("action-angle" variables) for all the Hamiltonian systems (2.2.5) and all the K-dV equations.

Great difficulties arise in attempts to generalize this method of integrating the K-dV equation to the periodic case. These will become clearer after we have derived the natural analogue to (2.2.1) for $\hat{T}$. We consider the bases (2.1.1) or (2.1.2) and compute the time derivatives by analogy with (2.2.2) at $x = x_0$ instead of $x = \infty$:

(2.2.8) $$\begin{cases} \dot{\varphi}_+ = A\varphi_+ + \lambda_{11}\varphi_+ + \lambda_{12}\varphi_-, \\ \dot{\varphi}_- = A\varphi_- + \lambda_{21}\varphi_+ + \lambda_{22}\varphi_- \end{cases}$$

in the basis (2.1.1), or

(2.2.8′) $$\begin{cases} \dot{c} = Ac + a_{11}c + a_{12}s, \\ \dot{s} = As + a_{21}c + a_{22}s \end{cases}$$

in the basis (2.1.2), where $\lambda_{11} = \lambda$, $\lambda_{12} = \mu$, $\lambda_{21} = \bar{\mu}$, $\lambda_{22} = \bar{\lambda}$ for real $k$ and all the $a_{ij}$ are real for real $E$. We have the matrix
$\Lambda = \begin{pmatrix} \lambda_{11} & \lambda_{12} \\ \lambda_{21} & \lambda_{22} \end{pmatrix}$, or $\Lambda = \begin{pmatrix} a_{11} & a_{12} \\ a_{21} & a_{22} \end{pmatrix}$, in the bases (2.1.1) and (2.1.2), respectively, the coefficients of which can be expressed in terms of $u(x_0)$, $u'(x_0)$, ... and $k$ by the following formulae:

(2.2.9) $$\begin{cases} (\dot{\varphi}_+)_{x=x_0} = 0 = (A\varphi_+)_{x=x_0} + \lambda_{11} + \lambda_{12} \\ \qquad (\varphi_+ = 1 \quad \text{for} \quad x = x_0), \\ \left(\frac{d}{dx}\dot{\varphi}_+\right)_{x=x_0} = 0 = \left(\frac{d}{dx} A\varphi_+\right)_{x=x_0} + ik(\lambda_{11} - \lambda_{12}). \end{cases}$$

For real $k$ we have

(2.2.10) $$\Lambda=\begin{pmatrix}\lambda & \mu \\ \bar{\mu} & \bar{\lambda}\end{pmatrix},\qquad \lambda=\lambda_{11},\qquad \mu=\lambda_{12},$$

where the trace Sp $\Lambda = 0$ or $\lambda + \bar{\lambda} = 0$, and the coefficients $\lambda$, $\mu$ are as in (2.2.9). The matrix $\Lambda = (a_{ij})$ for the basis (2.1.2) is obtained in exactly the same way. We note that in the basis (2.1.2) the coefficients $a_{ij}$ of $\Lambda$ are polynomials in $k^2 = E$, $u(x_0)$, $u'(x_0)$, ... The determinant det $\Lambda$ does not depend on the choice of basis and is a polynomial

(2.2.11) $$\det\Lambda = P_{2n+1}(E),$$

the operator $A = A_n$ being of degree $2n + 1$. To calculate the dynamics of the matrix $\hat{T}(x_0, E)$ in any of the bases we must compute (2.2.8) for $x = x_0 + T$, where $T$ is the period. For example, in the basis (2.1.1) and for real $k$ we obtain

(2.2.12) $$\begin{cases}[\dot{\varphi}_+]_{x=x_0+T}=\dot{a}+\dot{b}=[A(a\varphi_++b\varphi_-)]_{x=x_0+T}+\\ \qquad\qquad +\lambda(a+b)+\mu(\bar{b}+\bar{a}),\\ \left[\frac{d}{dx}\dot{\varphi}_+\right]_{x=x_0+T}=ik(\dot{a}-\dot{b})=\left[\frac{d}{dx}A(a\varphi_++b\varphi_-)\right]_{x=x_0+T}+\\ \qquad\qquad +\lambda ik(a-b)+\mu ik(\bar{b}-\bar{a}).\end{cases}$$

Bearing in mind that $A$ is real, we finally obtain without difficulty

(2.2.13) $$\dot{a}=\mu\bar{b}-b\bar{\mu},\qquad \dot{b}=(\lambda-\bar{\lambda})b+(a-\bar{a})\mu,$$

where

$$\Lambda=\begin{pmatrix}\lambda & \mu \\ \bar{\mu} & \bar{\lambda}\end{pmatrix}.$$

Clearly, the equations (2.2.13) are equivalent to the matrix equation

(2.2.14) $$\frac{\partial}{\partial t}\hat{T}=[\Lambda, \hat{T}],$$

where the matrix $\Lambda$ is defined by (2.2.8) and in the basis (2.1.2) depends polynomially on $E$, $u(x_0)$, $u'(x_0)$, ..., $u^{(2n)}(x_0)$. In the form (2.2.14) the equation holds, of course, in any basis. If at a point $x_0$ we have $u = u' = \ldots = 0$, then in the basis (2.1.1) we easily obtain

(2.2.15) $$\Lambda=\text{const}\begin{pmatrix}ik^{2n+1} & 0 \\ 0 & -ik^{2n+1}\end{pmatrix}.$$

Substituting (2.2.15) in (2.2.14) we clearly obtain the Gardner–Green–Kruskal–Miura formulae for rapidly decreasing potentials by setting $x_0 = \pm\infty$. In the periodic case, (3.2.14) can no longer be integrated. For the initial K-dV equation (2.2.14) the formula in the basis (2.1.2) was derived by another method by Marchenko [44], who went further in constructing a method of successive approximations for solving this equation.

We shall proceed in a different way. The matrix $\hat{T}$ depends on the parameters $x_0$, $t$, and $E$. We consider for $\hat{T}$ the pair of equations (2.2.14) and (2.1.5):

$$\frac{\partial}{\partial t}\hat{T}=[\Lambda, \hat{T}], \qquad \frac{\partial}{\partial x_0}\hat{T}=[Q, \hat{T}].$$

From the condition for compatibility of this pair of equations, $\frac{\partial}{\partial x_0}\frac{\partial}{\partial t}\hat{T}=\frac{\partial}{\partial t}\frac{\partial}{\partial x_0}\hat{T}$, we obtain

(2.2.16) $$\left[\frac{\partial\Lambda}{\partial x_0}-\frac{\partial Q}{\partial t}-[\Lambda, Q], \hat{T}\right]=0,$$

from which it follows, since the trace of the left-hand side vanishes, that

(2.2.17) $$\frac{\partial\Lambda}{\partial x_0}-\frac{\partial Q}{\partial t}=[\Lambda, Q].$$

Now (2.2.17) gives a new and very convenient commutation representation of the K-dV and all "higher order K-dV" equations by second order matrices that depend polynomially on $E$ (in the base (2.1.2)). An analogue of this commutation representation as well as analogues of (2.2.14) and (2.1.5) can also be obtained naturally for all other non-linear systems mentioned in Ch. 1 (see Ch. 3).

We now consider the "general K-dV equation"

(2.2.18) $$\dot{u}=\frac{\partial}{\partial x}\left(\sum_{i=0}^{n} c_i\frac{\delta I_{n-i}}{\delta u(x)}\right) \qquad (c_0=1)$$

and the equivalent Lax equation

(2.2.19) $$\dot{L}=[A, L], \text{ where } A=\sum_{i=0}^{n} c_iA_{n-i}.$$

The matrix $\Lambda$ constructed according to (2.2.8) depends additively on $A$, and we have the new representation (2.2.17)

(2.2.20) $$\frac{\partial\Lambda}{\partial x_0}-\frac{\partial Q}{\partial t}=[\Lambda, Q], \qquad \Lambda=\sum_{i=0}^{n} c_i\Lambda_{n-i}.$$

Suppose that we wish to find a stationary solution $\dot{u} = 0$ of the higher order K-dV equation (2.2.18). If $\dot{u} = 0$, then $\frac{\partial Q}{\partial t} = 0$ and we obtain

(2.2.21) $$\frac{d\Lambda}{dx_0}=[\Lambda, Q].$$

Thus, we have proved the following result.

COROLLARY 1 (see [38]). *The system of ordinary equations*

(2.2.22) $$\sum c_i\frac{\delta I_{n-i}}{\delta u(x)}=\text{const}$$

*is equivalent to the equation of Lax type* (2.2.21) *with second order matrices that are polynomially dependent on the extra parameter $E$ and are polynomial in $u(x)$, $u'(x)$, .... In particular, all the coefficients of the* *characteristic polynomial $P(E)$, where*

$$W^2 + P(E) = \det(W - \Lambda) = W^2 + \det \Lambda \ (\mathrm{Sp}\ \Lambda = 0),$$

*are polynomials in $u, u', u'', \ldots$ independent of $x$ (that is, they are integrals of the system (2.2.22)). The polynomial $P(E)$ itself is of degree $2n + 1$.*

We now prove the following beautiful and unexpected corollary of equation (2.2.14).

COROLLARY 2. *The roots of the polynomial $P(E) = 0$ are the complete set of boundaries of the forbidden and of the permitted zones of the Schrödinger operator $L$ with potential $u(x)$ satisfying the equation (2.2.22) for the stationary solutions of any higher order K-dV equation. In particular, all the stationary solutions of the K-dV equations are potentials whose number of zones does not exceed $n$.*

DERIVATION OF COROLLARY 2. From (2.2.14) $\dot{\hat{T}} = [\Lambda, \hat{T}]$ we obtain $\dot{\hat{T}} = 0$ or $[\Lambda, \hat{T}] = 0$. We consider the matrix element (in the base (2.1.1) $[\Lambda, \hat{T}]_{12} = (a - \bar{a})\mu + (\lambda - \bar{\lambda})b$ for real $k$. Since $\bar{\lambda} = -\lambda$, we have $2\mu a_{\mathrm{Im}} = 2\lambda b$. From this it follows that in all the non-degenerate levels (boundaries of the zones) for which $|b| \neq 0$

$$\left|\frac{a_{\mathrm{Im}}}{b}\right| = \left|\frac{\lambda}{\mu}\right|, \tag{2.2.23}$$

or $\det \Lambda = P(E) = |\lambda|^2 - |\mu|^2$ when $E = E_n$. By adding a constant to $u$ we can always achieve that all $E_n > 0$ and the $k_n = \sqrt{E_n}$ are real. Thus, Corollary 2 is proved. This proof is taken from [38]. Another derivation of Corollary 2 was indicated by Dubrovin in connection with a generalization of these results to matrix systems (see Ch. 3): since by the definition (2.1.6) the Bloch function $\psi_\pm(x, x_0, E)$ is an eigenfunction for $\hat{T}$, $\hat{T}\psi_\pm = e^{ip(E)}\psi_\pm$, it follows from $[\Lambda, \hat{T}] = 0$ that it is also an eigenvector for the matrix $\Lambda$. Since $\Lambda$ depends polynomially on $E$, the eigenvector $\psi_\pm$ is meromorphic on the Riemann surface $\Gamma$:

$$R(W, E) = \det(W - \Lambda) = 0. \tag{2.2.24}$$

In our case $\mathrm{Sp}\ \Lambda = 0$, and we have

$$R(W, E) = W^2 + P_{2n+1}(E) = 0, \tag{2.2.24'}$$

where $P_{2n+1}(E) = \det \Lambda$. The branch-points of $\Gamma$ are the boundaries of the zones, as indicated in Ch. 2, §1. It is important that this derivation easily generalizes not only to complex potentials, but also to almost-periodic potentials and to other linear operators.

A third derivation, which yields part of the result of Corollary 2, was obtained by Lax [50] simultaneously with, and independently of, [38], as was mentioned in the Introduction. In fact, following [50] we prove that every periodic stationary solution of any "higher order K-dV equation" has only finitely many forbidden zones in the spectrum of the operator

$L = -\frac{d^2}{dx^2} + u(x)$. For the sake of simplicity, we produce Lax's derivation for the original K-dV equation. Let $f_n$ be a non-degenerate periodic eigenfunction. We consider a normalization of the eigenfunction $f_n(x)$ instead of the bases (2.1.1) and (2.1.2) in which $Lf_n = E$, such that the development in time has the form

(2.2.25) $$\dot{f}_n = Af_n.$$

Since $Lf_n = E_n f_n$, (2.2.25) leads to a first order equation in $x$, which for $A = -4\frac{d^3}{dx^3} + 3(u\frac{d}{dx} + \frac{d}{dx}u)$ is as follows:

(2.2.25′) $$\dot{f}_n = (4E + 2u)f_{n_x} - u_x f_{n_x}.$$

The characteristics of the equation have the form

(2.2.26) $$\frac{dx}{dt} = -(4E + 2u),$$

and along the characteristics we have

(2.2.27) $$\frac{df_n}{dt} = u_x f_n.$$

Hence the zeros of $f_n$ are situated along the characteristics. For stationary solutions of the form $u(x - ct)$ (that is, $A = A_1 + cA_0$) we have

(2.2.28) $$f_n = f_n(x - ct), \quad \frac{dx}{dt} = c, \quad x = x_0 + ct.$$

Comparing (2.2.28) and (2.2.26), we obtain for the zeros of $f_n$:

(2.2.29) $$\frac{dx}{dt} = c = -(4E_n + 2u(x - ct)) = -4E_n - 2u(x_0).$$

For large values of $n$ the number $E_n$ is large and (2.2.29) cannot hold. This means that either the eigenfunction $f_n$ has no zeros, or that it cannot be non-degenerate. We know that eigenfunctions corresponding to high levels $E_n \to 0$ have more and more zeros. In consequence, all but finitely many levels are degenerate. A similar derivation for antiperiodic levels (period $2T$) shows that there are finitely many zones. It is easy to see that this derivation generalizes also to higher order K-dV equations. From this derivation it does not follow that the number of lacunae is $\leqslant n$ for the $n$th analogue of the K-dV equation (for the excluded case $n = 1$ see [50]). Moreover, this derivation, unlike Corollaries 1 and 2 above, does not give an algorithm for the integration of the stationary problem for higher order K-dV equations nor an algorithm for finding the boundaries of the zones.

We now pass on to the most recent work on stationary solutions of higher order K-dV equations. All the (2.2.22) are equations for extremals of the functional

(2.2.30) $$\delta\left(dI_{-1} + \sum_{i=0}^{n} c_i I_{n-i}\right) = 0 \quad (c_0 = 1),$$

where $I_{-1} = -\int u\,dx$. Therefore, all the equations (2.2.30) are Hamiltonian systems with $n$ degrees of freedom, depending on the $(n+1)$ constants $d, c_1, \ldots, c_n$. The coefficients of the polynomial $P_{2n+1}(E) = \det \Lambda$ are constructed in such a way that the leading one is a constant, then the next $n+1$ are formed from the parameters $(c_1, c_2, \ldots, c_n, d)$ and the last $n$, denoted by $J_1, \ldots, J_n$, give a set of polynomials that are algebraically independent integrals of the system (2.2.22). In fact, the integrals $J_1, \ldots, J_n$ are involutory and hence the system (2.2.22) is totally integrable, and its solutions, in principle, can be determined by the Liouville algorithm. (It will become clear later that the Hamiltonians $J_\alpha$ give a set of independent commuting systems, hence we do not prove this here.) Following the method applied above in the derivation of (2.2.17) from (2.2.14) and (2.1.5), we use the commutativity of all the higher order K-dV equations as dynamical systems in a function space. If we have two equations

$$\frac{\partial u}{\partial t_m} = \frac{\partial}{\partial x}\frac{\delta I_m}{\delta u(x)},$$

$$\frac{\partial u}{\partial t} = \frac{\partial}{\partial x}\sum_{i=0}^{n} c_i \frac{\delta I_{n-i}}{\delta u(x)},$$

then the solution is a function $u(x, t, t_m)$ by virtue of the commutativity of these systems. For the matrix $\hat{T}(x_0, E)$ we obtain

$$\frac{\partial \hat{T}}{\partial t_m} = [\Lambda_m, \hat{T}], \qquad \frac{\partial \hat{T}}{\partial t} = [\Lambda, \hat{T}], \tag{2.2.31}$$

where $\Lambda = \sum_{i=0}^{n} c_i \Lambda_{n-i}$. If $u(x)$ is a stationary solution, that is, $\frac{\partial u}{\partial t} = 0$ (so that $u$ is a solution of (2.2.22)), then

$$\frac{\partial \hat{T}}{\partial t} = \frac{\partial \Lambda_m}{\partial t} = 0.$$

In all cases we obtain from the compatibility of the two equations (2.2.31), by analogy with (2.2.17),

$$\frac{\partial \Lambda_m}{\partial t} - \frac{\partial \Lambda}{\partial t_m} = [\Lambda, \Lambda_m]. \tag{2.2.32}$$

If $\frac{\partial \hat{T}}{\partial t} = \frac{\partial \Lambda_m}{\partial t} = 0$, then

$$\frac{d\Lambda}{dt_m} = [\Lambda_m, \Lambda]. \tag{2.2.33}$$

When $m = 0$, (2.2.33) becomes (2.2.21) with $\Lambda_0 = Q$. Clearly, (2.2.33) is defined over the same phase space as (2.2.21) and gives a set of commuting Hamiltonians of dynamical systems on the phase space of the problem (2.2.21) (or on the set of stationary solutions of the higher order K-dV equations, given by (2.2.22)).

When $m = 1$, (2.2.33) determines the dynamics of finite-zone potentials under time evolution governed by the K-dV equation, which is described explicitly in the form of a finite-dimensional dynamical system, represented by a matrix equation for second order matrices polynomially dependent on the parameter $E$ (in the basis (2.1.2)). Here, the matrix $\Lambda$ replaces the Schrödinger operator $L$ and the characteristic polynomial of $\Lambda$ has the form $R(W, E) = \det (W - \Lambda)$, and the Riemann surface $R(W, E) = 0$ is also the "spectrum" (in the sense of Ch. 2, § 1) of the potential we are studying.

We exhibit the matrix $\Lambda_0 = Q$, $\Lambda_1$, $\Lambda_2$ in the basis (2.1.1) for real $k$, the corresponding polynomials $P_3(E)$, $P_5(E)$, and the integrals $J$ (for $n = 1$), $J_1$ and $J_2$ for $n = 2$. The matrix $\Lambda = \begin{pmatrix} \lambda & \mu \\ \bar{\mu} & \bar{\lambda} \end{pmatrix}$ has the following form:

(2.2.34) $$1)\ n=0;\quad \lambda = ik - \frac{iu}{2k},\quad \mu = \frac{iu}{2k},\quad R_1 = E - c,$$

and the potential has the form $u = c$.

(2.2.35) $$2)\ n=1;\quad \Lambda = \Lambda_1 + c\Lambda_0,$$

$$\lambda = \frac{i}{k}\left(-\frac{u''}{2} + u^2 - 4k^4 + ck^2 - c\frac{u}{2}\right),$$

$$\mu = u' + \frac{i}{k}\left(\frac{u''}{2} - u^2 - 2k^2 u + c\frac{u}{2}\right),$$

$$R_3(E) = |\lambda|^2 - |\mu|^2 = E^3 + \frac{1}{2}cE^2 + \frac{1}{16}(c^2 - 4d) + \frac{1}{16}(cd - J),$$

$$J = (u')^2 - (2u^3 + cu^2 + 2du);$$

the potential has the form

$$x - x_0 = \int \frac{du}{\sqrt{2u^3 + cu^2 + 2du + J}}.$$

(2.2.36) $$3)\ n=2;\ \Lambda = \Lambda_2 + c_2\Lambda,\qquad (c_1 = 0),$$

$$\lambda = \frac{i}{k}\left[\frac{1}{2}u^{\mathrm{IV}} - (4uu'' + 3(u')^2 - 3u^3) - 2u^2k^2 + \right.$$
$$\left. + 16k^6 + c_2k^2 - c_2\frac{u}{2}\right],$$

$$\mu = -u''' + 6uu' - 4u'k^2 + \frac{i}{k}\left[\frac{1}{2}u^{\mathrm{IV}} + 4uu'' + 3(u')^2 - \right.$$
$$\left. - 3u^3 + k^2(2uu'' - 4u^2) - 8uk^4 + c_2\frac{n}{2}\right],$$

$$P_5(E) = |\lambda|^2 - |\mu|^2 = E^5 + \frac{1}{4}c_2E^3 - \frac{1}{16}dE^2 + \frac{1}{32}J_1 + \frac{1}{4}c_2^2\Big)E + \frac{J_2}{2^8} + \frac{c_2 d}{2^7},$$

$$J_1 = p_1p_2 - \left(\frac{1}{2}q_2^2 + \frac{5}{2}q_1^2q_2 + \frac{5}{8}q_1^4\right) + c_2q_1^2 - dq_1,$$

$$J_2 = p_1^2 - 2q_1p_1p_2 + 2(q_2 - c_2)p_2^2 + q_1^5 + 2c_2q_1^3 + dq_1^2 - 4q_1q_2^2 + 4c_2q_1q_2 - 2dq_2.$$

The Hamiltonian system (2.2.22) for $c_1 = 0$ is given by the Hamiltonian $H = J_1$, and the canonical coordinates are:

(2.2.37) $$p_1 = q_2',\quad p_2 = u',\quad q_1 = u,\quad q_2 = -\frac{5}{2}u^2 + u''.$$

In particular, for the Lamé potential $3u(x)$, where $u(x) = 2\wp(x)$, $\wp(x)$ being the Weierstrass elliptic function, we can exhibit the boundaries of the zones: if $(\wp')^2 = 4\wp^3 - g_2\wp - g_3$ and $e_1, e_2, e_3$ are the roots of the polynomial $4x^3 - g_2x - g_3$ (all $e_i$ are real, $e_1 < e_2 < e_3$), then the boundaries of the zones are

(2.2.38) $E_1 = 3e_1, \quad E_2 = -\sqrt{3g_2}, \quad E_3 = 3e_2, \quad E_4 = \sqrt{3g_2}, \quad E_5 = 3e_3.$

Convenient coordinates $\gamma_1, \gamma_2$ on the level surfaces $J_1$ = const, and $J_2$ = const are given as follows (let $\sum_{i=1}^{5} E_i = 0$):

$$\begin{cases} 2(\gamma_1 + \gamma_2) = u = q_1, \\ \gamma_{1,2} = -\frac{1}{4}\left[q_1 \pm \sqrt{2q_2 - 8\sum E_iE_j}\right], \end{cases} \tag{2.2.39}$$

and all $p_1, p_2, q_1, q_2$ can be expressed in terms of $\gamma_1$ and $\gamma_2$. The interpretation of the coordinates $\gamma_1$ and $\gamma_2$, the formation of "angle variables" from them, and the completion of the integration of the equation (2.2.22) for $n = 2$ will be discussed in §3 (see (2.3.14)). In general, the real solutions of (2.2.22) are almost-periodic functions with a group of periods $T_1, \ldots, T_n$. Bounded solutions correspond to tori in the phase space, and the group of periods can be expressed in terms of the boundaries of the zones, or, what amounts to the same thing, in terms of the constants $(c_1, \ldots, c_n, d)$ and the integrals $J_1, \ldots, J_n$. Later, in §3, we shall give convenient formulae for the periods $T_1, \ldots, T_n$ and for the potentials themselves. The actual methods, to be developed in §3, are all connected with a very important circumstance, which has so far not been clear: the $n$-dimensional level surface $J_1 = a_1, \ldots, J_n = a_n$, when continued into the complex domain, is an Abelian variety (the complex torus $T^{2n}$), which is the Jacobian variety of the Riemann surface $\Gamma$ given by the equation $R(W, E) = 0$ in which $R(W, E) = \det(W - \Lambda) = W^2 + P_{2n+1}(E)$. This important property generalizes the natural case $n = 1$, where the Abelian varieties are defined as the complex solutions of the equation $u'' = \frac{\partial P_3(u)}{\partial u}$, $P_3$ being a polynomial of degree 3. The corresponding result when the number of lacunae is $n > 1$ is non-trivial and will be obtained by combining the methods of §3 with those of the present section.

We give one more useful application of the equations (2.1.15) and (2.2.17) for the matrices $\hat{T}$ and $\Lambda$. We recall the formula (2.1.19) for $\chi_R(x, E)$ (in the base (2.1.1)), $\chi_R(X, E) = \frac{k\sqrt{1-a_R^2}}{a_{\text{Im}} + b_{\text{Im}}}$. From (2.1.5) and (2.1.17) follow the general identities

$$(2.2.40)\qquad \begin{cases} 2\mu_R = -\dfrac{d}{dx_0}\left(\dfrac{\lambda_{\mathrm{Im}}+\mu_{\mathrm{Im}}}{k}\right), \\ 2b_R = -\dfrac{d}{dx_0}\left(\dfrac{a_{\mathrm{Im}}+b_{\mathrm{Im}}}{k}\right). \end{cases}$$

Comparing these with the form of $\chi_R$ and the condition $\dot{a}_R \equiv 0$, we obtain

$$(2.2.41)\qquad \chi_R = (\alpha\chi_R)', \text{ where } \alpha = \frac{\lambda_{\mathrm{Im}}+\mu_{\mathrm{Im}}}{k}.$$

For the K-dV equation we have $\Lambda = \Lambda_1$ and $\alpha = -2(2E + u)$. From (2.2.41) it follows that the quantity $I(k) = p(E) = \int_{x_0}^{x_0+T} \chi_R dx = \int_{x_0}^{x_0+T} \chi\, dx$ is conserved. We shall use (2.2.41) in §3 in the calculation of the dynamics of finite-zone potentials in terms of the parameters $\gamma_1, \ldots, \gamma_n$.

As $E \to \infty$, we have the expansion

$$I(k) \sim kT + \sum_{n\geqslant 0} \frac{1}{(2k)^{2n+1}} \int_{x_0}^{x_0+T} \chi_{2n+1}(x)\, dx,$$

where the integrals $I_{n-1} = \int_{x_0}^{x_0+T} \chi_{2n+1} dx$ are Hamiltonians of the higher order K-dV equations, $\dot{u} = \dfrac{\partial}{\partial x}\dfrac{\delta I_n}{\delta u(x)}$, $\dot{L} = [A_n, L]$. It is natural to introduce a generating function for this set of equations. We consider an operator $A_z$, which depends on the parameter $z$ and has the following property (by (2.1.17) and (2.1.15)):

$$(2.2.42)\qquad [A_z, L] = \frac{\partial}{\partial x}\frac{\delta p(z)}{\delta u(x)} = \frac{\partial}{\partial x}\left(-\frac{1}{2\chi_R(x, z)}\right),$$

where $\chi(x, z) = \chi_R + i\chi_I$, $\chi_I = \frac{1}{2}(\ln \chi_R)'$, $-i\chi' + \chi^2 + u - E = 0$. Then all the operators $A_n$ are obtained as coefficients of the expansion of $A_z$, as $z \to \infty$, in powers of $z^{-\frac{1}{2}}$. We can verify that $A_z$ has the form

$$(2.2.43)\qquad A_z = -\frac{1}{4}\left[\frac{1}{\chi_R(x, z)}\frac{d}{dx} - \frac{1}{2}\left(\frac{1}{\chi_R(x, z)}\right)'\right]\frac{1}{L-z}.$$

We now give a convenient algorithm for obtaining all the matrices $\Lambda_m$ (in the base (2.1.2)), which will also be very useful in Ch. 3 for the generalization to first order matrix operators. We consider the differential equation $\lambda' = [Q, \lambda]$, in which the matrix $Q$ in the base (2.1.2) has the form $Q = \begin{pmatrix} 0 & E-u \\ -1 & 0 \end{pmatrix}$. This equation has a unique solution as a formal series $\lambda = \begin{pmatrix} 0 & 1 \\ 0 & 0 \end{pmatrix} + \frac{\lambda_1}{E} + \frac{\lambda_2}{E^2} + \dots$; the coefficients $\lambda_n$ are determined from the recurrence relation $\left[\lambda_n, \begin{pmatrix} 0 & 1 \\ 0 & 0 \end{pmatrix}\right] = \lambda'_{n-1} + \left[\lambda_{n-1}, \begin{pmatrix} 0 & 0 \\ 1 & 0 \end{pmatrix}\right]$. Then the matrices $\Lambda_m$ have the form

$$\Lambda_m = \begin{pmatrix} 0 & 1 \\ 0 & 0 \end{pmatrix} E^m + \lambda_1 E^{m-1} + \dots + \lambda_m. \tag{2.2.44}$$

## §3. The inverse problem for periodic and almost-periodic (real and complex) finite-zone potentials. The connection with the theory of Abelian varieties

Starting from the Definitions 1 and 1′ of Ch. 2, §1, by a "finite-zone" potential we now understand a potential $u(x)$ for which the Bloch eigenfunction $\psi_\pm(x, x_0, E)$, normalized by (2.1.6) and (2.1.7), is meromorphic on the Riemann surface $\Gamma$:

$$\Gamma\colon\ W^2 = P_{2n+1}(E) = \prod_{i=1}^{2n+1} (E - E_i), \tag{2.3.1}$$

and has the asymptotic form

$$\psi_\pm \sim e^{\pm ik(x-x_0)}, \quad k^2 = E, \quad E \to \infty. \tag{2.3.2}$$

According to (2.1.13), we study on $\Gamma$ the function $\chi_R$,

$$\chi_R = \frac{1}{2i} W(\psi_+, \psi_-) = \frac{1}{2i}(\psi'_+\psi_- - \psi'_-\psi_+),$$

using the representation (2.1.14) and (2.1.16), (2.1.16′):

$$\psi_\pm(x, x_0, E) = \sqrt{\frac{\chi_R(x_0, E)}{\chi_R(x, E)}} \exp\left\{ i \int_{x_0}^{x} \chi_R(x, E)\, dx \right\},$$

where according to (2.1.19) $\chi_R = \dfrac{k\sqrt{1 - a_R^2}}{a_I + b_I}$ (basis (2.1.1), real $k$) or $\chi_R = \sqrt{1 - \frac{1}{4}(\alpha_{11} + \alpha_{22})^2 / \alpha_{21}}$ (basis (2.1.2), arbitrary $E$). We note that at degenerate points $E_n$ of the spectrum we have, according to (2.1.10), $\hat{T} = \pm \begin{pmatrix} 1 & 0 \\ 0 & 1 \end{pmatrix}$, $\alpha_{21} = 0$, and $\alpha_{21}(x_0, E)$ has a simple zero in $E$. By contrast, at non-degenerate points $E_n$ of the spectrum we have $\alpha_{21}(x_0, E_n) \not\equiv 0$. Let $\widetilde{E}_\alpha$ ($\alpha = 1, 2, 3, \dots$) be all the degenerate points of the (periodic and anti-periodic) spectrum. Since the quantity $(1 - a_R^2) = 1 - \frac{1}{4}(\alpha_{11} + \alpha_{22})^2$ has a two-fold zero at all points $\widetilde{E}_\alpha$ and a simple zero at the non-degenerate points $E_1, \dots, E_{2n+1}$, we see that

$$\sqrt{1 - a_R^2} = \sqrt{\prod_{i=1}^{2n+1} (E - E_i)} \cdot f(E), \tag{2.3.3}$$

where $f(E)$ is an entire function having only simple zeros at the points $\widetilde{E}_\alpha$.

Therefore, the quotient

$$\tilde{\alpha}_{21} = \frac{\alpha_{21}}{f(E)} \tag{2.3.4}$$

is an entire function of $E$. We have the expression

$$\chi_R = \frac{\sqrt{\prod_{i=1}^{2n+1} (E - E_i)}}{\tilde{\alpha}_{21}}. \tag{2.3.5}$$

As $E \to \infty$, the quantity $\chi_R(x_0, E)$ has the asymptotic form $\chi_R \sim k = \sqrt{E}$. Hence, by (2.3.5), the quantity $\tilde{\alpha}_{21}$ has the asymptotic form

$$\tilde{\alpha}_{21} \sim E^n \quad \text{as} \quad E \to \infty. \tag{2.3.6}$$

Since $\alpha_{21}(x_0, E)$ is an entire function, we see from (2.3.6) that it is, in fact, a polynomial:

$$\tilde{\alpha}_{21}(x_0, E) = \prod_{j=1}^{n} (E - \gamma_j(x)). \tag{2.3.7}$$

From this we obtain the final result.

COROLLARY 3. *The quantity* $\chi_R(x_0, E) = \frac{1}{2i} W(\psi_+, \psi_-)$ *has the following form*:

$$\chi_R(x, E) = \frac{\sqrt{\prod_{i=1}^{2n+1} (E - E_i)}}{\prod_{j=1}^{n} (E - \gamma_j(x))} = \frac{\sqrt{P(E)}}{P_n(x, E)}. \tag{2.3.8}$$

NOTE. The quantity $\tilde{\alpha}_{21} = \prod_{j=1}^{n} (E - \gamma_j(x))$ can be determined otherwise; it is a solution of the equation

$$-y''' + 4(u - E)y' + 2u'y = 0. \tag{2.3.9}$$

which is polynomial in $E$.

From this argument we can obtain another derivation of (2.3.8), since (2.3.8) is satisfied by the product $\varphi_1 \cdot \varphi_2$ of two solutions of the equation $L\varphi_\alpha = E\varphi_\alpha$, setting $\varphi_1 = \psi_+$, $\varphi_2 = \psi_-$ (the derivation in [41] and [47]).

Following the derivation in [40] and [46] we evaluate the product $\psi_+ \psi_-$ starting from (2.1.14):

$$\psi_+ \psi_- = \frac{\chi_R(x_0, E)}{\chi_R(x, E)} = \prod_{j=1}^{n} \frac{E - \gamma_j(x)}{E - \gamma_j(x_0)}. \tag{2.3.10}$$

Formally speaking, this derivation refers only to periodic potentials because it uses the translation matrix $\hat{T}$. However, the formulae (2.3.8) and (2.3.10) hold also in the almost-periodic case. To prove this we can proceed as follows: later, when we obtain the functions $\gamma_j(x)$, we have to verify that the function $\psi_\pm(x, x_0, E)$ defined by (2.1.14) really satisfies the equation $L\varphi = E\varphi$.

We now draw some conclusions from the results obtained.

1. *The poles of the Bloch function lie only on one sheet over the points $\gamma_j(x_0)$ if all the $\gamma_j(x_0)$ are pairwise distinct.*

PROOF. From (2.1.16), (2.3.8) we see that the poles of $\psi_\pm$ can lie only above points $\gamma_j(x_0)$ where $\chi_R$ has poles, since $\psi_\pm = c + i\chi(x_0, E)s$ and since $c$ and $s$ are entire functions of $E$. If there were poles on both sheets $(\gamma_j(x_0), \pm)$ then in the product $\psi_+\psi_- = \prod_j \frac{E-\gamma_j(x)}{E-\gamma_j(x_0)}$ we would have a double pole. This contradiction completes the proof. We note that for real periodic potentials the poles $\gamma_j(x_0)$ occur one each in the forbidden zones.

2. *Symmetric functions of $\gamma_1(x), \ldots, \gamma_n(x)$ can be expressed in terms of the potential $u(x)$ and its derivatives; in fact,*

$$u(x) = -2\sum_{j=1}^{n} \gamma_j(x) + \sum_{i=1}^{2n+1} E_i, \tag{2.3.11}$$

$$\sum \gamma_i\gamma_j = \frac{1}{8}(3u^2 - u'') + \frac{1}{2}\sum E_iE_j - \frac{3}{8}\left(\sum E_i\right)^2. \tag{2.3.11'}$$

PROOF. Starting from the asymptotic form of $\chi_R(x, k)$ as $k \to \infty$, we have (see (2.1.18)

$$\chi_R(x, k) \sim k + \sum_{n\geq 1} \frac{\chi_{2n+1}(x)}{(2k)^{2n+1}},$$

where $\chi_1(x) = -u(x), \ldots$ Comparing this fact with the form of $\chi_R$ (see (2.3.8)), we obtain the result.

3. *Each finite-zone potential $u(x)$ is a stationary solution of one of the higher order* K-dV equations (2.2.22).

PROOF. We start from (2.1.17):

$$\frac{\delta p(E)}{\delta u(x)} = \frac{\delta}{\delta u(x)} \int_{x_0}^{x_0+T} \chi_R(x, E)\, dx = -\frac{1}{2\chi_R(x, E)}$$

in the case of periodic potentials. From the form of $\chi_R$ it follows that amongst the coefficients of the expansion of $-\frac{1}{2\chi_R} = \frac{\delta p}{\delta u}$ in powers of $\frac{1}{\sqrt{E}}$ there are only finitely many linearly independent functions of $x$, and this leads to the relation

$$\sum_{m=-1}^{n} c_m \frac{\delta I_m}{\delta u(x)} = 0,$$

where $I_{m-1} = \int_{x_0}^{x_0+T} \chi_{2m+1}dx$ $(m \geqslant 0)$ are the coefficients of the expansion of $p(E)$, as $E \to \infty$, as follows from (2.1.18) and (2.1.15). From this the proof follows in the periodic case. The almost-periodic case is similar: the integrals over the period must be replaced by the mean value; all the formulae remain valid. For real periodic functions the result is suggested by standard uniqueness theorems (see [1]), from which it follows that the potentials form at most an $n$-dimensional family for a given spectrum (with $n$ forbidden zones); actually, by the results of Ch. 2, §2 we have an exactly $n$-dimensional family. However, this result is needed also for almost-periodic and for complex potentials.

4. *The set of quantities* $\gamma_1(x), \ldots, \gamma_n(x)$ *satisfies the system of equations*

$$\gamma_j' = \pm \frac{2i\sqrt{P_{2n+1}(\gamma_j)}}{\prod\limits_{k\neq j}(\gamma_k-\gamma_j)}. \tag{2.3.12}$$

Strictly speaking, these equations must be treated as equations in $x_0$ for the set of points $P_1(x_0), \ldots, P_n(x_0)$ on $\Gamma$, where $P_j(x_0) = (\gamma_j(x_0), \pm)$, and the poles of $\psi_\pm(x, x_0, E)$ lie at $P_j(x_0)$. Above each lacuna $[E_{2i}, E_{2i+1}] = l_i$ in the real case lies the cycle $a_i$ on $\Gamma$ that is obtained by fusing the two intervals $a_i = (l_i, +) \cup (l_i, -)$ at the end-points $(E_{2i}, +) = (E_{2i}, -)$ and $(E_{2i+1}, +) = (E_{2i+1}, -)$. The points $P_j(x_0)$ lie one each on the cycles $a_j$ (by Corollary 1 above) and more over these cycles as $x_0$ varies. The whole "phase point" $(P_1, \ldots, P_n)$ lies on the torus $T^n = S_1^1 \times \ldots \times S_n^1$ (the direct product of the cycles $a_j$), and (2.3.12) is actually written down for the motion of the point $(P_1, \ldots, P_n) \in T^n$.

PROOF OF (2.3.12). According to 1. above, the poles of $\psi_\pm$ lie only on one of the sheets $(\gamma_j, \pm)$ over $\gamma_j(x_0)$. On the other sheet there is no pole at $(\gamma_j, \pm)$. This means that in the formula (2.1.16) for $\psi_\pm$, or what is the same thing, in the quantity $\chi = \chi_R + i\chi_I = \chi_R + \frac{i}{2}\frac{d}{dx_0}\ln \chi_R$ (by (2.1.12)) the pole is cancelled for $E = \gamma_j$ and one choice of sign in the radical. From the form (2.3.8) of $\chi_R$ we obtain

$$\left(\frac{d}{dx}\frac{i}{2}\prod_{k=1}^{n}(E-\gamma_k)\right)_{E=\gamma_j} = \left(\sqrt{\prod_{k=1}^{2n+1}(E-E_k)}\right)_{E=\gamma_j}. \tag{2.3.13}$$

Solving (2.1.13) for $\gamma_j'$ we obtain (2.3.12). This proves the assertion.

We now analyse by way of illustration cases of one and two zones; the two-zone potentials were not previously known.

EXAMPLE 1. ($n = 1$). The equations (2.3.12) have the form

$$\gamma=\gamma_1, \quad \gamma' = \pm 2i\sqrt{\prod_{j=1}^{3}(\gamma-E_j)}, \quad u = -2\gamma + \sum_{i=1}^{3} E_i;$$

these formulae transform in an obvious way into (2.2.35).

EXAMPLE 2. ($n = 2$). The equations (2.3.12) for the parameters have the form

(2.3.14) $$\gamma_1' = \pm \frac{2i\sqrt{P_5(\gamma_1)}}{\gamma_1-\gamma_2}, \qquad \gamma_2' = \pm \frac{2i\sqrt{P_5(\gamma_2)}}{\gamma_2-\gamma_1},$$

where $-2(\gamma_1+\gamma_2)=u$, $\gamma_1\gamma_2 = \frac{1}{8}(3u^2-u'')+\frac{1}{2}\Sigma E_iE_j$. As was indicated in §2 (see (2.2.39), $\gamma_1$ and $\gamma_2$ are coordinates on the level surface of the two integrals $J_1$ and $J_2$ for the stationary solutions of the second order K-dV equation.

The equations (2.3.14) can be integrated by a change of variable

(2.3.15) $$d\tau = \frac{1}{\gamma_2-\gamma_1}\,dx.$$

For real potentials, where $E_2 \leqslant \gamma_1 \leqslant E_3$, $E_4 \leqslant \gamma_2 \leqslant E_5$, we introduce two functions $F_1(\tau)$ and $F_2(\tau)$, setting

(2.3.16) $$\tau = \int_{E_2}^{F_1} \frac{dq}{2\sqrt{P_5(q)}}, \qquad \tau = \int_{E_4}^{F_2} \frac{dq}{2\sqrt{P_5(q)}}.$$

We select an initial point $\tau_0$ so that

(2.3.17) $$\gamma_1(\tau) = F_1(\tau), \qquad \gamma_2(\tau) = F_2(\tau+\tau_0).$$

From (2.3.15) we have

(2.3.18) $$x-x_0 = \int_0^\tau (F_2(q+\tau_0)-F_1(q))\,dq.$$

The potential $u(x)$ has the form $u(x) = -2(\gamma_1+\gamma_2) + \sum_{i=1}^{5} E_i$, and in conjunction with (2.3.16), (2.3.17) and (2.3.18) we obtain our final expression for the two-zone almost periodic potentials:

$$u(x) = -2[F_1(\tau)+F_2(\tau+\tau_0)] + \sum_{i=1}^{5} E_i,$$

where

$$x-x_0 = \int_0^\tau (F_2(q+\tau_0)-F_1(q))\,dq.$$

In (2.2.38) values of the boundaries of the zones $E_i$ were exhibited for which, under a special choice of $x_0$ and $\tau_0$, the Lamé potential $3u(x)$ results, in which $u(x)$ is a one-zone potential (the Weierstrass $\wp$-function). Other potentials of this family (with the same spectrum as $3u(x)$) have the form (see [42])

(2.3.19) $$\begin{cases} v(x) = 2\wp(x-\beta_1)+2\wp(x-\beta_2)+2\wp(x-\beta_3), \\ \beta_1+\beta_2+\beta_3 = 0, \\ \beta_2-\beta_3 = \frac{1}{2}\wp^{-1}\left[-\wp(\beta_1-\beta_3)+\sqrt{g_2-3\wp^2(\beta_1-\beta_3)}\right]. \end{cases}$$

We now pass on to the connection with Abelian varieties. We have proved (see (2.3.11)) the formula $u(x) = -2\Sigma\gamma_j + 2\Sigma E_i$ for the potential

$u(x)$. The Bloch eigenfunction $\psi_{\pm}(x, x_0, E)$ has the set of poles $P_1(x_0), \ldots, P_n(x_0)$, which are points on the Riemann surface $\Gamma$, situated one each over the points $\gamma_1(x_0), \ldots, \gamma_n(x_0)$ of the $E$-plane. Strictly speaking, (2.3.11) is to be understood as follows: there is a canonical projection $\pi$: $\Gamma \to \mathbf{C}$ of $\Gamma$ onto the $E$-plane, and the potential $u(x)$ has the form

$$u(x) = -2 \sum_{j=1}^{n} \pi(P_j(x)) + \text{const.} \tag{2.3.20}$$

Essentially, the projection $\pi$ is a numerical function on $\Gamma$, invariant under interchange of the sheets (the canonical involution, which always occur on a hyperelliptic surface $\Gamma$). It is natural to define a numerical function $\sigma$ on the symmetrized set of $n$ points $(P_1, \ldots, P_n)$ such that

$$\sigma(P_1, \ldots, P_n) = \sum_{j=1}^{n} \pi(P_j).$$

The symmetrized sets $(P_1, \ldots, P_n)$ of points of $\Gamma$ form an algebraic variety, the symmetric power $\mathbf{S}^n(\Gamma)$), and $\sigma$ is an algebraic function on this variety. The set of points $(P_1, \ldots, P_n)$ varies when the parameter $x$ is changed, and the value of the function

$$-2\sigma(P_1(x), \ldots, P_n(x)) + \text{const}$$

is the potential $u(x)$, by (2.3.11). In classical algebraic geometry it has long been known that the symmetric power is birationally isomorphic to the $2n$-dimensional torus $\mathbf{J}(\Gamma)$, the Jacobian variety for $\Gamma$. Therefore, in particular, $\sigma$ is expressible in terms of multidimensional $\theta$-functions (of the Riemann $\theta$-function and its derivatives). The birational equivalence $\mathbf{S}^n(\Gamma) \to \mathbf{J}(\Gamma)$ is realized by the standard Abel map, which we describe below. We select a base of cycles on $\Gamma$:

$$a_1, \ldots, a_n, \ldots, b_1, \ldots, b_n, \tag{2.3.21}$$

where the intersection matrix has the form

$$a_i \circ a_j = b_i \circ b_j = 0, \qquad a_i \circ b_j = \delta_{ij} \tag{2.3.21'}$$

(the cycles $a_i$ for a real hyperelliptic surface $\Gamma$ were indicated above, they are the complete inverse images of the forbidden zones on $\Gamma$). We consider the basis of holomorphic differentials on $\Gamma$:

$$\Omega_k = \sum_{l=1}^{n} c_{kl} \frac{E^{l-1}\, dE}{\sqrt{P_{2n+1}(E)}} \qquad (k = 1, \ldots, n), \tag{2.3.22}$$

normalized by the conditions

$$\oint_{a_j} \Omega_k = 2\pi i \delta_{jk}. \tag{2.3.22'}$$

So we obtain the matrix

$$B_{kl} = \oint_{b_l} \Omega_k. \tag{2.3.23}$$

We know that this matrix is symmetric and has a real part of definite sign (the matrix $B_{kl}$ cannot be split into blocks, for $n = 2$ it is a complete set of constraints). The full $n \times 2n$ matrix

(2.3.24) $$\left(\oint_{a_j} \Omega_k, \oint_{b_j} \Omega_k\right) = (2\pi i\delta_{jk}, B_{jk})$$

defines $2n$ vectors in $n$-dimensional complex space $\mathbf{C}^n$, the integral linear combinations of which form a lattice that defines the torus $\mathbf{J}(\Gamma)$ as a Jacobian variety (the "Jacobian of $\Gamma$"). The Abel map $A$: $\mathbf{S}^n(\Gamma) \to \mathbf{J}(\Gamma)$ is defined as follows. We fix the points $P_1^0, \ldots, P_n^0$. We set

(2.3.25) $$A(P_1, \ldots, P_n) = (\eta_1, \ldots, \eta_n),$$

where

$$\eta_k = \sum_{j=1}^{n} \int_{P_j^0}^{P_j} \Omega_k,$$

and $\Omega_k$ is the basis (2.3.22) of holomorphic differentials. Clearly, the parameters $\eta_k$ are determined to within a vector of the lattice (2.3.24). This is the way the Abel mapping is constructed.

An important observation, mainly due to Akhiezer [10] consists in the fact that (translating the language of [10] into contemporary language) under the Abel map the set of zeros $\{P_1(x), \ldots, P_n(x)\}$ of the eigenfunction $\psi_\pm$ transforms into a straight line on the Jacobian variety (we shall show later that the eigenfunction $\mathscr{E}$, which Akhiezer constructed in his examples for the half-line is the same as the Bloch function $\psi_\pm(x, x_0, E)$ when the potential is even, $u(x) = u(-x)$, and $x_0 = 0$). We use the information on the zeros $P_j(x)$ of $\psi_\pm(x, x_0, E)$ lying over the points $\gamma_j(x)$, on the poles $P_j(x_0)$ lying over $\gamma_j(x_0)$, and on the asymptotic form $\psi_\pm \sim e^{\pm ik(x-x_0)}$, as $E \to \infty$. We consider the logarithmic differential

$$\omega = \left(\frac{d \ln \psi}{dE}\right) dE,$$

which has the following properties:

a) it has poles with residue $-1$ at the points $P_j(x_0)$;

b) it has poles with residue $+1$ at the points $P_j(x)$;

c) it has a pole of the second order at $E = \infty$, which in the local parameter $z$ has the form: $\omega \sim -i(x - x_0)\dfrac{dz}{z^2}$, where $z = \dfrac{1}{k} = \dfrac{1}{\sqrt{E}}$, since $\psi_\pm \sim e^{\pm ik(x-x_0)}$, as $E \to \infty$;

d) all the integrals over the cycles $\oint_{a_i} \omega$ and $\oint_{b_i} \omega$ are integer multiples of $2\pi i$, because $\psi_\pm$ is a single-valued function on $\Gamma$.

We claim that *the properties* a), b), c), d) *completely determine the function* $\psi_\pm(x, x_0, E)$ and that *as* $x$ *varies, the set of zeros* $(P_1(x), \ldots, P_n(x))$ *moves in a straight line on the Jacobian variety* $\mathbf{J}(\Gamma)$ *after the Abel map* $\mathbf{A}$.

PROOF. We introduce the differential $\Omega$ whose only singularity is a pole of the second order at $E = \infty$, and whose asymptotic form is $\Omega = -i\frac{dz}{z^2}$, as $E \to \infty$, where $z = \frac{1}{\sqrt{E}}$. We normalize $\Omega$ by the conditions

$$\oint_{a_j} \Omega = 0 \quad (j = 1, \ldots, n). \tag{2.3.26}$$

Then $\Omega$ has the form

$$\Omega = i\frac{E^n + \alpha_1 E^{n-1} + \ldots + \alpha_n}{2\sqrt{P_{2n+1}(E)}}\, dE,$$

where the coefficients $\alpha_j$ are determined by (2.3.26).

Let $\Omega^{PQ}$ be a differential on $\Gamma$ whose only singularities are a pole of residue $-1$ at $Q$ and a pole of residue $+1$ at $P$, normalized by the condition

$$\oint_{a_j} \Omega^{PQ} = 0 \quad (j = 1, \ldots, n). \tag{2.3.27}$$

In algebraic geometry it is known that if the $\Omega_j$ are normalized by (2.3.22′) and $\Omega^{PQ}$ by (2.3.27), then (for any path from $P$ to $Q$)

$$\oint_{b_j} \Omega^{PQ} = \int_Q^P \Omega_j, \tag{2.3.28}$$

where the cycles $(a_i, b_j)$ satisfy (2.3.21′). We now consider the differential $\omega = d_E \ln \psi$ and represent it in the form

$$\omega = (x - x_0)\,\Omega + \sum_{j=1}^{n} \Omega^{P_j(x) P_j(x_0)} + D, \tag{2.3.29}$$

where $D$ is the holomorphic differential $D = \sum_{i=1}^{n} \alpha_i \Omega_i$. From the condition $\oint_{a_j} \omega = 2\pi i m_j$; in which the $m_j$ are integers, we obtain, using the normalizations (2.3.22), (2.3.26) and (2.3.27),

$$2\pi i m_j = \Big(\sum_{k=1}^{n} \alpha_k \delta_{kj}\Big)\, 2\pi i \quad (j = 1, \ldots, n),$$

from which it follows that all the $\alpha_j$ are integers, $\alpha_j = m_j$. From the conditions $\oint_{b_j} \omega = 2\pi i n_j$, in which $n_j$ are the integers, we obtain, using (2.3.28)

$$2\pi i n_j = (x - x_0)\, U_j + \sum_{k=1}^{n} \int_{P_k(x_0)}^{P_k(x)} \Omega_j + \sum_{k=1}^{n} m_k B_{kj}, \tag{2.3.30}$$

where $U_j = -\oint_{b_j} \Omega$. From (2.3.30) it also follows that, to within a lattice vector, under variation of the parameter the set of points $(Q_1(x), \ldots, Q_n(x))$, describes a straight line on the torus $\mathbf{J}(\Gamma)$ whose gradient is defined by the vector $(U_1, \ldots, U_n)$, depending only on $\Gamma$.

(2.3.30) can be rewritten as

(2.3.30′)
$$\eta_j = \eta_j^0 + (x - x_0)U_j,$$

in which the $n_j$ are coordinates in $\mathbf{C}^n$, defined to within an element of the lattice (2.3.24). The restriction of the algebraic function $\sigma$ to the rectilinear winding (2.3.30′) also gives (in the real case) the potential $u(x)$, which is almost periodic with the group of periods $(T_1, \ldots, T_n)$, where

(2.3.31)
$$T_j^{-1} = \sum_{k=1}^{n} B^{jk} U_k \; .$$

(Here $B^{jk}$ is the inverse of the matrix of periods (2.3.23).) Under continuation into the complex $x$-domain, the potential becomes a meromorphic almost-periodic function with $2n$ periods $T_1, \ldots, T_n, T_1', \ldots, T_n'$, where the $T_j'$ along the imaginary axis have the form

(2.3.31′)
$$T_j' = \frac{2\pi i}{U_j}.$$

For complex $\Gamma$ the potential has $2n$ periods in the complex domain. For the potential $u(x)$ to be periodic in $x$, it is necessary and sufficient that the real periods $T_1, \ldots, T_n$ satisfy $n-1$ equations of the form

$$\sum_{j=1}^{n} n_{ij} T_j = 0 \qquad (i = 1, \ldots, n-1),$$

where the $n_{ij}$ are integers.

If the whole group of $2n$ complex periods $(T_1, \ldots, T_n, T_1', \ldots, T_n')$ reduces to two generators, then $u(x)$ is expressible in terms of elliptic functions (for this $2n-2$ integral linear relations are needed).

Thus, the Abel map integrates the equations (2.3.12) for the quantities $(\gamma_1, \ldots, \gamma_n)$. From our results so far, bearing in mind that $u(x)$ is uniquely determined by the set of initial points $[P_1(x_0), \ldots, P_n(x_0)]$ on $\Gamma$, we obtain the following theorem.

THEOREM 1) *The set of (real and complex) almost periodic finite-zone potentials with a given spectrum* $\Gamma$ *is canonically isomorphic to the Jacobian variety* $\mathbf{J}(\Gamma)$ *of* $\Gamma$, *which is a* $2n$*-dimensional Abelian variety, and this isomorphism is realized by the analytic operations described above. The group of periods is determined by the spectrum* $\Gamma$. 2) *The set of all complex solutions of the stationary problem for higher order K-dV equations is, to within a birational equivalence, for given constants* $(d, c_1, \ldots, c_n)$ *and levels of the commuting integrals* $J_1, \ldots, J_n$ *(or for a given spectrum) the Abelian variety* $\mathbf{J}(\Gamma)$, *where the Riemann surface* $\Gamma$ *is defined by the equation* $W^2 - P_{2n+1}(E) = 0$ (see (2.2.24)). *The affine part of* $\mathbf{J}(\Gamma)$ *is canonically embedded in the space* $\mathbf{C}^{2n}$.

As was shown above (see (2.3.20)), $u(x)$ has the form

(2.3.32) $u(x) = -2\sigma(\eta_1^0 + (x - x_0)U_1, \ldots, \eta_n^0 + (x - x_0)U_n) + \text{const},$

in which $\sigma$ is an algebraic function on the torus $\mathbf{J}(\Gamma)$ and is expressed in algebraic form in terms of the Riemann $\theta$-function and its derivatives, where the classical Riemann $\theta$-function is constructed from the lattice (2.3.24) in the standard form:

$$(2.3.33)\quad \theta(\eta_1, \ldots, \eta_n) = \sum_{m_1, \ldots, m_n} \exp\left\{\frac{1}{2}\sum_{j,k} B_{jk}m_jm_k + \sum_k m_k\eta_k\right\}$$

(the $m_1, \ldots, m_n$ are integers).

In [41] there is a convenient explicit formula expressing $\sigma$ in terms of the $\theta$-function (2.3.33). For $u(x)$ we have

$$(2.3.34)\quad u(x) = -2\frac{d^2}{dx^2}\ln\theta(\eta_1^0 + U_1(x - x_0) - K_1, \ldots, \eta_n^0 + U_n(x - x_0) - K_n) + C,$$

$$K_j = \frac{1}{2}\sum_{k=1}^n B_{kj} - \pi ij, \quad \eta_j^0 = \sum_k \int_\infty^{P_k(x_0)} \Omega_j;$$

where $C$ is a constant depending only on $\Gamma$. Let us prove this formula. We consider the function

$$(2.3.35)\quad F(P) = \theta(\eta(P) - \eta^0),$$

where $P \in \Gamma$ and $\eta_j(P) = \int_\infty^P \Omega_j$. The function $F(P)$ is single-valued on $\widetilde{\Gamma}$ cut along the cycles $(a_j, b_j)$ (2.3.21). By a result of Riemann, (see [56]), $F(P)$ has $n$ zeros $P_1, \ldots, P_n$ on $\Gamma$ (for almost all $\eta^0 \in \mathbf{J}(\Gamma)$, and the following relation holds on $\mathbf{J}(\Gamma)$:

$$(2.3.36)\quad A(P_1, \ldots, P_n) = \eta^0 - \vec{K},$$

where $K_j = \frac{1}{2}\sum_{k=1}^n B_{kj} - \pi_{ij}$ are the "Riemann constants". Therefore,

$$(2.3.37)\quad \frac{1}{2\pi i}\oint_{\partial\widetilde{\Gamma}} \pi_i(P)\, d\ln F(P) = \sum_{j=1}^n \pi(P_j) + \operatorname*{res}_{P=\infty} \pi(P)\, d\ln F(P) = \sigma(\eta^0 - \vec{K}) + \operatorname*{res}_{P=\infty} \pi(P)\, d\ln F(P).$$

Since the $\theta$-function (2.3.33) has the properties

$$\theta(\eta_1, \ldots, \eta_k + 2\pi i, \ldots, \eta_n) = \theta(\eta_1, \ldots, \eta_n),$$

$$\theta(\eta_1 + B_{1k}, \ldots, \eta_n + B_{nk}) = \exp\left(-\frac{B_{kk}}{2} - \eta_k\right)\theta(\eta_1, \ldots, \eta_n),$$

the integral (2.3.37) is

$$(2.3.37')\quad \frac{1}{2\pi i}\oint_{\partial\widetilde{\Gamma}} \pi(P)\, d\ln F(P) = \sum_{k=1}^n \oint_{a_k} \pi(P)\,\omega_k.$$

From (2.3.37) and (2.3.37′) we have

$$\sigma(\eta^0-\vec{K}) = -\operatorname{res}\pi(P)\,d\ln F+\sum_{k=1}^{n}\oint_{a_k}\pi(P)\,\omega_k. \tag{2.3.38}$$

To derive (2.3.34) from (2.3.38) we must calculate the residue res $\pi(P)d$ ln $F$ at $P = \infty$, where $\eta^0$ is chosen so that

$$\eta^0-\vec{K}=A(P_1(x),\ldots,P_n(x))=A(P_1(x_0),\ldots P_n(x_0))+\vec{U}\,x-x_0).$$

To do this it remains to observe that the quantities $U_j$ in (2.3.30) and $c_{kl}$ in (2.3.22) are connected by $U_j = c_{jn}$ (a consequence of the relation between the periods of the differentials on $\Gamma$; see [55]).

From the preceding results we have seen that the Bloch function $\psi_\pm$ and the potential $u$ itself are completely determined by the spectrum, (the Riemann surface $\Gamma$), and the "divisor" (the set of poles), that is, the numbers $\gamma_1(x_0), \ldots, \gamma_n(x_0)$ together with an indication of the sheet on which the poles lie. The zeros of $\psi_\pm$ also form a divisor $(\gamma_1(x), \sigma_1), (\gamma_2(x), \sigma_2), \ldots, (\gamma_n(x), \sigma_n)$, where $\sigma_j = \pm$ labels the sheet in which the zero lies. If the potential is real, then all the $\gamma_j$ lie in the lacunae $E_{2j} \leqslant \gamma_j \leqslant E_{2j+1}$, the Riemann surface is real, and it makes sense to talk of the "upper sheet" (+) and the "lower sheet" (−) for real values of $E$ over the lacunae (the positive and negative values of the root $\sqrt{-P_{2n+1}}$). If the sign of the sheet is +, then $\psi_+(x, x_0, E) \to 0$, as $x \to +\infty$, for $E$ inside the lacuna; if the sign of the sheet is − when $E$ is in the lacuna, then $\psi_-(x, x_0, E) \to 0$, as $x \to -\infty$. Hence the function $\psi_\pm(\tau, x_0, E) = f_j(\tau)$, $E = \gamma_j(x)$, belongs to the discrete (non-degenerate) spectrum of the Sturm-Liouville problem on the half-line

$$\tau \geqslant x, \qquad f(x)=0, \qquad f(+\infty)=0,$$
$$\tau \leqslant x, \qquad f(x)=0, \qquad f(-\infty)=0,$$

depending on the sign (±) of the sheet on which the zero $P_j(x)$ of $\psi(x, x_0, E)$ lies, and the set $\{\gamma_j\}$ comprises all the non-degenerate levels of these spectra. A similar though rather simpler situation arose in the work of Shabat [11]: he suggested a method of studying non-reflecting potentials by means of "conditional" eigenvalues $\gamma_j$ on two half-lines for which he derived in [11] an equation analogous to (2.3.5) (but, of course, in this case the Riemann surface degenerates and there is, in general, no Abelian variety; equations of the type (2.3.5) in this degenerate case were derived by completely different methods). In our case, when $x$ varies, the set of points $(\gamma_j, \pm)$ describes the cycles $a_j$, and the sign of the sheet changes when passing through a branch-point.

We consider now the case of even potentials, $u(x) = u(-x)$, with $x_0 = 0$. It is easy to verify that when $x \mapsto -x$ the sheets of $\Gamma$ are exchanged. Hence, for the equation $u(x) = u(-x)$ to hold it is necessary and sufficient that the poles $\gamma_j(0)$ are invariant under this exchange, that is, occur at branch-points. Moreover, when $x_0 = 0$, the function $\chi(x_0, E)$ has the form

$$\chi(0, E) = \chi_R(0, E),$$

since

$$\chi_I(0, E) = -\frac{1}{2}\left[\frac{d}{dx} \ln \prod_{j=1}^{n} (E-\gamma_j)\right]_{x=0} = 0,$$

because $\gamma_j'(0) = 0$; this follows from (2.3.5) and the fact that $\gamma_j(0)$ lies at a branch-point. Therefore, $\psi_\pm(x, 0, E)$ in this case has the form

$$\psi_\pm(x, 0, E) = c + i\frac{\sqrt{P_{2n+1}(E)}}{\prod_{j=1}^{n}(E-\gamma_j(0))}s,$$

where the $\gamma_j(0)$ lie at the branch-points (ends of the lacunae). This gives the Akhiezer formula of [10] for the function $\mathscr{E}$ in case the $\gamma_j(0)$ are taken as the lower boundaries of the lacunae. The point of the derivation is the fact that for a given zone structure there are only finitely many even potentials.

We conclude this section by showing that the parameters $(\eta_1, \ldots, \eta_n)$ on the torus $\mathbf{J}(\Gamma)$ give "angle" variables, canonically conjugate to the "action" variables, which, for the Hamiltonian systems (2.2.22), can be taken as the integrals $J_1, \ldots, J_n$, the last $n$ coefficients of the polynomial $P_{2n+1} = \det \Lambda$ (see §2). The Poisson brackets here form the constant non-singular matrix

$$\begin{cases} [J_k, \eta_j] = a_{kj}, \\ [J_k, J_s] = [\eta_k, \eta_s] = 0. \end{cases} \tag{2.3.39}$$

It turns out (see §4) that the parameters $\eta_k$ on $\mathbf{J}(\Gamma)$ also give the set of "angles" for the time evolution by the K-dV and higher order K-dV equations: by the higher order K-dV equations, all the derivatives $\dot{\eta}_k$ are constant, and by virtue of the operator $\frac{\partial}{\partial x}$, which is connected with the time dynamics, the Poisson brackets all vanish: $[\eta_k, \eta_s] = 0$.

## §4. Applications. The time dynamics of finite-zone potentials according to the K-dV equations. The universal fibering of Jacobian varieties (the hyperelliptic case).

We study the time evolution according to any of the higher order K-dV equations. In Ch. 2, §2 we have derived the equation (2.2.33) for the time evolution, which for $m = 1$ coincides with the usual K-dV equation:

$$\frac{d\Lambda}{dt_m} = [\Lambda, \Lambda_m], \tag{2.4.1}$$

where $\Lambda = \sum_{i=0}^{n} c_i \Lambda_{n-i}$, (2.4.1) holds for the set of solutions of (2.2.22)

$$\sum c_i \frac{\delta I_{n-i}}{\delta u(x)} = d,$$

and the algorithm for finding the matrices in the basis (2.1.2) is given by (2.2.44).

Starting from the results of §3, we now find formulae for the time dynamics in terms of the parameters $\gamma_j$. We use (2.2.41), which in the basis (2.1.1) has the form $\frac{\partial \chi_R}{\partial t} = \frac{\partial}{\partial x}(\alpha\chi_R)$, where $\alpha = \frac{\lambda_I + \mu_I}{k}$ and $\alpha = -2(u+2E)$ for the initial K-dV equation ($m = 1$). Using the form

$$\chi_R = \frac{\sqrt{P_{2n+1}(E)}}{P_n(E, x)},$$

we obtain

$$\frac{\partial P_n}{\partial t} = \frac{\partial \alpha}{\partial x} P_n - \alpha \frac{\partial P_n}{\partial x}, \tag{2.4.2}$$

or, substituting $E = \gamma_j$ in (2.4.2) after solving for $\dot{\gamma}_j$ and using (2.3.5) for $\gamma_j'$

$$\dot{\gamma}_j = \pm \frac{2i\alpha_j \sqrt{P_{2n+1}(\gamma_j)}}{\prod_{k\neq j} (\gamma_k - \gamma_j)}, \tag{2.4.3}$$

where $\alpha_j = -2(u + 2E)|_{E=\gamma_j(x)}$ for the original K-dV equation, and by (2.3.11) $u = -2(\sum_{j=1}^{n} \gamma_j) + \sum_{i=1}^{2n+1} E_i$; this gives the final formula for $\dot{\gamma}_j$.

EXAMPLE. Consider $n = 2$ and the original K-dV equation with $\alpha = -2(u + 2E)$. We obtain

$$\begin{cases} \dot{\gamma}_1 = \pm \dfrac{8i\left(\gamma_2 - \frac{1}{2}\sum_{i=1}^{5} E_i\right)\sqrt{P_5(\gamma_1)}}{\gamma_1 - \gamma_2}, \\ \dot{\gamma}_2 = \pm \dfrac{8i\left(\gamma_1 - \frac{1}{2}\sum_{i=1}^{5} E_i\right)\sqrt{P_5(\gamma_2)}}{\gamma_2 - \gamma_1}. \end{cases} \tag{2.4.3'}$$

As in (2.3.15), we introduce the parameter $\tau$ given by $d\tau = \frac{dx}{\gamma_2 - \gamma_1}$, and obtain

$$\begin{cases} \dfrac{d\gamma_1}{\sqrt{-P_5(\gamma_1)}} = \pm 8\left(\gamma_2 - \frac{1}{2}\sum E_i\right) d\tau, \\ \dfrac{d\gamma_2}{\sqrt{-P_5(\gamma_2)}} = \pm 8\left(\gamma_1 - \frac{1}{2}\sum E_i\right) d\tau. \end{cases} \tag{2.4.3''}$$

Suppose, for simplicity, that $\sum_{i=1}^{5} E_i = 0$. Introducing the parameter $w$ where $8(\gamma_2\gamma_1)d\tau = dw$, we then have

$$(2.4.4)\qquad \frac{\gamma_1\, d\gamma_1}{\sqrt{-P_5(\gamma_1)}} = dw, \qquad \frac{\gamma_2\, d\gamma_2}{\sqrt{-P_5(\gamma_2)}} = dw,$$

where $dw$ is a first order differential on $\Gamma$. Now (2.4.4) and thus also (2.3.3) can be integrated in the obvious way. Comparing the result with the formulae (2.3.11) for the potential, we find that $\frac{\partial}{\partial t}\tau_0 = 4$, $\frac{\partial}{\partial t}x_0 = 4(F_1(\tau_0) - \frac{1}{2}\Sigma E_i)$, and the potential has the form

$$(2.4.5)\qquad u(x, t) = -2[F_1(\tau(x - x_0(t))) + F_2(\tau(x - x_0(t)) + \tau_0(t))] + \sum E_i,$$

where the hyperelliptic functions $F_1$ and $F_2$ are defined in Ch. 2, §3 (see (2.3.16).

If $u(x, 0)$ is the Ince (Lamé) potential, where $u(x, 0) = 6\wp(x)$ with the boundaries of the zones (2.2.38) ($\wp(x)$ is the Weierstrass elliptic function), then we have

$$(2.4.6)\qquad u(x, t) = 2\wp(x - \beta_1(t)) + 2\wp(x - \beta_2(t)) + 2\wp(x - \beta_3(t)),$$

where

$$\beta_1 + \beta_2 + \beta_3 = 0, \qquad t = \int_0^{\beta_1 - \beta_2} \frac{dz}{\sqrt{12(g_2 - 3\wp^2(z))}},$$

$$\beta_2 - \beta_3 = \frac{1}{2}\wp^{-1}[-\wp(\beta_1 - \beta_3) + \sqrt{g_2 - 3\wp^2(\beta_1 - \beta_3)}].$$

It turns out that the Abel map (2.3.25) integrates (2.4.3) for K-dV equation and all its higher order analogues, and the parameters $\eta_k$ have constant derivatives by virtue of all these equations. We can calculate the derivatives

$$(2.4.7)\qquad \dot{\eta}_k = W_k^{(m)}$$

by the $m$th K-dV equation:

$$(2.4.8)\qquad \frac{\partial}{\partial t}u = \frac{\partial}{\partial x}\frac{\delta I_m}{\delta u(x)}.$$

The idea of this calculation, by analogy with §3, consists in the fact that under time development the eigenfunction $\psi_\pm(x, x_0, E)$ has the asymptotic form, as $E \to \infty$ (for a suitable normalization)

$$(2.4.9)\qquad \psi \sim \exp[ik(x - x_0) + ik^{2n+1}(t - t_0)].$$

If (2.4.9) can be established (this involves certain difficulties), then, by analogy with §3, we obtain the following result: let $\omega_m$ be a differential on $\Gamma$ such that at $E = \infty$ it has a pole of order $2m + 2$ (for the $m$th K-dV equation (2.4.8), as $E \to \infty$)

$$(2.4.10)\qquad \omega_m = \frac{dz}{z^{2m+2}} + \text{a regular part}$$

where $z = \frac{1}{\sqrt{E}} = \frac{1}{k}$, and normalized by the conditions

(2.4.10′) $$\oint_{a_j} \omega_m = 0 \quad (j = 1, \ldots, n).$$

Arguing as in §3 in "the case of the differential $d_E \ln \psi$, we have the required result, namely

(2.4.11) $$\dot{\eta}_k = W_k^{(m)}, \quad W_k^{(m)} = \oint_{b_k} \omega_m.$$

Substituting in the formula (2.3.34), which expresses the potential $u(x, t)$ in terms of the Riemann $\theta$-function, we obtain the K-dV dynamics in the form

(2.4.12) $$u(x, t) = -2\frac{d^2}{dx^2} \ln \theta ((x - x_0) U_1 + (t - t_0) W_1^{(1)} + \\ + \eta_1^0 - K_1, \ldots, (x - x_0) U_n + (t - t_0) W_n^{(1)} + \eta_n^0 - K_n) + C.$$

So we have obtained the following result.

THEOREM. *The time development of finite-zone potentials according to the* K-dV *equation or any of its higher order analogues is described by* (2.4.3), (2.4.5), (2.4.11), (2.4.12) *and represents a motion on a torus along a rectilinear winding on the variety of all potentials with the given spectrum* $\Gamma$, *which is isomorphic to the Jacobian variety* $\mathbf{J}(\Gamma)$, *the complex torus* $\mathbf{T}^{2n}$. *In particular, the rectilinear structure on* $\mathbf{J}(\Gamma)$ *is determined by "higher order K-dV" equations, written in the form of a set of commuting polynomial Hamiltonian systems with n degrees of freedom and the Hamiltonians* $J_1, \ldots, J_n$ *in the realization of* §2.

In the case of a real Riemann surface $\Gamma$ (or a real bounded potential) the motion is on the torus $\mathbf{T}^n$, which can be visualized in the form of a direct product of the cycles $a_j$, the inverse images of the lacunae on $\Gamma$. This motion is conditionally periodic in time with a set of $n$ real and $n$ imaginary periods, where the periods are expressed in terms of the integrals $W_k^{(m)}$ of the differentials $\omega_m$ (for the $m$th analogue of the K-dV equation) over the cycles $b_k$ and the lattice matrix $B_{jl}$.

Since the higher order K-dV equations are rectilinear on $\mathbf{J}(\Gamma)$, we see that on this Abelian variety the law of addition of points, by means of motions along these systems holds. In the realization of §2 of the varieties $\mathbf{J}(\Gamma)$ of finite-zone potentials by the equations $J_1 = J_1^0, \ldots, J_n = J_n^0$, all the higher order K-dV equations were realized as Hamiltonian dynamical systems in a phase space with $n$ degrees of freedom, depending on the $n + 1$ constants $(d, c_1, \ldots, c_n)$ (if $\Sigma E_i = 0$, then $c_1 = 0$), and the Hamiltonians $J_\alpha (\alpha = 1, \ldots, n)$ of these systems are polynomials in the phase variables, having vanishing Poisson brackets $[J_\alpha, J_\beta] = 0$. The Jacobian $\mathbf{J}(\Gamma)$ is defined in $\mathbf{C}^{2n}$ by the equations $J_1 = J_1^0, \ldots, J_n = J_n^0$, and the Riemann surface $\Gamma$ has the form

$$R(W, E) = W^2 - P_{2n+1} = 0,$$

where

$$P_{2n+1}(E) = -\det\Lambda,\quad \Lambda = \Lambda_n + \sum_{i=1}^{n} c_i\Lambda_{n-i};$$

for $n = 2$, according to (2.2.36), we have the Abelian varieties $\mathbf{J}(\Gamma)$:

$$J_1 = p_1p_2 - \left(\frac{1}{2}q_2^2 + \frac{5}{2}q_1^2q_2 + \frac{5}{8}q_1^4\right) + c_2q_1^2 - dq_1,$$

$$J_2 = p_1^2 - 2q_1p_1p_2 + 2(q_2 - c_2)p_2^2 + q_1^5 + 2c_2q_1^3 + dq_1^2 - 4q_1q_2^2 + 4c_2q_1q_2 - 2dq_2;$$

$$-\det\Lambda = P_5(E) = E^5 + \frac{1}{4}c_2E^3 - \frac{1}{16}dE^2 + \left(\frac{1}{32}J_1 + \frac{1}{4}c_2^2\right)E + J_2/2^8 + c_2d/2^7,$$

where $c_1 = \Sigma E_i = 0$, $q_1 = u$, $q_2 = -\frac{5}{2}u^2 + u''$, $p_1 = q_2'$, $p_2 = u'$.

Isomorphic Abelian varieties are obtained when the Riemann surfaces are isomorphic, that is, when the roots $(E_i)$ of the polynomial $P_{2n+1}(E)$ differ only by a factor $\lambda$ if $\Sigma E_i = c_1 = 0$. After taking account of this equivalence, there remain altogether $2n + 2$ isomorphic surfaces $\Gamma$($n$ being the genus) because one of the $2n + 2$ branch-points is distinguished and is situated at $E = \infty$. We recall that hyperelliptic Riemann surfaces are characterized uniquely by the roots of a polynomial $Q_{2n+2}(E)$ to within a common fractionally-linear transformation and an arbitrary permutation of the roots. We place one branch-point $E_{2n+2} = \infty$ at infinity, and also assume that $\Sigma E_i = 0$; the remaining $2n$ numbers can be simultaneously multiplied by one and the same number $\lambda$. Furthermore, all the coefficients of the polynomial $\det\Lambda = P_{2n+1}(E)$ can be expressed symmetrically in terms of $E_1, \ldots, E_{2n+1}$. The only remaining non-symmetry in our constructions is the selection of the branch-points $E_{2n+2} = \infty$.

Let $v \in \mathbf{V}$ be a base point of the variety of moduli of hyperelliptic curves $\Gamma$ defined by the set $v = (E_1, \ldots, E_{2n+2})$ to within a common fractionally-linear transformation and a permutation; let $\widetilde{\mathbf{V}} \xrightarrow{2n+2} \mathbf{V}$ be our covering associated with the selection of the branch-point $E_{2n+2} = \infty$. Over each point of $\mathbf{V}$ there is the Jacobian variety $\mathbf{J}(\Gamma)$ of the corresponding curve $\Gamma$, and we have the universal fibering of Jacobian varieties $\mathbf{M} \xrightarrow{\mathbf{J}(\Gamma)} \mathbf{V}$ and its $(2n + 2)$-sheeted covering $\widetilde{\mathbf{M}} \xrightarrow{\mathbf{J}(\Gamma)} \widetilde{\mathbf{V}}$, where $\widetilde{v} \in \widetilde{\mathbf{V}}$ is determined by the set $\widetilde{v} = (E_1, \ldots, E_{2n+1})$ to within a permutation and multiplication: $(E_1, \ldots, E_{2n+1}) \sim \lambda(E_1, \ldots, E_{2n+1})$, $\Sigma E_i = 0$. This variety $\widetilde{\mathbf{M}}$ can be computed in our constructions as follows: in $\mathbf{C}^{3n}$ the coordinates are $z_1, \ldots, z_{3n}$, where $z_i = q_i$, $z_{n+i} = p_i$, $i \leqslant n$; $z_{2n+1} = d$, $z_{2n+i} = c_i$, $2 \leqslant i \leqslant n$, and the symplectic form $\Omega = \sum_i dp_i \wedge dq_i$. All the Abelian varieties are given by the equations

$$\begin{cases} z_j = \text{const}, & j \geqslant 2n+1, \\ J_\alpha = \text{const}, & (\alpha = 1, \ldots, n), \end{cases}$$

and thus lie in the spaces $\mathbf{C}^{2n}$. The universal fibering splits into a family of fiberings of each $\mathbf{C}^{2n}\{z_j = \text{const}, j > 2n\}$ by level surfaces of all the polynomials $J_\alpha$, depending on the remaining $n$ coordinates $z_{2n+1}, \ldots, z_{3n}$ and on the parameters. An algorithm for calculating the polynomials $J_\alpha$ was described in §2. The law of addition on the Abelian varieties and all the one-parameter subgroups are given by Hamiltonian systems with the Hamiltonians $J_\alpha$. The group of multiplications of the roots $E_i$ by $\lambda$ acts by multiplying the coefficients of the polynomial $P_{2n+1}(E) = \det \Lambda$ by the corresponding power of $\lambda$. We can choose $\lambda$ by requiring that $c_2 = 1$.

Thus, we obtain the following result:
The manifold $\widetilde{\mathbf{M}}$, the space of the universal fibering of the Jacobian varieties $\mathbf{J}(\Gamma)$ with the distinguished branch-point $E_{2n+2} = \infty$, is a rational variety; this universal fibering with the fibre $\mathbf{J}(\Gamma)$ splits into a family of fiberings with rational fibering spaces of dimension $2n$, fibred by the polynomials $J_\alpha$ (an algorithm to calculate them was given in §2). On affine parts of the fibering space $\mathbf{C}^{2n}$ there is a symplectic form, and the Poisson brackets of all pairs of polynomials $J_\alpha$ vanish. The Abelian varieties are complex solutions of this commuting set of Hamiltonian systems.

We mention that the variety $\mathbf{M}$ itself is probably also rational, but here we have only proved its unirationality. In the general (non-hyperelliptic) case we can also develop an analogous method, using operators of higher order instead of the Schrödinger operator. For an approach to these problems, see Ch. 3, §2.

# CHAPTER 3

## GENERALIZATIONS. DISCRETE SYSTEMS AND MATRIX OPERATORS OF FIRST ORDER

### §1. The periodic problem for the Toda chain and the "K-dV difference equation"

As we have said in Ch. 1, §5, Manakov [35] and Flaschka [33], [34] found an $L - A$ pair for the Toda chain, proved that the Henon integrals [32] are involutory and integrated completely the Cauchy problem for the Toda chain with rapidly decreasing initial conditions $c_n \to \text{const}$, $v_n \to 0$, as $|n| \to \infty$, by the method of scattering theory. Furthermore, with the operator $L$ under the condition $v_n \equiv 0$ another physically interesting system is associated, the "difference K-dV equation", discovered in [35], for which Manakov first found the integrals and then integrated the system by the same method as in the rapidly decreasing case. Following unpublished work of S. P. Novikov, we consider here the periodic case for both these systems, by a method similar to that of Ch. 2. The operator $L$ has the form indicated in Ch. 1, §5, and the equation $L\psi_n = E\psi_n$ is

(3.1.1) $$(E-v_n)\psi_n = i\sqrt{c_n}\,\psi_{n-1} - i\sqrt{c_{n+1}}\,\psi_{n+1}.$$

After the change $\psi_n \mapsto i^n \psi_n$ we obtain

(3.1.1′) $$(E-v_n)\psi_n = \sqrt{c_n}\,\psi_{n-1} + \sqrt{c_{n+1}}\,\psi_{n+1}.$$

The Wronskian for the operator (3.1.1′) has the form

(3.1.2) $$W_n(\varphi, \psi) = (-1)^n \sqrt{c_{n+1}}\,(\psi_{n+1}\varphi_n - \psi_n\varphi_{n+1})$$

and does not depend on $n$. We use here an analogue $c(n, n_0, E)$, $s(n, n_0, E)$ to the basis (2.1.2) of Ch. 2, where

(3.1.3) $$\begin{cases} c_{n_0} = 1, & c_{n_0+1} = 0, \\ s_{n_0} = 0, & s_{n_0+1} = 1, \end{cases}$$

and $W(c, s) = \sqrt{c_{n_0+1}}(-1)^{n_0+1}$. In the periodic case the matrix of a translation by a period is defined in the usual way, (in the basis 3.1.2):

$$\hat{T} = \begin{pmatrix} \alpha_{11} & \alpha_{12} \\ \alpha_{21} & \alpha_{22} \end{pmatrix},$$

where the $\alpha_{ij}$ are real for real $E$ and $\det \hat{T} = 1$ (we assume that the period $N$ is even). In exactly the same way the Bloch eigenfunctions $\psi_\pm(n, n_0, E)$ are defined where $\psi_{n_0} = 1$, $\hat{T}\psi_\pm = e^{\pm ip(E)}\,\psi_\pm$; they are meromorphic on the Riemann surface $\Gamma$ and have branch-points at the boundaries of the zones. It is important to note that the operator $L$ (see (3.1.1) and (3.1.1′) has altogether finitely many forbidden and solution zones, and the neighbourhood of $E = \infty$ is always a forbidden zone. Thus, $\Gamma$ is always of finite genus and is defined by the equation

(3.1.4) $$y^2 = P_{2n+2}(E) = \sum_{i=1}^{2n+2} (E-E_i),$$

where the $E_i$ are the boundaries of the zones. We introduce the notation:

(3.1.5) $$\begin{cases} \psi_\pm(n, n_0, E) = \exp\left(\sum_{n_0}^{n-1} \Delta_n\right), \\ \chi^\pm(n, E) = e^{\Delta n}. \end{cases}$$

From (3.1.1) we obtain the following equation (an analogue to the Riccati equation):

(3.1.6) $$E - v_n = \sqrt{c_n}\,e^{-\Delta_{n-1}} + \sqrt{c_{n+1}}\,e^{\Delta_n}.$$

In the solution zones, where $p(E)$ is real, we obtain from (3.1.6)

(3.1.7) $$\begin{cases} \chi = \chi_{Re} + i\chi_{\mathrm{Im}}, & \psi_+ = \overline{\psi}_-, \\ \chi = \chi^+ = \overline{\chi^-}, & \Delta_n = \Delta_{n_R} + i\Delta_{n_I}, \\ \Delta_{n_R} = -\frac{1}{2}\ln\{[\sqrt{c_{n+1}}\,\chi_{\mathrm{Im}}(n+1, E)] - [\sqrt{c_n}\,\chi_{\mathrm{Im}}(n, E)]\}. \end{cases}$$

From (3.1.7) it follows that (in the solution zones)

(3.1.8) $$\psi_\pm(n, n_0, E) = \sqrt{\frac{\sqrt{c_{n+1}}\,\chi_{\mathrm{Im}}(n, E)}{\sqrt{c_n}\,\chi_{\mathrm{Im}}(n_0, E)}}\exp\left\{i\sum_{n_0}^{n-1} \Delta_{n_I}\right\}.$$

Further, from the definition of $W$ we have

$$(3.1.8') \qquad \begin{cases} W(\psi_+, \psi_-) = 2i\chi_{\mathrm{Im}}(n_0, E)\sqrt{c_{n_0+1}}, \\ \psi_\pm = c + \chi(n_0, E)\, s, \quad \chi = \chi^+ \quad \text{or} \quad \chi = \chi^-. \end{cases}$$

As in Ch. 2, the function $\chi(n_0, E)$ can be expressed in terms of $\hat{T}$, and $\chi_{\mathrm{Im}}(n, E)$ has the same form as in (2.1.19):

$$(3.1.9) \qquad \chi_{\mathrm{Im}}(n_0, E) = \frac{\sqrt{1 - \frac{1}{4}(\mathrm{Sp}\,\hat{T})^2}}{\alpha_{21}}.$$

As in Ch. 2, §3, the function $2i\chi_{\mathrm{Im}}(n_0, E)\sqrt{c_{n_0+1}}$ can be continued from the solution zones to all complex values of $E$ by the formula (3.1.9) and coincides with the Wronskian $W(\psi_+, \psi_-)$. As $E \to \infty$, it has one of two asymptotic forms (we recall that the Riemann surface $\Gamma$ given by (3.1.4) has two sheets $\pm$ over the point $E = \infty$, and $z = \frac{1}{E}$ is a local parameter on each of them)

$$(3.1.10) \qquad \begin{cases} \chi^+(n, E) \to \dfrac{E}{\sqrt{c_{n+1}}}\left(1 + O\left(\dfrac{1}{E}\right)\right), \\ \chi^-(n, E) \to \dfrac{\sqrt{c_{n+1}}}{E}\left(1 + O\left(\dfrac{1}{E}\right)\right) \end{cases}$$

on the two sheets of $\Gamma$, as $E \to \infty$. Hence, for $W(\psi_+, \psi_-) = \chi^+ - \chi^-$ we have the asymptotic form, as $E \to \infty$:

$$(3.1.10') \qquad W \to 2E + O(1).$$

Exactly as in Ch. 2, §3, from the asymptotic form (3.1.10′) we obtain

$$(3.1.11) \qquad \frac{1}{2} W(\psi_+, \psi_-) = \frac{\sqrt{\prod_{i=1}^{2m+2}(E - E_i)}}{\prod_{j=1}^{m}(E - \gamma_j(n_0))}.$$

From (3.1.11) it follows that in the solution zones

$$(3.1.11') \qquad \chi_{\mathrm{Im}}(n_0, E) = \frac{1}{2\sqrt{c_{n_0+1}}} W(\psi_+, \psi_-).$$

Comparing with the formulae (3.1.7), which are applicable within the solution zones, we obtain

$$(3.1.12) \qquad \chi^\pm(n, E) = \frac{\sqrt{R} + \sqrt{R + 4\Pi_n \Pi_{n+1} c_{n+1}}}{2\sqrt{c_{n+1}}\,\Pi_n},$$

where $R(E) = \prod_{j=1}^{2m+2}(E - E_j)$, $\Pi_n = \prod_{j=1}^{m}(E - \gamma_j(n))$. However, this expression cannot yet be regarded as the final result for $\chi^\pm(n_0, E)$. In fact, formally speaking (3.1.12) is algebraic in a Riemann surface $\Gamma'$, covering $\Gamma$ doubly. We know that the quantity $\chi^\pm(n_0, E)$ is algebraic on $\Gamma$ and has poles of first order over $\gamma_j(n_0)$ at only one of the inverse images of this point. This has an important consequence: the expression $R + 4c_{n+1}\Pi_n\Pi_{n+1}$ under the root sign must be a perfect square of some polynomial of degree $m + 1$:

(3.1.13) $$R+4c_{n+1}\Pi_n\Pi_{n+1}=\prod_{k=1}^{m+1}(E-\alpha_k(n))^2.$$

The conditions (3.1.13) lead to a full set of relations of the form

(3.1.13′) $$\begin{cases}\gamma_j(n+1)=f_j(\gamma_1(n),\ldots,\gamma_m(n)),\\ c_{n+1}=c_{n+1}(\gamma_1(n),\ldots,\gamma_m(n)),\end{cases}$$

giving a difference analogue of the Dubrovin equations for the $\gamma_j(n)$, and an analogue, which we shall need, of the "trace identities" for expressing $c_{n+1}$ in terms of $\gamma_1(n),\ldots,\gamma_m(n)$.

Of course, from the asymptotic form of $\chi^+(n, E)$, as $E\to\infty$, we can obtain the usual trace identities as in Ch. 2; for from (3.1.6) we have

(3.1.14) $$\chi^+(n,E)\sim\frac{E}{\sqrt{c_{n+1}}}\left(1-\frac{v_n}{E}-\frac{c_n}{E^2}+o\left(\frac{1}{E^3}\right)\right).$$

Expanding (3.1.12), as $E\to\infty$, we obtain

(3.1.15) $$\begin{cases}v_n=-\sum_{j=1}^{m}\gamma_j(n)+\frac{\sigma_1}{2},\\ c_n+c_{n+1}+v_n^2-\frac{\sigma_1v_n}{2}=\sum_{j\neq k}\gamma_j(n)\gamma_k(n)+\frac{\sigma_1^2}{8}-\frac{\sigma_2}{2}.\end{cases}$$

Thus, the usual trace formulae are insufficient to calculate $c_{n+1}$, and we obtain the missing expressions from (3.1.13).

Here are the simplest examples.

$m = 1$. For the sake of convenience, let $\sigma_1 = \sum_{i=1}^{4}E_i = 0$. Then from (3.1.13) we have

(3.1.16) $$\begin{cases}\sigma_2+4c_{n+1}=-2\alpha_n^2,\\ \sigma_3-4c_{n+1}(\gamma_{1n}+\gamma_{1n+1})=0,\\ \sigma_4+4c_{n+1}(\gamma_{1n}\gamma_{1n+1})=\alpha_n^4.\end{cases}$$

$m = 2$. Let $v_n\equiv 0$ (symmetric spectrum). Then

$$R(E)=\prod_{i=1}^{3}(E+E_i)(E-E_i),$$

$$\gamma_{1,n}+\gamma_{2,n}\equiv 0,\quad \sigma_1=\sigma_3=\sigma_5=0.$$

From (3.1.13) we obtain

(3.1.17) $$\begin{cases}\sigma_4-4c_{n+1}(\gamma_{1n}^2+\gamma_{1n+1}^2)=\frac{1}{4}(\sigma_2+4c_{n+1})^2,\\ 4c_{n+1}\gamma_{1n}^2\gamma_{1n+1}^2=-\sigma_6.\end{cases}$$

In these examples we can easily obtain from (3.1.16) and (3.1.17) the Dubrovin equation and an expression for $c_n$.

Continuing to follow the scheme of Ch. 2, §3, we consider the analytic properties of the function $\psi_\pm(n, n_0, E)$ on $\Gamma$:

1) $\psi_\pm$ has zeros $P_j(n) = [\gamma_j(n), \pm]$ and poles $P_j(n_0) = [\gamma_j(n_0), \pm]$, which for real $v_n$ and $c_n > 0$ lie one each in the lacunae on one of the sheets of $\Gamma$.

2) As $E \to \infty$, we have on both sheets of $\Gamma$

$$\psi_\pm \sim E^{\pm(n-n_0)}\cdot\text{const}.$$

Hence the differential $d_E \ln \psi$ has poles of the first order with residue +1 at $P_j(n)$ and with residue −1 at $P_j(n_0)$; at $E = \infty$ it has poles with residues $\pm(n-n_0)$ on the sheets (±). The integrals of $d_E \ln \psi$ over all the cycles are integer multiples of $2\pi i$. If

$$\Omega_l = \sum_{k=0}^{n-1} c_{kl} \frac{E^k\, dE}{\sqrt{R(E)}}$$

are holomorphic differentials, normalized by the conditions

$$\oint_{a_j} \Omega_l = 2\pi i \delta_{jl},$$

where the $a_j$ are cycles over the lacunae, then by analogy with the arguments of §3 we have

$$\sum_{j=1}^{n} \int_{P_j(n_0)}^{P_j(n)} \Omega_l = \left(2\pi i \oint_{b_l} \Omega\right)(n-n_0), \tag{3.1.18}$$

where $\Omega = \Omega^{\infty_+,\infty_-}$ is a differential having poles at the points $(\infty_+, \infty_-)$ with residues $(+1, -1)$, respectively, normalized by the conditions $\oint_{a_j} \Omega = 0$. Moreover, we have $\oint_{b_j} \Omega = \int_{\infty_+}^{\infty_-} \Omega_j$, as in §3, for the differentials $\Omega^{PQ}$, since the cycles $b_j$ are conjugate to $a_j$. In the real case we obtain

$$\oint_{b_j} \Omega = 2 \int_{E_{2n+2}}^{\infty} \Omega_j = U_j. \tag{3.1.19}$$

From (3.1.18) and (3.1.19) we obtain, as in §3, using the trace identity (3.1.15), an analogue of the Matveev–Its formula for $v_n$ in terms of the restriction of the $\theta$-function to the rectilinear winding in the direction $U_j$ on the Jacobian variety $\mathbf{J}(\Gamma)$:

$$v_n = -\frac{d}{dn} \ln \frac{\theta\left((n-n_0)\vec{U} + \eta^0_+ - \vec{K}\right)}{\theta\left((n-n_0)\vec{U} + \eta^0_- - \vec{K}\right)} + \text{const}; \quad (\eta^0_\pm)_j = \sum_k \int_{\infty_\pm}^{P_k(n_0)} \Omega_j. \tag{3.1.20}$$

Using the technique of Appendix 3, we can easily express the sum $c_n + c_{n+1}$ in terms of the Riemann $\theta$-function on the basis of the trace identities (3.1.15). However, we cannot obtain a convenient expression for

$c_n$. In principle, we can perform all the calculations starting from (3.1.13). Incidentally, under the condition $v_n \equiv 0$ the spectrum of $L$ is symmetric:

$$R(E) = \prod_{i=1}^{m+1} (E - E_i)(E + E_i). \tag{3.1.21}$$

The distribution of the zeros $\gamma_j(n)$ and the poles $\gamma_j(n_0)$ in the lacunae is also symmetric.

We now turn to the non-linear systems associated with $L$. Their Hamiltonians $I_k$ can be obtained, as indicated in [35], from the expansion of $\Delta_n(E) = \ln \chi(n, E)$ in terms of $E^{-1}$, as $E \to \infty$, which follows from (3.1.6) and (3.1.10):

$$\begin{aligned} \Delta_n^+(E) &\sim \ln E - \ln \sqrt{c_{n+1}} + \ln(1 + O(1/E)) = \\ &= \ln \frac{E}{\sqrt{c_{n+1}}} + \sum_{q \geq 1} b_{qn}/E^q = \\ &= \ln(E/\sqrt{c_{n+1}}) - \frac{v_n}{E} - \frac{c_n + \frac{v_n^2}{2}}{E^2} - \frac{\frac{v_n^3}{3} + v_n c_n + c_n v_{n-1}}{E^3} + \dots \end{aligned} \tag{3.1.22}$$

All the integrals $I_q = \sum_{n=n_0}^{n_0+N} b_{qn}$ are conserved in time by each of the systems, and all these systems commute; by definition, $c_n = e^{x_n - x_{n-1}}$.

The canonical coordinates are $(x_n, v_n)$. The first of these systems has the form

$$\begin{cases} \text{a)} \ \dot{x}_n = 1, \quad \dot{v}_n = 0, \quad I_1 = \sum v_n; \\ \text{b)} \ \dot{x}_n = v_n, \quad \dot{v}_n = c_{n+1} - c_n, \quad I_2 = \sum \left(\frac{v_n^2}{2} + c_n\right); \\ \text{c)} \ \dot{x}_n = v_n^2 + c_{n+1} + c_n, \quad \dot{v} = (v_{n+1} + v_n) c_{n+1} - (v_n + v_{n-1}) c_n, \\ \qquad I_3 = \sum \left(\frac{v_n^3}{3} + v_n c_n + c_n v_{n-1}\right). \end{cases} \tag{3.1.23}$$

For the system (3.1.23), c) the variety $v_n \equiv 0$ is invariant, and on it the system has the form of the "K-dV difference equation"

$$\dot{c}_n = c_n(c_{n+1} - c_{n-1}). \tag{3.1.23'}$$

To find important families of solutions of these equations we have to solve the stationary problem for linear combinations of these systems (that is, to look for the singular points of the Hamiltonians $H = \Sigma \lambda_j I_j$), which determines "potentials" $(v_n, c_n)$ with finitely many lacunae. In particular, the stationary solutions of the Toda chain (3.1.23), b) have the form of "0-zone" operators

$$\begin{cases} v_n = \dot{x}_n = \text{const}, & H = I_2 + \lambda I_1, \\ \dot{v}_n = 0 = c_{n+1} - c_n & (c_n \equiv \text{const}). \end{cases} \tag{3.1.24}$$

We obtain the "1-zone" operators $L$ from the stationary points of the Hamiltonian $H$, where

(3.1.25) $$H = I_3 + \lambda I_2 + \mu I_1$$

or

$$v_n^2 + c_{n+1} + c_n + \lambda v_n + \mu = 0,$$
$$(v_{n+1} + v_n)c_{n+1} + (v_n + v_{n-1})c_n + \lambda(c_{n+1} - c_n) = 0.$$

Clearly, the change of variables $v_n \to v_n$ + const, $c_n \to c_n$ allows us to assume that $\lambda \equiv 0$. However, (3.1.25) are difference equations and are difficult to solve (even in the one-lacuna case). To find an analytic form of the 1-zone operators $L$ we may proceed by one of two methods.

METHOD 1. We use the algebraic-geometric formula for $v_n$ (see (3.1.20) and then compute $c_n$, starting from (3.1.16).

METHOD 2. We use the fact that the time dynamics of the Toda chain for the one-zone potentials $(v_n, c_n)$ is a "simple wave" $\{v(n-ct), c(n-\alpha t)\}$ for all $t$, and we calculate the dynamics in $t$ instead of the difference equation in $n$.

After the Abel map, as in Ch. 3, §4, the time dynamics of all these systems becomes linear.

EXAMPLE 1. Let $m = 1$; using (3.1.16) we obtain ($\Sigma E_i = \sigma_1 = 0$)

(3.1.26)
$$\begin{cases} \gamma_n + \gamma_{n+1} = \dfrac{\sigma_3}{L(\gamma_n)}[2\gamma_n^2 + \sigma_2 \pm \sqrt{P_4(\gamma_n)}], \\ \gamma_n + \gamma_{n-1} = \dfrac{\sigma_3}{L(\gamma_n)}[2\gamma_n^2 + \sigma_2 \mp \sqrt{P_4(\gamma_n)}], \\ P_4(E) = \prod_{i=1}^{4}(E - E_i), \qquad L(\gamma_n) = 4\sigma_4 - \sigma_2^2 + 4\sigma_3\gamma_n, \\ \dot{v}_n = \dot{\gamma}_n = c_{n+1} - c_n = \dfrac{\sigma_3}{4}\left(\dfrac{1}{\gamma_{n+1}+\gamma_n} - \dfrac{1}{\gamma_n + \gamma_{n-1}}\right) = \\ \qquad = \sqrt{P_4(\gamma_n)}, \qquad \sigma_3 \neq 0. \end{cases}$$

If $\sigma_3 = 0$, then we have

(3.1.26′)
$$\begin{cases} \gamma_{n+1} + \gamma_n = 0, \\ \dot{\gamma}_n = c_{n+1} - c_n = \sqrt{P_4(\gamma_n)}. \end{cases}$$

Thus, for $m = 1$ the functions $\gamma_n(t) = \gamma(n - \alpha t)$ have the form of the elliptic function

$$t - t_0 = \int^{\gamma_n} \frac{d\tau}{\sqrt{P_4(\tau)}}.$$

According to (3.1.16) we find for the coefficients $c_n$ and $v_n$

(3.1.27)
$$\begin{cases} v_n = \gamma_n \qquad (\sigma_1 = 0), \\ 4c_{n+1} = \dfrac{\sigma_3}{\gamma_n + \gamma_{n+1}} \qquad (\sigma_3 \neq 0), \\ (\sigma_2 + 4c_{n+1})^2 + 4c_{n+1} - 4\sigma_4\gamma_n^2 = 0 \qquad (\sigma_3 = 0), \end{cases}$$

where $\gamma_n + \gamma_{n+1}$ are determined from (3.1.26).

EXAMPLE 2. Let $m = 2$ and $v_n \equiv 0$.

We now consider the symmetric spectrum with $\sigma_1 = \sigma_3 = \sigma_5 = 0$. We have $\gamma_{1n} + \gamma_{2n} = 0$ since $v_n = \Sigma\gamma_{jn} \equiv 0$; from the trace identities (3.1.15) we obtain

$$-(c_{n+1}+c_n)=\gamma_{1n}^2+\frac{1}{2}\left(\sum E_i^2\right). \tag{3.1.28}$$

We denote $\gamma_{1n}^2$ by $\gamma_n$. Using (3.1.17) we obtain

$$\begin{cases} \gamma_n=\frac{t_n}{2}\pm\sqrt{\frac{P_4(c_n)}{16^2c_n^2}}, \\ \gamma_{n-1}=\frac{t_n}{2}\mp\sqrt{\frac{P_4(c_n)}{16^2c_n^2}}, \\ \gamma_n-\gamma_{n-1}=\frac{2}{16c_n}\sqrt{P_4(c_n)}, \\ P_4(x)=[4\sigma_4-(\sigma_2+4x)^2]^2+16\cdot 4\sigma_6 x, \\ t_n=-\frac{(\sigma_2+4c_n)^2}{16c_n}+\frac{4\sigma_4}{16c_n}. \end{cases} \tag{3.1.29}$$

We now use the K-dV difference equation

$$\dot{c}_n = c_n(c_{n+1} - c_{n-1}).$$

Together with (3.1.29) we have

$$(\ln c_n^*)^{\cdot} = \gamma_n - \gamma_{n-1}. \tag{3.1.30}$$

Using the expression for $\gamma_n$ and $\gamma_{n-1}$ in terms of $c_n$ we obtain

$$\dot{c}_n=\frac{1}{8}\sqrt{P_4(c_n)}, \tag{3.1.31}$$

where $P_4(c_n) = [4\sigma_4 - (\sigma_2 + 4c_n)^2]^2 + 16\cdot 4\sigma_6 c_n$.

Finally we note the difference analogues to certain other results of Ch. 2.

By analogy with Ch. 2, §2, we define the matrices $Q(n_0, E)$ and $\Lambda(n_0, E)$ for a given non-linear system in the basis (3.1.3). Then

$$\hat{T}(n_0+1)=Q(n_0)\hat{T}Q^{-1}(n_0),$$

$$\frac{d}{dt}c(n, n_0, E)=\lambda_{11}c+\lambda_{12}s+Ac,$$

$$\frac{d}{dt}s(n, n_0, E)=\lambda_{21}c+\lambda_{22}s+As,$$

$$\dot{L}=[A, L], \qquad \dot{T}=[\Lambda, T],$$

$$\Lambda=\begin{pmatrix}\lambda_{11} & \lambda_{12}\\ \lambda_{21} & \lambda_{22}\end{pmatrix}, \quad Q=\begin{pmatrix}q_{11} & q_{12}\\ q_{21} & q_{22}\end{pmatrix}.$$

The spectrum of $L$ such that $\dot{L} = [A, L]$ is defined on the Riemann surface $\Gamma$:

$$\det(y - \Lambda(E)) = 0,$$

or, if Sp $\Lambda = 0$, then $y^2 = \det\Lambda = P_{2n+2}(E)$. As before, the coefficients of $\det\Lambda = P_{2n+2}(E)$ give "integrals" of the difference equation for the stationary problem. They can be used, in particular, to embed the Jacobian variety $\mathbf{J}(\Gamma)$ in a projective space, by analogy with Ch. 2. From the compatibility condition for the equations

$$\dot{T} = [\Lambda, T] \text{ and } T_{n+1} = QT_nQ^{-1}$$

we obtain

$$\frac{dQ(n)}{dt} = \Lambda(n+1)\,Q(n) - Q(n)\,\Lambda(n).$$

These equations are equivalent to the original non-linear system. For the stationary problem, then

$$\Lambda(n+1) = Q(n)\,\Lambda(n)\,Q^{-1}(n),$$

$$\frac{d}{dt}Q(n) \equiv 0.$$

In conclusion we mention that our arguments require that $L$ is "local" (and that quantities of Wronskian kind are conserved). It is important for us that, irrespective of the value of the period, a basis of eigenfunctions can be determined (for a given $E$) by values in a set of neighbouring points, in which the normalization of the Bloch function $\psi_{\pm}$ can also be uniquely determined.

## §2. First order matrix operators and their associated non-linear systems

We give here an account (with sketches of the proofs) of some very recent results of Dubrovin and Its as applied to non-linear systems associated with first order linear differential matrix operators. Its [52] has studied in detail the case of a two-dimensional matrix operator and its associated non-linear Schrödinger equation and modified K-dV equation, by comparing trace formulae (see Appendix 3 of the present survey), and for this case he has obtained convenient explicit formulae in terms of $\theta$-functions. The case of general matrix operators was analyzed by Dubrovin [51] (in our account we essentially follow [51]). Our aims were as follows.

1. The generalization of the methods of the authors to more than two-dimensional matrix operators meets certain difficulties. However, a number of physically interesting non-linear equations give rise to such operators; for example, for three-dimensional operators as was pointed out by Zakharov and Manakov in [27], the equation for the interaction of wave packets in non-linear media (see Ch. 1, §4 of the present survey).

2. The application to the theory of Abelian varieties according to the scheme of [43] (see Ch. 2, §4, of this survey). Here we can describe explicitly the universal fibering of Jacobian varieties that are not necessarily hyperelliptic.

3. As Tyurin has indicated, an interesting topic is the applicability of our technique to the classical problem of unirationality not only of the space of the universal fibering of Jacobian varieties, but also of the base of this fibering, that is, the unirationality of the space of moduli of algebraic curves. We show that the algebraic curves $\Gamma$ (see below) obtained within the framework of our construction form a necessarily unirational family, but

the codimension of this family in the space of moduli is always non-zero.

We consider an $n$-dimensional linear differential first order matrix operator $L = \frac{d}{dx} + U(x)$, where $U(x) = (u_i^j(x))$ is an $n \times n$ matrix, periodically dependent on $x$ with period $T$, and with zero diagonal elements $u_i^i \equiv 0$ $(i = 1, \ldots, n)$. We pose the eigenvalue problem for $L$:

$$L\psi = EA\psi, \qquad \psi(x + T, E) = e^{p(E)}\psi(x, E). \tag{3.2.1}$$

Here $E$ is a complex parameter and $A$ is a constant diagonal matrix $A = (a_i\delta_i^j)$, $\Sigma a_i = 0$. It is convenient to consider at once the family of such problems parametrized by the matrices $A$.

Let **A** be the $(n-1)$-dimensional space of complex diagonal $n \times n$ matrices of trace zero. In what follows we consider functions of $n-1$ variables, parametrized by the matrices $A$; to each matrix $A \in \mathbf{A}$ there corresponds a variable $x_A$, where $x_{A+B} = x_A + x_B$, $x_{\lambda A} = \lambda x_A$; the set of these $(n-1)$ variables is denoted by $X$. Let $V = V(X)$ be an $n \times n$ matrix depending on $X$. For each matrix $A \in \mathbf{A}$ we construct an operator $L_A$ depending on the parameter $E$:

$$L_A = \frac{\partial}{\partial x_A} + [A, V(X)] - EA. \tag{3.2.2}$$

The matrix $V$ is called *the potential of the operator* $L_A$, which acts on functions of $x_A$ and depends parametrically on the remaining variables $x_B (B \in \mathbf{A})$. We require that for different $A \in \mathbf{A}$ all the operators $L_A$ commute among themselves:

$$\begin{cases} [L_A, L_B] = 0 \Leftrightarrow \dfrac{\partial Q_B}{\partial x_A} - \dfrac{\partial Q_A}{\partial x_B} = [Q_A, Q_B], \\ Q_A = [A, V] - EA, \quad Q_B = [B, V] - EB. \end{cases} \tag{3.2.3}$$

The commutativity condition (3.2.3) is equivalent to the following non-linear equation for $V(X)$:

$$\left[A, \frac{\partial V}{\partial x_B}\right] - \left[B, \frac{\partial V}{\partial x_A}\right] = [[A, V], [B, V]]. \tag{3.2.4}$$

The formulae (3.2.3) show that the equation (3.2.4) admits a commutation representation by $n \times n$ matrices that are polynomially (linearly) dependent on the parameter $E$, In the sense of Ch. 2, §2 (see (2.2.20)). Varying the matrices $A$ and $B$, we obtain a system of $n-2$ equations for the matrix $V(X)$ depending on the $(n-1)$-fold argument, so that knowing the dependence of $V$ on one variable $x_A$, we can determine the dependence on any other $x_B (B \in \mathbf{A})$ by solving the Cauchy problem for (3.2.4). In what follows, we always assume that $V(X)$ is a solution of the system (3.2.4) where $A$ and $B$ range over all the diagonal matrices **A**.

We now construct a set of commuting dynamical systems on the variety of matrices $V(X)$ that are solutions of the system (3.2.4). It is easy to see that the equation $\frac{\partial}{\partial x_A}\lambda_A = [\lambda_A, Q_A]$ has a unique solution in the form of a formal series in $1/E$, beginning with $A$

(3.2.5) $$\lambda_A=\lambda_{0,A}+\frac{\lambda_{1,A}}{E}+\frac{\lambda_{2,A}}{E^2}+\ldots; \qquad \lambda_{0,A}=A.$$

If $V(X)$ is a solution of (3.2.4), then $\lambda_A$ satisfies the following equation:

$$\frac{\partial}{\partial x_B}\lambda_A=[\lambda_A, Q_B].$$

The matrix elements of $\lambda_{k,A}$ are polynomials in the elements of $V, \partial V/\partial x_A, \partial^2 V/\partial x_A^2 \ldots$ with constant coefficients, depending on $A$. Suppose now that $A_1, \ldots, A_{N+1} \in \mathbf{A}$.

DEFINITION 1. *The Nth equation of K-dV type* is

(3.2.6) $$[A, \dot{V}-(\lambda_{A_1, N+1}+\lambda_{A_2, N}+\ldots+\lambda_{A_{N+1}, 1})]=0,$$

where the matrices $\lambda_{A_k, N-k+2}$ are defined by the algorithm (3.2.5) and $V = V(X, t)$ is a solution of (3.2.4).

The equation (3.2.6) admits a commutation representation by $n \times n$ matrices polynomially dependent on $E$, in the sense of Ch. 2, §2:

(3.2.7) $$\frac{\partial\Lambda}{\partial x_A}-\frac{\partial Q_A}{\partial t}=[\Lambda, Q_A].$$

Here

(3.2.7′) $$\Lambda=\Lambda_{A_1, N}(E)+\ldots+\Lambda_{A_{N+1}, 0}(E),$$

where

(3.2.7″) $$\Lambda_{A, K}=AE^k+\lambda_{1, A}E^{k-1}+\ldots+\lambda_{k, A} \qquad (A\in\mathbf{A}).$$

It is easy to see that $\Lambda_{1,A} = -Q_A$, therefore (3.2.4) is the first equation of K-dV type in the sense of Definition 1.

Let $V = V(X)$ be a periodic function of the variable $x_A$ with period $T$ (for simplicity we assume in the subsequent formulae that all the diagonal elements of $A$ are pairwise distinct). Then we have the eigenvalue problem for $L_A$:

(3.2.8) $$L_A\psi(x_A, E) = 0, \qquad \psi(x_A + T, E) = e^{p(E)}\psi(x, E).$$

By analogy with Definition 1 in Ch. 2, §1, we make the following definition:

DEFINITION 2. The potential $V$ is said to be *finite-zoned* for the operator $L_A$ if the eigenfunctions of the problem (3.2.8) are meromorphic on a Riemann surface $\Gamma$ of finite genus, which gives an $n$-fold covering of the $E$-plane. The surface $\Gamma$ is then called the *spectrum* of $L_A$.

If $V$ is almost-periodic as a function of $x_A$, then Definition 2 is modified as follows:

DEFINITION 2′. The potential $V$ is said to be *finite-zoned* for the operator $L_A$ if $L_A$ for all $E$ has an eigenfunction $\psi(x_A, E)$ that is meromorphic on a Riemann surface $\Gamma$ of finite genus that gives an $n$-fold covering of the $E$-plane, with the "boundary condition" of (3.2.8) replaced by the following: the group of periods of the logarithmic derivatives of the coordinates of $\psi(x_A, E)$ is the same as the group of periods of $V$. The surface $\Gamma$ is called the *spectrum* of $L_A$.

We consider the stationary solutions of (3.2.6) (that is, those not depending on the time $t$). (We recall that $V(X)$ is a solution of the system (3.2.4).) To find them, as in Ch. 2, §2, we have a commutation representation of Lax type on matrices polynomially dependent on $E$:

$$\frac{\partial \Lambda}{\partial x_A} = [\Lambda, Q_A], \tag{3.2.9}$$

where the matrix $\Lambda$ is defined by (3.2.7). We consider the Riemann surface $\Gamma$ of the algebraic function $W = W(E)$, where $W(E)$ is given by the equation

$$R(W, E) = \det | W \cdot 1 - \Lambda | = 0. \tag{3.2.10}$$

Actually, $\Gamma$ is a complex algebraic curve in two-dimensional complex space with the coordinates $W$ and $E$, defined by (3.2.10). Since for an arbitrary $E$ there are $n$ values of $W(E)$, $\Gamma$ gives an $n$-fold covering of the $E$-plane. The infinitely distant part of $\Gamma$ consists of $n$ ordered points $\{1\}, \ldots, \{n\}$; the order is determined by the conditions $W(E) \sim a_{1i}E^N$ ($E \to \infty$) in a neighbourhood of $\{i\}$ (where $a_{1i}$ is the $i$th diagonal element of $A_1$).

Let $\pi$ be the factor group of the subgroup of scalar matrices in the group of all non-singular diagonal $n \times n$ matrices. Then $\pi$ acts on the potentials $V$ by the following rule:

$$V \to \pi^{-1} V \pi.$$

THEOREM 1. *The stationary equation* (3.2.9) *is a totally integrable Hamiltonian system with* $N \frac{n(n-1)}{2}$ *degrees of freedom, and the coefficients of the polynomial* $R(W, E)$ *give a complete set of commuting polynomial integrals of* (3.2.9).

2. *The potential* $V$ *is finite-zoned for all the operators* $L_A$; *their spectrum is the Riemann surface* $\Gamma$ *defined by* (3.2.10).

3. *The set of finite-zoned potentials* $V$ *with a given spectrum* $\Gamma$ *is the space of the principal* $\pi$*-fibering over the Jacobian variety* $J(\Gamma)$ *of* $\Gamma$.

PROOF. As in Ch. 2, §2, we see that $\Gamma$, defined by (3.2.10), does not depend on the variable $X$ and is invariant under all the dynamical systems of the form (3.2.6), that is, the coefficients of $R(W, E)$ are integrals of (3.2.9). Later we shall show that the systems (3.2.9) are Hamiltonian, hence the first part of the theorem follows by analogy with Ch. 2, §2.

In the solution space of the equation $L_A f(x, E) = 0$ we introduce the basis of solutions $c_1(x, y, E), \ldots, c_n(x, y, E)$ (denoting the variable $x_A$ simply by $x$ and regarding $A$ as fixed) such that $c_i^j(y, y, E) = \delta_i^j$ (where $y$ is a parameter). Let $\hat{T}(y, E)$ be the translation matrix for $L_A$ in the basis $c_1(x, y, E), \ldots, c_n(x, y, E)$ if $V$ is periodic in $x$. If $V$ evolves in time $t$ according to the K-dV type equation (3.2.6), then the time derivative for $\hat{T}$ has the form

$$\frac{\partial}{\partial t}\hat{T}=[\hat{T},\ \Lambda], \tag{3.2.11}$$

with $\Lambda$ defined by (3.2.7′). Since we look for stationary solutions of (3.2.6), the matrices $T$ and $\Lambda$ commute:

$$\hat{T}\Lambda = \Lambda\hat{T}. \tag{3.2.11′}$$

Let $\psi(x, E)$ be an eigenfunction of the problem (3.2.8) ($x = x_A$), normalized by fixing the value of one coordinate when $x = y$, for example, $\psi^1(y, E) = 1$.

We note that

$$\psi(x, E)=\sum_j \psi^j(y, E)\, c_j(x, y, E). \tag{3.2.12}$$

Here $(\psi^1(y, E), \ldots, \psi^n(y, E))$ is an eigenvector of $\hat{T}(y, E)$, hence also of $\Lambda(y, E)$, by (3.2.11). Consequently, the coordinates $\psi^j(y, E)$ can be expressed rationally (with the normalization taken into account) in terms of the elements of the matrix $W(E)\cdot 1 - \Lambda(y, E)$, that is, they are algebraic functions on the Riemann surface $\Gamma$ defined by (3.2.10). Therefore, by (3.2.12) $\psi(x, E)$ can be continued as a meromorphic function on the Riemann surface $\Gamma \setminus \infty$ (that is, away from the "infinitely distant part" of $\Gamma$). So we have proved that $L_A$ is finite-zoned with spectrum $\Gamma$. If $V$ is an almost-periodic solution of (3.2.9), then we define the eigenfunction $\psi$ by

$$\psi(x, E)=\sum_j \xi^j(y, E)\, c_j(x, y, E), \tag{3.2.12′}$$

where $(\xi^1(y, E), \ldots, \xi^n(y, E))$ is an eigenvector of $\Lambda(y, E)$ with the eigenvalue $W(E)$. The definition (3.2.12) is not contradictory because according to (3.2.9) $L_A$ commutes with the operator of multiplication by $\Lambda$. The finite-zone property is subsequently proved as in the periodic case (compare with the second method of proof of Corollary 2 in Ch. 2, §2).

We construct the matrix-valued function $\Psi(x, y, P)$, where $P$ is a point of $\Gamma$. Let $E$ be a point that is not a branch-point, so that for the given $E$ the problem (3.2.8) has exactly $n$ linearly independent arbitrarily ordered eigenfunctions $\psi_1(x, E), \ldots, \psi_n(x, E)$. We form from their coordinates a matrix $\psi_i^j(x, E)$. Let $\varphi_i^j(x, E)$ be its inverse, which exists because $\psi_1, \ldots, \psi_n$ are linearly independent. If $P \in \Gamma$, $P = (E, k)$, where $k$ labels the sheet, then we set

$$\Psi_i^j(x, y, P) = \psi_k^j(x, E)\cdot\varphi_i^k(y, E). \tag{3.2.13}$$

This definition does not depend on the original ordering of the eigenfunctions $\psi_1, \ldots, \psi_n$ nor on their normalization. The function $\Psi(x, y, P)$ becomes meromorphic on the Riemann surface $\Gamma \setminus \infty$.

We define the operation $Tr_P$, which is important in what follows, of taking the trace of a function defined on $\Gamma$. Let $\varphi = \varphi(P)$ be such a function, $P \in \Gamma$, that is, $P$ is a pair $(E, k)$ where $k$ labels the sheet. If $E$ is not a branch-point, there are exactly $n$ points $(E, 1), \ldots, (E, n)$ on $\Gamma$ over $E$. We then set

$$(Tr_P\varphi)(E) = \varphi((E, 1)) + \ldots + \varphi((E, n)). \tag{3.2.14}$$

Now $Tr_P\varphi$ is a single-valued function on the $E$-plane. In particular, if

$$G(x, y, E) = \begin{cases} Tr_P \Psi(x, y, P) & (x \leqslant y), \\ 0 & (x > y), \end{cases}$$

then $G(x, y, E)$ is the Green's matrix of $L_A$. Let $g(x, P) = \Psi(x, x, P)$. Then $g(x, P)$ has the following important properties.

a) The group of periods of $g(x, P)$ is the same as that of the potential $V$.

b) It is algebraic on $\Gamma$.

c) It gives the "spectral decomposition" for the matrix $\Lambda(x, E)$ that is, $g^2 = g$, $g(x, (E, k))\cdot g(x, (E, l)) = 0$ for $k \neq l$ (where $k$, $l$ label the sheets), $Tr_P g(x, P) = 1$, $Tr_P W\cdot g(x, P) = \Lambda(E, x)$.

d) $\dfrac{\partial g}{\partial x_3} = [g, Q_B]$ for any $B \in A$.

e) The variational derivative of the functional $p(E)\,\{V\}$ defined in (3.2.8) (in the periodic case) has the form

$$\frac{\delta p(E)}{\delta v_i^j(x)} = -(a_i - a_j)\, g_i^j \quad \text{(see (2.1.17))}.$$

f) As $P \to \{k\}$, $g(x, P)$ has an expansion of the form

$$g(x, P) = g_0 + \frac{g_1}{E} + \frac{g_2}{E^2} + \ldots,$$

where

$$g^j_{0_i} = \delta^k_i \cdot \delta^j_k; \qquad g^j_{1_i} = -\delta^k_i v^j_i + v^j_i \delta^j_k;$$

$$g^j_{2_i} = \delta^k_i \left[ \frac{v_i^{j\prime}}{a_i - a_j} + \frac{1}{a_i - a_j} \sum_s v_i^s (a_s - a_j)\, v_s^i \right] +$$
$$+ \left[ -\frac{v_i^{j\prime}}{a_i - a_j} + \frac{1}{a_i - a_j} \sum_s v_i^s (a_i - a_s) v_s^j \right] \delta^j_k - v_i^k v_k^j + \delta^k_i \left( \sum_s v_i^s v_s^j \right) \delta^j_k.$$

Here the dash denotes differentiation with respect to $x = x_A$.

From c) and e) it follows immediately that the systems (3.2.9) are Hamiltonian. Let us find the zeros and poles of the matrix elements $g_i^j(x, P)$, which are algebraic on $\Gamma$. From c) it follows that the poles of $g(x, P)$ are precisely the branch-points of $\Gamma$, that is, the points at which the different branches of the algebraic function $W(E)$ merge. We denote the set of branch-points by the symbol $\mathcal{D}_w$. There are $Nn(n-1)$ of them. From the expansion f) it follows that at the infinitely distant part of $\Gamma$

all the $g_i^j$ have zeros and their disposition is as follows: for $i \neq j$, $g_i^j(x, P)$ has double zeros at all the points $P = \{k\}$ $(k \neq i, j)$ and simple zeros at the points $P = \{i\}$, $P = \{j\}$; $g_i^i(x, P)$ has double zeros at all the points $P = \{k\}$, $(k \neq i)$, and $g_i^i(x, P) = 1$ at $P = \{i\}$. There is a convenient notation to describe the distribution of zeros and poles of a function defined on $\Gamma$. Let $\varphi(P)$ be a function on $\Gamma$ having zeros of multiplicity $n_1$ at $P_1$, $n_2$ at $P_2, \ldots$, and poles of multiplicity $m_1$ at $Q_1$, $m_2$ at $Q_2, \ldots$. We express this by saying that $\varphi(P)$ has the divisor $D = n_1P_1 + n_2P_2 + \ldots - m_1Q_1 - m_2Q_2 - \ldots$ on $\Gamma$. It is clear that the whole divisor $D$ of $\varphi(P)$, that is, the set of zeros and poles of $\varphi(P)$ with their multiplicities taken into account, can be decomposed into the difference $D = D_+ - D_-$, where $D_+ = n_1P_1 + n_2P_2 + \ldots$ is the set of zeros (the divisor of zeros), and $D_- = m_1Q_1 + m_2Q_2 + \ldots$ the set of poles of $\varphi(P)$ (the divisor of poles). The degree of the divisor $D$ is defined to be

$$\deg D = \sum_i n_i - \sum_j m_j = \deg D_+ - \deg D_-.$$

If $\varphi(P)$ is an algebraic function on $\Gamma$, that is meromorphic everywhere on $\Gamma$, then $\deg D = 0$ (the number of zeros, counting multiplicities, is equal to the number of poles, counting multiplicities). In particular, the result on the zeros and poles of $g_i^j(x, P)$ on $\Gamma$ can be expressed concisely as:

$$\text{(3.2.15)} \quad \text{divisor } (g_i^j(x, P)) = \sum_{k \neq i, j} 2\{k\} + \{i\} + \{j\} - \mathcal{D}_w + \ldots =$$
$$= 2\Sigma - \{i\} - \{j\} - \mathcal{D}_w + \ldots,$$

where $\Sigma = \sum_k \{k\}$, and the dots denote the unknown divisor of the zeros of $g_i^j(x, P)$ in the finite part of $\Gamma$. To find this we turn to a study of the analytic properties of the matrix $\Psi(x, y, P)$ defined by (3.2.13). We note the formula $\Psi(x, y, P) = c(x, y, E(P))\, g(y, P)$ similar to (3.2.12). Therefore, the poles of the matrix elements $\Psi(x, y, P)$ lie only at the branch-points (that is, the divisor of poles of $\Psi_i^j(x, y, P)$ is equal to $\mathcal{D}_w$ for all $i$ and $j$). The matrix $\Psi(x, y, P)$ is of rank 1; its columns are eigenfunctions of $L_A$, which acts on the variable $x$, and differ only in normalization, and the rows are eigenfunctions of the adjoint operator $L_A^*$, which acts on $y$ and is defined as follows:

$$L_A^* = \frac{\partial}{\partial y} - Q_A^T \quad (T \text{ denotes the transpose}).$$

Therefore the relations

$$\text{(3.2.16)} \qquad \frac{\Psi_i^k(x, y, P)}{\Psi_j^k(x, y, P)} = \frac{g_i^k(y, P)}{g_j^k(y, P)}$$

and

$$\text{(3.2.16')} \qquad \frac{\Psi_k^i(x, y, P)}{\Psi_k^j(x, y, P)} = \frac{g_k^i(x, P)}{g_k^j(x, P)} \quad \text{do not depend on } k.$$

From (3.2.16) and (3.2.16′) it follows that the zeros of $\Psi_i^j(x, y, P)$ split into two parts: zeros depending on $j$ and on $x$, and zeros depending on $i$ and on $y$. We express this in the following way:

(3.2.17) $\quad \textit{divisor of the zeros of } (\Psi_i^j(x, y, P)) = d_i(y) + d^j(x),$

where the divisors $d_i(x)$ and $d^j(y)$ have the form

(3.2.17′) $\quad d_i(y) = P_{1i}(y) + \dots, \qquad d^j(x) = Q_1^j(x) + \dots$

Thus, we define the divisor of $g_i^j(x, P)$ as

(3.2.18) $\quad \textit{divisor } (g_i^j(x, P)) = 2\Sigma - \{i\} - \{j\} - \mathcal{D}_w + d_i(x) + d^j(x).$

The function $g_i^j(x, P)$ is algebraic on $\Gamma$, hence the degree of its divisor is zero (the number of zeros is equal to the number of poles). Therefore, $\deg\,[d_i(x) + d^j(x)] = 2[N\frac{n(n-1)}{2} - (n-1)]$. Since the operators $L_A$ and $L_A^*$ are entirely of equal standing, we have the important relation

(3.2.19) $\quad \deg d_i(x) = \deg d^j(x) = N\frac{n(n-1)}{2} - (n-1) = \text{genus}\,(\Gamma) \equiv p.$

Hence $d_i(x) = P_{1i}(x) + \dots + P_{pi}(x)$, $d^j(x) = Q_1^j(x) + \dots + Q_p^j(x)$. The divisors $d_i(x)$ and $d^j(x)$ can therefore be regarded as points of the $p$th symmetric power $S^p\Gamma$ of $\Gamma$ (see the definition in Ch. 2, §3). We recall that the Abelian map $\mathfrak{A}$ from the $k$th symmetric power of $\Gamma$ into its Jacobian variety

(3.2.20) $\quad \mathfrak{A}\colon S^k\Gamma \to \mathbf{J}(\Gamma),$

is almost everywhere one-to-one if $k = p = $ genus $(\Gamma)$. (More accurately, it is a birational equivalence, see (2.3.25).) The map $\mathfrak{A}$ has the following property: if $D = D_+ - D_-$ is the divisor of the zeros and of the poles of $\varphi(P)$, algebraic on $\Gamma$ then (by the classical theorem of Abel)

(3.2.21) $\quad \mathfrak{A}(D_+) = \mathfrak{A}(D_-).$

We know that $g_i^j(x, P)$ is algebraic on $\Gamma$ and has a divisor of the form (3.2.8). Now, bearing (3.2.21) in mind we obtain a system of linear equations on $\mathbf{J}(\Gamma)$ for the quantities $\mathfrak{A}(d_i(x))$, $\mathfrak{A}(d^j(x))$:

(3.2.22) $\quad \mathfrak{A}(d_i(x)) + \mathfrak{A}(d^j(x)) = \mathfrak{A}(\mathcal{D}_w) - \mathfrak{A}(2\Sigma - \{i\} - \{j\}).$

It is easy to see that the specification of any one of the unknowns, for example, of $\mathfrak{A}(d^1(x))$, determines all the remaining $\mathfrak{A}(d_i(x))$, $\mathfrak{A}(d^j(x))$. Thus, we have constructed a correspondence

(3.2.23) $\quad V \to \eta, \qquad \eta \in \mathbf{J}(\Gamma),$

where $V$ is a solution of (3.2.9) and the specification of $\eta \in \mathbf{J}(\Gamma)$ uniquely determines the distribution of the zeros of $g(x, P)$ (for fixed $x$) on $\Gamma$.

We now show that, as $x$ varies, the point $\eta$ on $\mathbf{J}(\Gamma)$ moves in a straight line, that is,

(3.2.24) $$\eta(x) = \eta(y) + (x - y)\vec{U};$$

where $\vec{U}$ is a constant vector. We consider the function $\widetilde{\Psi}(x, y, P)$ where

(3.2.25) $$\widetilde{\Psi}^j(x, y, P) = \Psi_i^j(x, y, P)/g_i^j(y, P)$$

does not depend on $i$. Then $\widetilde{\Psi}^j(x, y, P)$ has on $\Gamma \setminus \infty$, zeros at the points of $d^j(x)$ and poles at the points of $d^j(y)$, and as $P \to \{k\}$, it has the asymptotic form $\exp\{\alpha_k E(x-y)\}$. The latter assertion follows from the formula

(3.2.26) $$\widetilde{\Psi}^j(x, y, P) = \exp\left\{\int_y^x \chi^j(\xi, P)\,d\xi\right\},$$

where

$$\chi^j(\xi, P) = a_j E - \sum_s u_s^j(\xi)\,\frac{g_i^s(\xi, P)}{g_i^j(\xi, P)},$$

and from the expansions for $g_i^j(\xi, P)$ given in f). Hence we find ourselves in a situation similar to that of Ch. 2, §3, in the proof of (2.3.30) and (2.3.30′). Reasoning in exactly the same way, we obtain the relation on $\mathbf{J}(\Gamma)$

(3.2.27) $$\mathfrak{A}(d^j(x)) - \mathfrak{A}(d^j(y)) = (x-y)\vec{U},$$

which is equivalent to (3.2.4). We shall choose the vector $\vec{U}$ below.

We have seen that the specification of $\eta$ on $\mathbf{J}(\Gamma)$ determines the position of the zeros of all the functions $\Psi_i^j(x, y, P)$ for arbitrary $x$ and $y$. It is easy to see that the "phases" $\chi^j(x, P)$ are likewise determined for all $\xi$. It remains to define the "amplitudes" of the functions $\Psi_i^j(x, y, P)$, that is, the functions $g_i^j(y, P)$. We note that $g_i^i(y, P)$ is uniquely determined by its zeros, and the functions $g_i^j(y, P)$, $i \neq j$, are uniquely determined to within a constant factor: that is, for a given point $\eta \in \mathbf{J}(\Gamma)$ we can construct a matrix $\widetilde{g}(y, P)$, which is, in general, distinct from our $g(y, P)$:

$$\widetilde{g}_i^j(y, P) = \varepsilon_i^j g_i^j(y, P),$$

where $\varepsilon_i^j$ = constant. If we require that $\widetilde{g}^2 = \widetilde{g}$, then we find that $\varepsilon_i^j = \varepsilon_i/\varepsilon_j$, $\varepsilon_1, \ldots, \varepsilon_n \neq 0$, that is, the non-uniqueness in the definition of the "amplitudes" $g_i^j(y, P)$ of the functions $\Psi_i(x, y, P)$ lies in the action of the group $\pi$. The same non-uniqueness occurs also in the definition of the potential $V$. It is noteworthy that the action of $\pi$ on the variety of potentials $V$ commutes with all the dynamical systems of the form (3.2.6).

We give an explicit construction of $V$. We define the following functions on $\mathbf{J}(\Gamma)$: let $\eta \in \mathbf{J}(\Gamma)$. We construct for $\eta$ the set of divisors $d_i$, $d^j$ of degree $p$ on $\Gamma$ such that $\mathfrak{A}(d^1) = \eta$, and the remaining $\mathfrak{A}(d_i)$,

$\mathfrak{A}(d_j)$ are determined by the system of linear equations (3.2.22). Let $h_i^j(P)$ be an algebraic function on $\Gamma$ whose divisor of zeros and poles has the form

$$2\Sigma - \{i\} - \{j\} + d_i + d^j - \mathcal{D}_w$$

(compare with (3.2.18)). We normalize $h_i^j$ as follows: for $i \neq j$ we require $h_i^j(P) \sim -\frac{1}{E}$, as $P \to \{i\}$; for $i = j$ we require that $h_i^j(\{i\}) = 1$. Let $\sigma_i^{jk}(\eta)$ be the coefficient of $\frac{1}{E^2}$ in the expansion of $h_i^j(P)$, as $P \to \{k\}$. Also, let $\rho_i^k(\eta)$ be the coefficient of $1\backslash E^2$ in the expansion of $h_i^i(P)$, as $P \to \{k\}$. Then

$$v_i^j(x) = v_i^j(x_0) \exp\left\{\int_{x_0}^{x} \sum a_k \sigma_i^{jk}\, dx\right\}, \tag{3.2.28}$$

where the integration is in the direction of $\vec{U}$, and the constants $v_i^j(x_0)$ satisfy the relations

$$\text{(3.2.28}'\text{)} \qquad \left\{ \begin{aligned} &\frac{v_i^k(x_0)\, v_k^j(x_0)}{v_i^j(x_0)} = -\sigma_i^{jk}(\eta) \quad \text{when } i \neq j, \\ &v_i^k(x_0)\, v_k^i(x_0) = -\rho_i^k(\eta). \end{aligned} \right.$$

As independent variables we can take, for example, the parameters $v_i^1(x_0)$, $i = 2, \ldots, n$.

We note now that the above arguments carry over trivially to the calculation of the time dependence on the variety of potentials $V$ that are solutions of the Hamiltonian system (3.2.9). We need merely replace the operators $L_A$ everywhere by $\frac{\partial}{\partial \tau} + \tilde{\Lambda}$, where $\tilde{\Lambda}$ is another matrix of the form (3.2.7) and the definition of $g$ is independent of the operator in question ($g$ gives a "spectral decomposition" for $\Lambda$). We write down the law of time evolution of a point $\eta$ on $\mathbf{J}(\Gamma)$ for standard equations of K-dV type

$$[A, \dot{V} - \lambda_{N+1, B}] = 0 \tag{3.2.29}$$

(the remaining ones are linear combinations of these). Then

$$\eta(\tau) - \eta(\tau_0) = \vec{W}\cdot(\tau - \tau_0). \tag{3.2.30}$$

The vector $\vec{W}$ can be found as follows. Let $\omega_{N,i}$ be an Abelian differential of the second kind with an $(N+1)$-fold pole at $P = \{i\}$, normalized by the conditions

$$\oint_{\alpha_j} \omega_{N,i} = 0, \qquad (j = 1, \ldots, p). \tag{3.2.31}$$

Here $\alpha_i$, $\beta_i$ $(i = 1, \ldots, p)$ is the set of cycles on $\Gamma$ with the intersection matrix of the form (2.3.21′).

Let

(3.2.31′)
$$\oint_{\beta_j} \omega_{N,i} = U_{ji}.$$

Then the vector $\vec{W}$ in (3.2.30) has the form

(3.2.32)
$$W_j = \sum_i b_i U_{ji}.$$

In particular, we have shown that the potential $V$ is finite-zoned for all the operators $\frac{\partial}{\partial \tau} + \widetilde{\Lambda}$, where $\widetilde{\Lambda}$ is of the form (3.2.7). This completes the proof of the theorem.

EXAMPLE. Let $n = 2$. $V = \begin{pmatrix} 0 & v_+ \\ v_- & 0 \end{pmatrix}$, $A = \begin{pmatrix} 1 & 0 \\ 0 & -1 \end{pmatrix}$; we consider the $(N + 1)$th equation of K-dV type. The matrix $\Lambda_{N+1}(E)$ has the form

$$\Lambda_{N+1} = \begin{pmatrix} Q & P_+ \\ P_- & -Q \end{pmatrix}, \qquad \deg Q = N+1, \qquad \deg P_\pm = N,$$

$$P_\pm = \mp 2v_\pm \prod_{j=1}^{N} (E - \gamma_j^\pm), \qquad Q = E^{N+1} + \dots$$

The Riemann surface $\Gamma$ has the form $W^2 - R(E) = 0$, where $R(E) = Q^2 + P_+P_-$, $\deg R = 2N + 2$. The projection $g$ has the form

$$g = \begin{pmatrix} \dfrac{Q+\sqrt{R}}{2\sqrt{R}} & \dfrac{P_+}{2\sqrt{R}} \\ \dfrac{P_-}{2\sqrt{R}} & \dfrac{-Q+\sqrt{R}}{2\sqrt{R}} \end{pmatrix}.$$

The divisor $d_2 + d^1$ is the complete inverse image of the points $E = \gamma_1^+, \dots, E = \gamma_N^+$ on $\Gamma$. Thus, $d^1$ comprises the $N$ points $P_1, \dots, P_N$ lying over the points $E = \gamma_1^+, \dots, E = \gamma_N^+$. Let $\Sigma E_i = 0$, where $R(E) = \prod_{i=1}^{2N+2} (E - E_i)$. Then

(3.2.33)
$$\begin{cases} v_+(x) = c \exp\left\{ -2 \int_{x_0}^{x} \sum \gamma_i^+ \, dx \right\}, \\ v_-(x) = \dfrac{\rho(x)}{v_+(x)}. \end{cases}$$

The algebraic function $\rho = v_+ v_-$ has the form

(3.2.33′)
$$\rho = \frac{1}{2} \det \begin{vmatrix} \gamma_1^{N+1} + \sqrt{R(\gamma_1)} & \gamma_1^{N-2} & \dots & 1 \\ \vdots & & & \vdots \\ \gamma_N^{N+1} + \sqrt{R(\gamma_N)} & \gamma_N^{N-2} & \dots & 1 \end{vmatrix} \left( \prod_{i<j} (\gamma_i - \gamma_j) \right)^{-1} - \frac{1}{4} \sum_{i<j} E_i E_j \qquad (\gamma_i = \gamma_i^+).$$

We now write down the $\pi$-fibering over $\mathbf{J}(\Gamma)$ referred to in the theorem. To do this we rewrite (3.2.33) in the form

$$(3.2.34)\qquad \begin{cases} v_+ = v_+(\eta, \eta_0, c) = c \exp\left\{\int\limits_{\eta_0}^{\eta} \omega\right\}, \\ v_- = v_-(\eta, \eta_0, c) = \dfrac{\rho(\eta)}{v_+(\eta, \eta_0, c)}. \end{cases}$$

Here $c$ is the coordinate in the fibre of $\pi$ (for $n = 2$, $\pi$ is one-dimensional), $\eta, \eta_0 \in \mathbf{J}(\Gamma)$, and $c$ together with the initial point $\eta_0$ determine the potential $V$ in accordance with the theorem, and the dependence of $V$ on $\eta$ includes the dependence of the potential on a displacement along the trajectories of any one of the dynamical systems of the form (3.2.6); also,

$$\omega = d \ln v_+ = \frac{v'_+}{v_+} dx + \sum_i{}' \frac{\partial v_+/\partial t_i}{v_+} dt_i$$

is a closed meromorphic differential on $\mathbf{J}(\Gamma)$. The periods of $\omega$ on the torus are given by the transition functions of the required $\pi$-fibering:

$c \to c \exp \oint \omega$ for a circuit of $\eta_0$ along a cycle in $\mathbf{J}(\Gamma)$. Here $\omega$ has the form

$$(3.2.34')\qquad \omega = -2\left(\sum \gamma_i\right) dx + \left(4 \sum_{i<j} \gamma_i\gamma_j - 2 \sum E_iE_j\right) dt_1 + \dots$$

In concluding this section we mention that, by analogy with Ch. 2, §3, we can prove a converse theorem: any finite-zoned potential (in the sense of Definitions 2 or 2′) of $L_A$ is a solution of a stationary equation of K-dV type (that is, of the form (3.2.9)). The proof is based on the properties of $g(x, P)$ described above.

Simultaneously with [51] and [52], the papers [60] and [61] were completed, in which matrix operators are studied by the methods of Marchenko [44], [45].

## APPENDIX 1

### NON-REFLECTIVE POTENTIALS AGAINST THE BACKGROUND OF FINITE-ZONE POTENTIALS. THEIR ALGEBRAIC-GEOMETRIC AXIOMATICS

As was shown in Ch. 2, §2, every periodic or almost-periodic stationary solution of any higher order K-dV equation is a finite-zone potential. However, these equations also have degenerate "separatrix" solutions. In [38] it was shown that rapidly decreasing separatrix solutions are non-reflecting potentials $u(x)$, associated with the so-called "multisoliton" solutions of the K-dV equation, which have been studied in detail, for example in [20]. The general separatrix solutions of any of the higher order K-dV equations are, as shown in [38], degenerate limits of conditionally periodic potentials, when some of the group of periods $T_1, \dots, T_n$ tend to $\infty$.

By virtue of the K-dV equation it is natural to call the dynamics of such potentials "multisoliton solutions against a finite-zone background". A study of the inverse scattering problem for potentials of this kind was made by Krichever [49]; some months earlier for multisoliton potentials against the background of one-zone potentials the problem was solved by another method in [48]. We give here an account of the main arguments of [49], including the algebraic-geometric axiomatics of the class of non-reflective potentials against a finite-zone background. We begin by explaining the axiomatics. For convenience we recall some elementary concepts of algebraic geometry. By a divisor we mean a set of points $P_i$ on a Riemann surface $\Gamma$ of finite genus with multiplicities $k_i$, formally written as a sum:

$$\text{divisor } D = \sum_i k_i P_i.$$

By the divisor $(f)$ of a function we mean the divisor of its zeros and poles, where the multiplicities of the zeros are positive and those of the poles are negative. The divisor $(-D)$ has the form $\Sigma - k_i P_i$.

Formal addition of divisors is defined in the obvious way. A divisor whose multiplicities are all positive, $k_i \geqslant 0$, is said to be *effective*. We say that an effective divisor is greater than zero: $D \geqslant 0$.

DEFINITION. An algebraic function $f$ on $\Gamma$ is said to belong to the space $L(D)$ if $(f) + D \geqslant 0$. We define the *degree of the divisor* $D$, denoted by $n(D)$, to be the sum of the multiplicities of its points. We have the standard (Riemann) estimate

$$\dim L(D) \geqslant n(D) - g + 1 \quad (g = \text{genus } \Gamma),$$

and equality holds for divisors of degree $n(D) > 2g - 2$. Now let $\Gamma'$ be a Riemann surface, doubly covering $\Gamma$ with branch-points $P_1, \ldots, P_n$, forming the "branching divisor" $\sum_{j=1}^{m} P_j$, $\Gamma \xrightarrow{\pi} \Gamma'$. We define an involution $T$ on $\Gamma$, which interchanges the sheets of the covering, with $TP_J = P_j$. For any divisor $D$ we define the "conjugate divisor" $D^+ = T(D)$.

We also assume that an algebraic function $E(P)$ is given on $\Gamma'$, with simple poles at $Q_1, \ldots, Q_s$, whose sum forms the "divisor of poles" $D_\infty = -\sum_{\alpha=1}^{s} Q_\alpha$, and that all the poles of $E$ lie at branch-points $Q_\alpha = P_{J_\alpha}$. We denote by $\widetilde{E} = \pi^* E$ the lifting of $E$ to $\Gamma$.

DEFINITION. We say that a potential $u(x)$ on the interval $[a, b]$ has correct algebraic-geometric properties if there exists an eigenfunction $\psi(x, P)$, $x \in [a, b]$ $(P \in \Gamma)$ such that

A) it satisfies the Sturm-Liouville equation

$$-\psi''(x, P) + u(x)\psi(x, P) = \widetilde{E}(P)\psi(x, P);$$

B) it is meromorphic everywhere on $\Gamma$, except at the poles of $\widetilde{E}$ at the points $\pi^{-1}(D_\infty)$ and the poles of $\psi(x, P)$ do not depend on $x$;

C) near poles of $\widetilde{E}$ the following asymptotic form holds:

$$\psi(x, P) \sim \text{const}\, e^{i\sqrt{\widetilde{E}(P)}(x-x_0)}$$

(that is, $\psi e^{-i\sqrt{\widetilde{E}(P)}(x-x_0)}$ is regular at the poles of $\widetilde{E}$). The properties A) and C) are natural requirements in the axiomatics for this class of potentials. As for B), it is a natural generalization of properties of the Bloch function. We recall that the Bloch function $\psi$ is meromorphic on a Riemann surface $\Gamma$, doubly covering a rational (trivial) surface, the $E$-plane, on which $E$ has a pole at infinity (by definition). The two-sheeted character of the covering is a natural requirement because the Sturm-Liouville equation is of the second order, and we have a basis of solutions when there are two sheets (see Ch. 2, §1, and Ch. 3, §2). Apart from $\psi$ we also consider the function $\psi^+ = T^*\psi$ (the sheets are exchanged) and their Wronskian

$$W(\psi, \psi^+) = \psi\psi^{+\prime} - \psi^+\psi',$$

which is constant when condition A) is satisfied (that is, does not depend on $x$).

We now come to a statement of the "scattering data" problem on the Riemann surface $\Gamma$ defining the potential $u(x)$.

EXAMPLE 1. In the case of a rapidly decreasing non-reflecting potential (see Ch. 2, §1) a complete set of scattering data on the rational surface $\Gamma$ with the parameter $k = \sqrt{E}$ was the collection of points $i\varkappa_1, \ldots, i\varkappa_s$ (in the upper half-plane $\operatorname{Im} k > 0$ for real $u(x)$, or on the upper sheet of the surface $k = \sqrt{E}$ for real $E$), and the set of numbers $c_1, \ldots, c_s$. The numbers $i\varkappa_1, \ldots, i\varkappa_s$ determine the position of the discrete levels $E_j = -\varkappa_j^2$.

The eigenfunctions $f(x, k)$ and $g(x, k)$ defined by the conditions

$$f_\pm(x, k) \to e^{\pm ikx} \quad (x \to +\infty),$$
$$g_\pm(x, k) \to e^{\pm ikx} \quad (x \to -\infty),$$

do not satisfy the conditions A), B), C). By analogy with the Bloch function (see Ch. 2), we introduce a new function $\psi(x, x_0, k)$ proportional to $f(x, k)$ by a factor depending rationally on $k$, and such that $\psi(x, x_0, k) \equiv 1$ when $x = x_0$. Then $\psi(x, x_0, k)$ satisfies A), B), C). Any point $P$ on $\Gamma$ is determined by the parameter $k = \sqrt{E}$, and the function $E(P)$ has the form $E = k^2$. It is easy to verify that the zeros $P_j(x)$ of $\psi$ lie over points $\gamma_j(x)$ and the poles $P_j(x_0)$ over points $\gamma_j(x_0)$ ($j = 1, \ldots, N$) on $\Gamma$. The "divisor of the poles" has the form

$$D = \sum_{j=1}^{N} P_j(x_0).$$

The Wronskian $W(\psi, \psi^+)$ vanishes at the branch-point $k = 0$ and at the points of the discrete spectrum

$$Q^{\pm} = \pm i\varkappa_1, \ldots, Q_n^{\pm} = \pm i\varkappa_n, \qquad E_\alpha = -\varkappa_\alpha^2 \qquad (\alpha = 1, \ldots, n).$$

Clearly, $n = N$. The set of points

$$[P_1(x_0), \ldots, P_N(x_0), +i\varkappa_1, \ldots, +i\varkappa_N],$$

where the $P_j$ lie over $\gamma_j$, uniquely determine the potential $u(x)$. Equations in $x$ for the quantities $\gamma_j(x)$ for rapidly decreasing non-reflecting potentials were first found by Shabat [11], in the language of "conditional eigenvalues". At the branch-points $k = 0$ and at all the points $k_j = i\varkappa_j$ the following equations hold:

$$\psi(x, x_0, i\varkappa_j) \equiv \psi^+(x, x_0, i\varkappa_j)$$

or

$$W(\psi, \psi^+) = 0, \qquad k = i\varkappa_j.$$

We call the pair of divisors

$$D = \sum_j P_j(x_0) \quad \text{and} \quad d = \sum_j Q_j^+$$

a "complete set of scattering data".

EXAMPLE 2. Suppose that we are given a non-degenerate $N$-zone periodic or almost-periodic potential (real or complex) with the Bloch eigenfunction $\psi^{\pm}(x, x_0, E)$, as defined in Ch. 2. The divisor of the poles has the form

$$D = \sum_{j=1}^{N} P_j(x_0),$$

where the $P_j(x_0)$ lie on $\Gamma$ over the points $\gamma_j(x_0)$ of the $E$-plane. The function $\psi$ satisfies the conditions A), B), C), and its Wronskian has the form

$$W(\psi, \psi^+) = 2i \frac{\sqrt{R(E)}}{\prod_j (E - \gamma_j(x_0))}, \qquad R(E) = \prod_{i=1}^{2N+1} (E - E_i).$$

The Wronskian vanishes only at the branch-points $E_i$, which are completely determined by $\Gamma$ and the involution $T$ interchanging the sheets. In this case the divisor of poles $D$ completely determines the potential $u(x)$, by the results of Ch. 2.

We now pass on to the general case of potentials with correct algebraic-geometric properties.

Let $\psi(x, P)$ be a function on $\Gamma$ satisfying B) and C), and let $D$ be its divisor of poles. Let $d = \Sigma\lambda_i\varkappa_i$ be another effective divisor (where the $\lambda_i$ are numbers and the $\varkappa_i$ are points on $\Gamma$).

DEFINITION. We say that the pair of effective divisors $(D, d)$ is compatible if the following conditions are satisfied:

$$n(d) = \sum \lambda_i = \dim L_-(\Delta) - 1,$$
$$\dim[L(\Delta - d) \cap L_-(\Delta)] = 1,$$

where the effective divisor $\Delta$ has the form

$$\Delta = D + D^{+} - D_{\infty},$$

$D^{+} = T(D)$, $D_{\infty}$ is the divisor of poles of the function $E$ on $\Gamma$, and $L_{-}(\Delta)$ is the subspace of rational functions $f$ on $\Gamma$ that change sign under the interchange $T$ of sheets and such that $(f) + \Delta \geqslant 0$.

Let us explain this definition. The Wronskian $W(\psi, \psi^{+})$ belongs to $L_{-}(\Delta)$; for it has poles of the first order at the points of $D_{\infty}$, it can have poles at the points of the divisors $D$ and $D^{+}$, by definition, and finally, it changes sign when the sheets $T$ are interchanged. The divisor $d$ arises from those zeros of $W(\psi, \psi^{+})$ that do not lie at branch-points of $\Gamma$. Now we have the following proposition.

PROPOSITION 1. *If $\psi$ satisfies the requirements* B) and C) *above, and if the divisor $D$ of its poles and the divisor $d$ of the zeros of the Wronskian $W(\psi, \psi^{+})$ form a compatible pair* $(D, d)$, *then $\psi$ satisfies a Sturm-Liouville equation for a certain potential $u(x)$. Conversely, if $\psi$ satisfies the requirements* A), B), C) *and if $D$ is the divisor of its poles, then there exists a $d$ such that the pair* $(D, d)$ *is compatible.*

The proof of the direct assertion follows from the fact that the Wronskian $W(\psi, \psi^{+})$ belongs to the space $L(\Delta - d) \cap L_{-}(\Delta)$, which is one-dimensional according to the compatibility condition for $(D, d)$.

Thus, as $x$ varies, the Wronskian as a function of $P$ is simply multiplied by a constant $c(x)$. However, it clearly follows from Condition C) that $c(x) \equiv 1$. Hence $dW/dx \equiv 0$ from which the direct assertion follows easily. The expression $\frac{\psi''}{\psi} + E$ does not depend on $E$, as follows from its properties. The proof of the converse assertion follows easily from the definition of a compatible pair. If the divisor $d$ has the form

$$d = \sum_{i} \lambda_i \varkappa_i,$$

then at the points $\varkappa_i$ the following identities hold (identically in $x$):

$$(\psi \equiv \psi^{+})_{\varkappa_i}, \ \ldots, \ \left(\frac{d^q}{dz^q}\psi \equiv \frac{d^q}{dz^q}\psi^{+}\right)_{\varkappa_i}, \ \ldots$$

$$(q = 0, \ldots, \lambda_i - 1),$$

where $z$ is a local parameter on $\Gamma$ near $\varkappa_i$.

In [49] the following theorem is proved.

THEOREM. *If the inverse scattering problem is soluble for any compatible pair of effective divisors* $(D, d)$ *(that is, if there exist $u(x)$ and $\psi(x, P)$ with the correct algebraic-geometric properties), then the Riemann surface $\Gamma'$ is rational, the function $E$ on it has exactly one pole at $E = \infty$ and $\Gamma$ is hyperelliptic and covers the $E$-plane twice.* (*All the potentials $u(x)$ with correct algebraic-geometric properties in this case satisfy one of the higher order K-dV equations.*)

We do not prove this theorem here (it follows in a straightforward way from some very simple algebraic-geometric facts). The theorem completes the axiomatics of the class of potentials satisfying equations of K-dV type. In fact, all the potentials of this class can be defined as meromorphic almost-periodic functions on the whole complex $x$-plane with a group of $2n$ periods $T_1, \ldots, T_n, T'_1, \ldots, T'_n$, where any part of the periods can degenerate and become infinite. The number of poles of $\psi(x, P)$ (the degree of the divisor $n(D)$) must not be smaller than the genus of $\Gamma$:

$$N = n(D) \geqslant g,$$

a degree of $n(D)$ for this class of potentials is $N - g$. We now give an explicit construction of this class of potentials.

In the first place, we can use the equations of Ch. 2, §2, for the poles $\gamma_j(x_0)$ in the variable $x_0$ (or the zeros $\gamma_j(x)$) of the Bloch function $\psi_\pm(x, x_0, E)$, which can be defined naturally in the given case, for example, by means of the formula

$$\psi_\pm(x, x_0, E) = c(x, x_0, E) + i\chi(x_0, E)s(x, x_0, E),$$

where

$$\chi(x_0, E) = \frac{\varphi(E)\sqrt{R(E)} - \frac{i}{2}\frac{d}{dx_0}\prod_{j=1}^{N}(E - \gamma_j(x_0))}{\prod_{j=1}^{N}(E - \gamma_j(x_0))},$$

$R(E) = \prod_{i=1}^{2n+1}(E - E_i)$ is a polynomial defining $\Gamma$, and

$$\varphi(E) = \prod_{j=1}^{q}(E - i\varkappa_j)^{\lambda_j}$$

is the polynomial defined by $d$.

$$d = \sum_{j=1}^{q}\lambda_j Q_j,$$

where the $Q_j$ lie above the points $i\varkappa_j$ on the $E$-plane. One can imagine that $R(E)\cdot[\varphi(E)]^2$ is obtained as a result of degeneration of a more complicated Riemann surface for which all the roots $i\varkappa_j$ are of multiplicity $2\lambda_j$ (we may assume that $\lambda_j = 1$). For the poles $\gamma_j(x_0)$ we obtain the equation similar to (2.3.12)

$$\frac{d\gamma_j}{dx_0} = 2i\,\frac{\varphi(\gamma_j)\sqrt{R(\gamma_j)}}{\prod_{k\neq j}(\gamma_j - \gamma_k)}.$$

An equation like (2.4.3) is obtained for the time dynamics under the K-dV equation and its higher order analogues by the formal change

$$\sqrt{R(\gamma_j)} \mapsto \sqrt{R(\gamma_j)}\,[\varphi(\gamma_j)]^2.$$

For a Riemann surface of genus $g = 0$ we can solve these equations in the obvious way and obtain rapidly decreasing non-reflecting potentials and multisoliton solutions for the K-dV equations (see [11], [8], [20]).

For a Riemann surface $\Gamma$ of genus $g = 1$ the equations can be solved without difficulty in terms of elliptic functions. We do not perform this integration, because the resulting formulae are already published in [48], [49].

Of course, for all $g \geqslant 2$ all these equations can be solved in terms of hyperelliptic functions on the Riemann surface. We draw attention to the following circumstance: if $\Omega_k$ are differentials of the first kind on $\Gamma$, normalized as in Ch.2, §3, then we have the Abel map $\mathfrak{A}$ :

$$\eta_k = \sum_{j=1}^{N} \int_{P_j(x_0)}^{P_j(x)} \Omega_k \qquad (k=1, \ldots, n),$$

(where $n$ is the genus of $\Gamma$ and $N$ is the number of poles $P_j(x_0)$ and zeros $P_j(x)$ of $\psi_\pm(x, x_0, E)$). As before, the parameters $\eta_k$ on the torus (Jacobian variety) $\mathbf{J}(\Gamma)$ are such that

$$\frac{d\eta_k}{dx} = U_k = \text{const}, \qquad \frac{d\eta_k}{dt_m} = W_k^m = \text{const}$$

(by virtue of all higher K-dV equations of order $m$). Moreover, the points of the divisor $d = \Sigma\lambda_j Q_j$ depend neither on $x$ nor on the time. However, there remain $(N - n)$ unknown parameters whose dynamics is not contained in $\mathbf{J}(\Gamma)$. The Abel map here has the form

$$\mathbf{S}^N(\Gamma) \xrightarrow{A} \mathbf{J}(\Gamma),$$

where the inverse image $A^{-1}(\eta_1, \ldots, \eta_n)$ of a point is the complex projective space $\mathbf{CP}^{N-n}$. In [49] the following results are obtained.

Let $N = n + k$ and let $P_1(x_0), \ldots, P_{N+k}(x_0)$ be the poles of $\psi(x, P)$, $D = P_1 + \ldots + P_{n+k}$ and let $d = \varkappa_1 + \ldots + \varkappa_k$ be that half of of the zeros of the Wronskian $W(\psi, \psi^+)$ that do not lie at branch-points of the surface

$$y_2 = \prod_{i=1}^{2n+1} (E - E_i)$$

and chosen on the upper sheet. According to the results above, the pair $(D, d)$ uniquely determines the potential $u(x)$ with the eigenfunction $\psi(x, P)$. Suppose that $u_i(x)$ are the $n$-zone potentials on $\Gamma$ given by the divisors of the poles

$$[P_1(x_0), \ldots, P_{n-1}(x_0), P_{n+i}(x_0)] \leftrightarrow u_i(x),$$

and $\psi_i(x, P)$ the corresponding Bloch eigenfunctions.

PROPOSITION 2. *The eigenfunction $\psi$ for $u(x)$ can be represented in the form*

(A.1.1) $$\psi(x, P) = \sum_{i=1}^{k} a_i(x)\psi_i(x, P),$$

*where the $a_i(x)$ do not depend on the point of the Riemann surface (on E) and can be determined from the system of equations*

(A.1.2) $$\begin{cases} \sum_{i=1}^{k} a_i(x)[\psi_i(x, \varkappa_s) - \psi_i^+(x, \varkappa_s)] = 0 \quad (s = 1, \ldots, k), \\ \sum_{i=1}^{k} a_i(x) = 1. \end{cases}$$

Now let $K(x) = \int_{x_0}^{x} u(x)dx,\ K_i(x) = \int_{x_0}^{x} u_i(x)dx.$

PROPOSITION 3.

$$K(x) = \sum_{i=1}^{k} a_i(x) K_i(x).$$

We also introduce the "monosoliton potentials against the background of $n$-zone potentials". We can show that the integral $K(x)$ of a multisoliton potential against the background of an $n$-zone potential can be expressed rationally in terms of the similar integrals $K_{ij}(x)$ of the "monosoliton potentials against the background of $n$-zone potentials" and of the $n$-zone potentials $K_i(x)$ themselves. This rational representation can be thought of as a non-linear analogue of the superposition of the monosoliton solutions against the background of $n$-zone solutions with a given Riemann surface $\Gamma$. For real bounded potentials $u(x)$ associated with $\Gamma$: $y^2 = \prod_{j=1}^{2n+1}(E - E_j)$, where $-\infty < E_1 < \ldots < E_{2n+1} < \infty$, the points $\varkappa_j$ lie in the intervals $E_{2k} < \varkappa_s < E_{2k+1}$ $(-\infty < \varkappa_s < E_1)$, and the poles $\gamma_j(x_0)$ lie one each in the intervals obtained (complementary to the solution zones and the points $\varkappa_j$.) Then $\psi_i^+(x, \varkappa_s) \to 0$ and $\psi(x, \varkappa_s) \to 0$, as $x \to \pm\infty$, as follows from Proposition 2.

COROLLARY. *Under these hypotheses about the distribution of the poles $\gamma_j(x_0)$ and of the points $\varkappa_j$ the potential $u(x)$ is smooth and bounded in $x$, and as $x \to \pm\infty$, the potential tends exponentially to the finite-zone potential $u_\pm(x)$, where $u_+(x)$ is given on $\Gamma$ by the divisor of the poles of the Bloch function, which is equivalent to $D - d$, while $u_-(x)$ is given by a divisor equivalent to $D - d^+$* (we recall that any divisor of degree $n$ is equivalent to an effective divisor, that is, to a sum of $n$ distinct points; to say that two divisors $D_1$ and $D_2$ are equivalent means that the difference $D_1 - D_2$ is the complete divisor of the zeros and poles $(f)$ of an algebraic function $f$ on $\Gamma$).

For a surface of genus $n = 1$ the difference of the two potentials $u_{\pm}(x)$ reduces to a phase shift, as indicated in [48].

In the general case this "displacement" of the potential on the torus $\mathbf{J}(\Gamma)$ is given by the divisor $(d - d^{+})$, as follows from the corollary; the corresponding displacement vector in the parameter $\eta_k$ on the Jacobian variety is

$$\eta_q^{-\infty,+\infty} = \sum_{j=1}^{k} \int_{\varkappa_j^{+}}^{\varkappa_j^{-}} \Omega_q \quad (q = 1, \ldots, n),$$

$$d = \varkappa_1^{+} + \ldots + \varkappa_k^{+}, \qquad d^{+} = \varkappa_1^{-} + \ldots + \varkappa_k^{-},$$

where the points $\varkappa_j^{\mp}$ lie over $\varkappa_j$ of the $E$-plane on the upper and lower sheets, respectively, and the integral is over "half" of a cycle $a_s$ joining the points $\varkappa_j^{\mp}$, in the lacuna in which $\varkappa_j$ lies. Here the $\Omega_q$, normalized as in Ch. 2, §3, form a basis of differentials of the first kind.

For monosoliton potentials against the background of $n$-zone potentials with a single point $\varkappa_1 = \varkappa$, we obtain from Proposition 3

$$K = \frac{\varphi_2}{\varphi_2 - \varphi_1} K_1(x) + \frac{\varphi_1}{\varphi_1 - \varphi_2} K_2(x),$$

where

$$\varphi_i = \psi_i(x, \varkappa) - \psi_i^{+}(x, \varkappa),$$

$$K = \int u\, dx, \qquad K_i = \int u_i\, dx.$$

For the eigenfunctions $\psi_i$ we can use a modification of Its' formula [54]:

$$\text{(A.1.3)} \quad \psi_i(x, P) = e^{(x-x_0)\int\Omega} \frac{\theta(\eta(P) + (x - x_0)\vec{U} + \eta^{0i} - \vec{K})}{\theta(\eta(P) + \eta^{0i} - \vec{K})}$$

(for the definition of the differentials here, see Ch. 2, §3), where

$$\eta_j(P) = \int_{\infty}^{P} \Omega_j,$$

$$\eta_j^{0i} = \sum_{k=1}^{n-1} \int_{\infty}^{P_k(x_0)} \Omega_j + \int_{\infty}^{P_{n+i}(x_0)} \Omega_j.$$

The time dynamics is included automatically. (A.1.3) remains true at all times $t$, and for the eigenfunctions $\psi_i(x, t, P)$, which depend on the time by virtue of the K-dV equation, we have

(A.1.4)

$$\psi_i(x, t, P) = e^{(x-x_0)\int\Omega + i(t-t_0)\int\omega_2} \frac{\theta(\eta(P) + (x - x_0)\vec{U} + (t - t_0)\vec{W}^{(1)} + \eta^{0i} - \vec{K})}{\theta(\eta(P) + \eta^{0i} - \vec{K})},$$

in which the differential $\omega_2$ has zero $a$-periods and a singularity $dz/z^4$, as

$E \to \infty$, where $z = 1/\sqrt{E}$.

# APPENDIX 2

## ANOTHER METHOD OF OBTAINING SOME THEOREMS IN CH. 2, §2

Quite recently (at the beginning of 1975) a preprint of a new paper by Lax has reached us, in which he essentially develops the results of his earlier paper [50], which was mentioned in the Introduction and was discussed in Ch. 2, §2. Although the actual results are contained in the previously published paper [38], the proofs differ substantially from the methods of [38]. We quote the basic arguments from Lax's preprint, whose main results are as follows.[1]

1) A stronger form is given of the result of [50] on the spectrum of a Schrödinger operator with periodic potential $u(x)$ satisfying any one of the stationary higher order K-dV equations (2.2.22). Stated in our language, although the theorem in [50] that these potentials are finite-zoned, is non-constructive it is now also proved that the number of forbidden zones does not exceed $n$ (true, this proof is also non-constructive, and in Lax's paper there is no analogue of the algorithm for finding the boundaries of the zones which is described in [38] and in Ch. 2, §2 of this survey. This will be clear from the derivation below).

2) Polynomial integrals are found of the stationary K-dV equations (2.2.22) by another method than that of [38]. As stated already, the boundaries of the zones have so far not been expressed in terms of these integrals.[2]

3) It is proved that the simple eigenvalues of the operator $L$ are commutative (or involutory), and also that all the integrals $p(E_j)$ are involutory, where $p(E)$ is the Bloch dispersion law. The fact that the discrete eigenvalues of a rapidly decreasing potential are involutory was already established in [21]; for the periodic case this was also well known after [21]; a rigorous proof was published, for example, in [66]. As for the function $p(E)$, it is known that it is determined by the spectrum of the periodic problem (including degenerate levels), since the trace of the translation matrix $a_R = \frac{1}{2}\,\mathrm{Sp}\,\hat{T}$ is an entire function of order $\frac{1}{2}$ and is completely determined by the levels of the periodic problem $a_R = 1$ including their multiplicities. (In classical terminology, an eigenvalue $e^{ip(E)}$ of the translation matrix $\hat{T}$ is called a "Floquet exponent".) Since for almost any small perturbation $u + \delta u$ the eigenvalues become non-degenerate and the functional $p(E)$ is

---

[1] Lax's paper has now been published [63].

[2] A construction of these integrals was also achieved simultaneously and independently in the survey [62] – see the "Concluding remarks" at the end of this paper.

smooth, the vanishing of the Poisson brackets $[p(E), p(E')] \equiv 0$ follows formally from [66]. This was pointed out by Faddeev (see the Introduction and [38], §1).

PROOF OF 1). We consider the higher order K-dV equation and its operator representation

$$\dot{u} = \frac{\partial}{\partial x}\left(\sum c_i \frac{\delta I_{n-i}}{\delta u(x)}\right), \qquad \dot{L} = [A, L],$$

where $A = \Sigma c_i A_{n-i}$. We know that $A$ is a skew-symmetric operator. If $\dot{u} = 0$ and $\dot{L} = 0$, then $[L, A] = 0$. Hence, if $L\psi_j = \lambda_j \psi_j$ and $\lambda_j$ is a non-degenerate level of the periodic problem, then $A\psi_j = \mu_j \psi_j$. This follows from elementary algebra. From the skew symmetry it follows that $\mu_j$ is purely imaginary. From the reality of the operator and of $\psi_j$ it follows that $\mu_j$ is real. Therefore, $\mu_j = 0$. Hence, all the non-degenerate periodic eigenfunctions $\psi_j$ of $L$ satisfy $A\psi_j = 0$. Since the order of $A$ is $2n + 1$, we see that there can be no more than $2n + 1$ non-degenerate levels $\lambda_j$ for $L$, because all the $\psi_j$ must be linearly independent. Thus, 1) is proved. The proof is non-constructive: it is not clear so far how to determine the disposition of the levels $\lambda_j$ by this method.

PROOF OF 2). We use the result of [21], [22] that the integrals $I_m = \int_0^T P_m(u, u', \ldots, u^{(m)})dx$ are involutory. The Poisson brackets has the form

$$\text{(A.2.1)} \qquad [I_n, I_m] = \int_0^T \left(\frac{\delta I_m}{\delta u(x)} \frac{d}{dx} \frac{\delta I_n}{\delta u(x)}\right) dx,$$

and we know that $[I_n, I_m] = 0$. From the vanishing of this integral for an arbitrary periodic function $u(x)$ it follows, clearly, that

$$\text{(A.2.2)} \qquad \frac{\delta I_m}{\delta u(x)} \frac{d}{dx} \frac{\delta I_n}{\delta u(x)} = \frac{d}{dx} J_{mn}(u, u', \ldots),$$

where $J_{mn}$ is a polynomial in $u, u', u'' \ldots$. Obviously,

$$J_{mn} + J_{nm} = \frac{\delta I_n}{\delta u(x)} \frac{\delta I_m}{\delta u(x)}.$$

We consider the stationary equation

$$\text{(A.2.3)} \qquad \sum_{i=0}^{n} c_i \frac{\delta I_{n-i}}{\delta u(x)} = d, \qquad \text{or} \qquad \sum_{i=0}^{n+1} c_i \frac{\delta I_{n-i}}{\delta u(x)} = 0,$$

where $d = c_{n+i}$, $I_{-1} = -\int_0^T u\,dx$. Multiplying this equation by $\frac{d}{dx} \frac{\delta I_m}{\delta u(x)}$ and using (A.2.2), we obtain

(A.2.4) $$\frac{d}{dx}\Big(\sum_{i=0}^{n+1} c_i J_{n-i,\, m}\Big)=0.$$

Let

(A.2.4′) $$\widetilde{I}_m=\sum_{i=0}^{n+1} c_i J_{n-i,\, m} \qquad (m=0, \ldots, n-1).$$

By (A.2.4) all the polynomials $\widetilde{I}_m$ are integrals of (2.2.22), and 2) is proved. It is easy to prove that the integrals $\widetilde{I}_m$ are algebraically independent. By a direct calculation it can be shown that the $\widetilde{I}_m$ are also conserved in time under the higher order K-dV equations (this is done in the preprint).

Of course, from this it follows more or less that these integrals $\widetilde{I}_m$ can be expressed in terms of the boundaries of the zones of the potential. Therefore, by the results of Ch. 2, §2, they can be expressed in terms of the constants $c_0 = 1, c_1, \ldots, c_n, c_{n+1}$ and the integrals $J_\alpha$. However this expression is not clear so far, and Lax gives no indication on how to calculate the boundaries of the zones in terms of the integrals $\widetilde{I}_m$ and the constants $c_0, \ldots, c_{n+1} (c_0 = 1, c_{n+1} = d)$ (see [65]).

We note, finally, that it follows trivially from these results that the set of periodic potentials for a given spectrum is an $n$-dimensional real algebraic variety in $R^{2n}$ whose bounded connected components are isomorphic to the tori $T^n$. The complexification of these varieties and almost-periodic potentials are not discussed in Lax's paper. In his exposition Lax starts out from the following problem: how does one minimize the functional $I_n$ under given constraints $\{I_k = a_k\}$ $(k < n)$ over the class of periodic functions with a given period?

## APPENDIX 3

### ON THE USE OF LINEAR AND NON-LINEAR TRACE FORMULAE FOR THE INTEGRATION OF EQUATIONS OF K-dV TYPE AND THE EXPRESSION OF THE BLOCH SOLUTION OF THE SCHRÖDINGER EQUATION IN TERMS OF A θ-FUNCTION

In the mathematical literature, beginning with the papers [13] and [14] of Gel'fand, Levitan, and Dikii which were completed in 1953–1955, there are many publications dealing with the derivation of various identities for sums of powers of the eigenvalues of linear differential operators, in other words, trace formulae.

Formulae of this kind which we call non-linear trace formulae, have already appeared in the main part of this survey. In fact, we have established the relation

(A.3.1) $$\sum_{i=1}^{n} \gamma_i(x, t)=\frac{-u(x, t)}{2}+\frac{1}{2}\sum_{j=1}^{2n+1} E_j,$$

and from (2.3.11′) it follows at once that

$$\text{(A.3.2)} \qquad \sum_{i=1}^{n} \gamma_i^2(x, t) = \frac{u_{xx}(x, t)}{4} - \frac{u^2(x, t)}{2} + \frac{1}{2}\sum_{j=1}^{2n+1} E_j.$$

The computations leading to (A.3.2) can easily be generalized to higher order trace formulae; in particular, for the sums of the cubes of the quantities $\gamma_i$ we easily obtain the expression

$$\sum_{i=1}^{n} \gamma_i^3 = -\frac{u^3}{2} + \frac{3}{4} u u_{xx} + \frac{15}{32} u_x^2 - \frac{3}{32} u_{xxxx} + \frac{1}{2}\sum_{i=1}^{n} E_i^3.$$

We note that, to within the evaluation of the constants $\frac{1}{2}\sum_{i=1}^{n} E_i^k$, these formulae follow directly from the results of Dikii [14], as was explained in [54]. They can be obtained even more simply (again to within the constant $\frac{1}{2}\sum_{i=1}^{n} E_i^k$) from the fact mentioned earlier in this survey, that the polynomial

$$P = \prod_i (E - \gamma_i(x, t))$$

satisfies the differential equation

$$-P_{xxx} + 4P_x u(x, t) + 2u_x P = 4EP_x.$$

We emphasize that the derivation of the trace formulae in our situation is connected with the Riemann surface $\Gamma$ of the function $\sqrt{\prod_{i=1}^{2n+1}(E - E_i)}$ and holds for arbitrary complex values of the $E_i$. A foundation for trace formulae for arbitrary boundaries of zones can also be obtained using the following theorem of Its [54], which is of independent interest.

THEOREM. *The function*

$$\psi(x, t, E) = e^{i\omega(E)x}\frac{\theta(A(E) + x\vec{U} + t\vec{W} + l(0, 0))\,\theta(t\vec{W} + l(0, 0))}{\theta(A(E) + t\vec{W} + l(0, 0))\,\theta(x\vec{U} + t\vec{W} + l(0, 0))},$$

*in which*

$$\omega(E) = \int_{E_{2n+1}}^{E} \frac{z^n + b_1 z^{n-1} + \dots b_n}{2\sqrt{\prod_{j=1}^{2n+1}(z - E_j)}}\,dz, \qquad \oint_{a_k} d\omega(E) = 0 \qquad (k = 1, \dots, n),$$

$$(A(E))_k = \int_{\infty}^{E} \Omega_k \qquad (k = 1, \dots, n),$$

*satisfies the Schrödinger equation with the potential* (2.3.34). *Here* $\psi(x, t, E)$ *and* $\psi(x, t, E^*)$, $E^* = TE$ *(where T is the involutory auto-*

*morphism of* $\Gamma$ *interchanging the sheets) form a fundamental system of solutions of the Schrödinger equation.*

However, the Riemann surface $\Gamma$ generates a series of trace formulae of another kind, namely linear trace formulae. They are obtained by integrating the form $E^m \, d\ln F$, where $F$ is defined by (2.3.35), over the boundary $\partial\widetilde{\Gamma}$ of $\widetilde{\Gamma}$, which is obtained by cutting $\Gamma$ along all the basic cycles $a_i$ and $b_i$, and they have the form

$$\sum_{k=1}^{n} \gamma_k^m(x,t) = \sum_{k=1}^{n} \oint_{a_k} E^m \Omega_k - \operatorname*{res}_{E=\infty} \{E^m d\ln F\}. \tag{A.3.3}$$

The evaluation of the residue in the first of these formulae has already been discussed and led to (2.3.34). For $m = 2$ the evaluation of the residue and a subsequent differentiation with respect to $x$ give

$$\frac{d}{dx}\sum_{k=1}^{n} \gamma_k^2(x,t) = -\frac{u_{xxx}}{12} - \frac{u_t}{6}. \tag{A.3.4}$$

It was observed by one of the present authors in [53] and [54] that a comparison between linear and non-linear trace formulae can be taken as a basis for integration of non-linear equations connected with Riemann surfaces.

This idea was used for the integration of the non-linear Schrödinger equation and the modified Korteweg-de Vries equation (see [52]).

When applied to the Korteweg-de Vries equation it is particularly effective: by differentiating the right-hand side of (A.3.2) and comparing the resulting identity with (A.3.4), we see that the function $u(x)$ (originally defined as the sum $u = -2\sum \gamma_i + \sum E_i$) satisfies the K-dV equation, a calculation of the residue on the right-hand side of (A.3.3) for $m = 1$ gives rise to an explicit representation for $u$. To integrate the higher order analogues of the K-dV equation within this framework we must be justified in writing out explicitly the corresponding Jacobi problem, that is, selecting the corresponding direction (the vector $\vec{W}$) on the Jacobian of the Riemann surface. It turns out that the form of the corresponding vector $\vec{W}$ is uniquely determined by the condition that the right-hand sides of (A.3.3) after the evaluation of the residue and subsequent differentiation with respect to $x$, has the form $Gu$, where $G$ is a linear differential operator with constant coefficients. Therefore, the form of the corresponding vectors $\vec{W}$ is determined by the asymptotic form of the expression $\dfrac{\varphi_k(E)}{\sqrt{R(E)}}$, as $E \to \infty$, where $\omega_k = \dfrac{\varphi_k(E)}{\sqrt{R(E)}}\, dE$. A detailed description of this method of integrating the higher order K-dV equations is given in [54]. We quote here only the formula for $\vec{W}$ that describes the solution of the second K-dV equation:

$$W_j = 32i\left\{c_{j3} + \frac{1}{2}cc_{j2} + c_{j1}\left[\frac{3}{8}c^2 - \frac{1}{2}\sum_{i>j} E_iE_j\right]\right\}, \qquad c = \sum_{j=1}^{n} E_j.$$

We recall that the second K-dV equation has the form

$$u_t = u_x^{(5)} - 20u_xu_{xx} - 10uu_{xxx} - 30u^2u_x.$$

We ought to mention also that in this "trace" approach to the integration of non-linear equations we can use non-linear trace formulae, different in form and origin, which we call dynamic. The source of these formulae are the equations (2.4.3); for example, for the K-dV equation it follows directly from (2.4.3) that

$$\frac{\partial\gamma_j(x,t)}{\partial t} = 2(u+2\gamma_j)\frac{\partial\gamma_j}{\partial x},$$

from which we obtain by summation over $j$ the equation

$$\text{(A.3.5)} \qquad \frac{d}{dx}\sum_{j=1}^{n}\gamma_j^2(x,t) = \frac{1}{2}uu_x - \frac{1}{4}u_t.$$

Comparing (A.3.5) and (A.3.4) we again confirm that $u$ is a solution of the K-dV equation, and this argument makes no use of the Schrödinger equation, but deals all the time with objects directly connected with the Riemann surface.

Its has shown that when applied to hyperelliptic surfaces generated by a polynomial of even order $P(E) = \prod_{j=1}^{2n}(E - E_j)$, this approach leads to the integration of a series of non-linear evolutionary systems, generalizing the non-linear Schrödinger equation and modified K-dV equation, which were discussed above. Here are some examples of such systems:

$$\begin{cases} iu_t + u_{xx} - 2vu = 0, \\ iv_t - v_{xx} + 2\left(v\frac{u_x}{u}\right)_x = 0, \end{cases}$$

$$\begin{cases} v_z - 6vv_x + v_{xxx} - 3\left[\frac{u_x}{u}\left(v_x - \frac{u_x}{u}v\right)\right]_x = 0, \\ u_z - 6u_xv + u_{xxx} = 0, \end{cases}$$

$$\begin{cases} if_t + f_{xx} - 2iff_x - 2iv_x = 0, \\ iv_t - v_{xx} - 2i(fv)_x = 0, \end{cases}$$

$$\begin{cases} v_z - 6vv_x + v_{xxx} - 3[f(fv - iv_x)]_x = 0, \\ f_z - 3f^2f_x + f_{xxx} - \frac{3}{2}i(f^2)_{xx} - 6(vf)_x = 0. \end{cases}$$

## CONCLUDING REMARKS[1]

1. In the summer of 1975 L. D. Faddeev brought to the attention of the authors new preprints which he had obtained in May in the USA. A paper of McKean and van Moerbeke "On Hill's equation" contains proofs of a number of the results quoted in Ch. 2, §3 of this survey. Apparently the authors of this preprint were not familiar with the papers [39], [42], [43], published in 1974. Incidentally, the text contains a historical inaccuracy: the fact that the Lamé potential $n(n+1)\,\wp(x)$ is finite-zoned was established not in 1975 but many years before (see, for example, our Introduction).

A number of preprints (of Kac and van Moerbeke) deals with periodic problems for the Toda chain and a discrete version of the K-dV equation. All these papers were written in 1975; the first, which contains only a small part of the results, has appeared in print recently (Proc. Nat. Acad. Sci. USA 72 (1975), 1627; Adv. in Math. **16** (1975), 160 ff.). In subsequent preprints some of the results of Ch. 3, §1 of this survey, which have not previously been published, are obtained independently. The authors evidently are not familiar with the paper of Manakov [35] on discrete systems, in which an $L-A$ pair of operators for the discrete K-dV equation was first found (see also Ch. 1, §5 and Ch. 3, §1).

In addition, we have received from Japan a preprint (by E. Date and S. Tanaka "Exact solutions for the periodic Toda lattice"), in which the method of the authors is also applied to the periodic problem for the Toda chain.

2. I. M. Gel'fand informs us that the survey of Gel'fand and Dikii "Asymptotic behaviour of the resolvent of Sturm-Liouville equations and the algebra of the Korteweg-de Vries equations", Uspekhi Mat. Nauk 30:5 (1975), 67–100 = Russian Math. Surveys **30**:5 (1975), 77–113, contains the construction of the integrals for stationary problems for higher order K-dV equations of which an account is given in Appendix 3 of this survey from Lax's recent preprint. They have found this construction independently and at the same time as Lax. Moreover, Gel'fand and Dikii also prove that these integrals are involutory, having subjected them to a concrete analysis (this result is not in Lax's paper). Without knowing the results of Gel'fand and Dikii, Novikov and Bogoyavlenskii have also proved that these integrals are involutory, as a consequence of a very simple and quite general theorem on the interrelation of the Hamiltonian formalisms of non-stationary and stationary equations. (Attention to this connection had already been drawn in [38], but it had not been formulated accurately.) We give here a statement of this theorem (see [64]). We recall that according to Gardner, Hamiltonian systems on a function space have the form

[1] This addendum was received by the Editors on 24 September 1975.

$$(1)\qquad \frac{\partial u}{\partial t}=\dot{u}=\frac{d}{dx}\frac{\delta I}{\delta u(x)},$$

where

$$I=\int P(u,u',u'',\ldots,u^{(n)})dx,\quad u'=\frac{\partial u}{\partial x},$$

and $I$ is the Hamiltonian. We assume that $P$ does not depend explicitly on $x$. We consider the system formally without making precise the nature of the function space, which can even turn out to consist of increasing functions.

For the two functionals

$$I=\int P\,dx,\quad J=\int Q\,dx$$

the Poisson bracket has the form

$$(2)\qquad \langle I,J\rangle=\int\left(\frac{d}{dx}\frac{\delta J}{\delta u(x)}\right)\frac{\delta I}{\delta u(x)}dx,$$

and the Hamiltonian $\langle I, J\rangle$ formally defines the commutator of the two flows (on any class of functions on a straight line). If the flows commute, $\langle I, J\rangle = 0$, then we have, by definition,

$$\left(\frac{d}{dx}\frac{\delta J}{\delta u(x)}\right)\frac{\delta I}{\delta u(x)}=\frac{dV(I,J)}{dx},$$

where $V = V(u, u', \ldots)$. We consider the stationary problem $\dfrac{\delta I}{\delta u(x)} = 0$ (or const) which has a Hamiltonian form in the canonical coordinates (see [38], Example 2 for $n = 2$)

$$(3)\qquad \begin{cases} q_1=u,\quad q_2=u',\quad q_3=u'',\ \ldots,\quad q_n=u^{(n-1)}\\ p_1=\dfrac{\partial P}{\partial u'}-\left(\dfrac{\partial P}{\partial u''}\right)'+\left(\dfrac{\partial P}{\partial u'''}\right)''-\ldots\pm\left(\dfrac{\partial P}{\partial u^{(n)}}\right)^{(n-1)},\\ p_2=\dfrac{\partial P}{\partial u''}-\left(\dfrac{\partial P}{\partial u'''}\right)''+\left(\dfrac{\partial P}{\partial u^{(4)}}\right)'''-\ldots\pm\left(\dfrac{\partial P}{\partial u^{(n)}}\right)^{(n-2)},\\ \cdots\cdots\cdots\cdots\cdots\cdots\cdots\cdots\\ p_n=\dfrac{\partial P}{\partial u^{(n)}} \end{cases}$$

with the Hamiltonian

$$(4)\qquad H(p,q)=P-u'p_1-u''p_2-\ \ldots\ -u^{(n)}p_n,$$

provided that they can all be expressed by the coordinates $(p, q)$. This is in text-books on the calculus of variations with higher order derivatives.

If $I = \int P(u, u', \ldots, u^{(n)})dx$, $J = \int Q(u, u', \ldots, u^{(m)})dx$, and $\langle I, J\rangle = 0$, then on the set of fixed points of the flow (1) the functional $J$ defines the flow $\dot{u} = \dfrac{d}{dx}\dfrac{\delta J}{\delta u(x)}$ as a finite-dimensional dynamical system, for example, as a higher order K–dV equation in [38]. Under these conditions we have the following general and fairly simple result:

THEOREM. *If $m < n$ and $\langle I, J \rangle = 0$, then the flow $\dot{u} = \frac{d}{dx} \frac{\delta J}{\delta u(x)}$ on the set of fixed points of the flow* (1) *is a one-parameter family of finite-dimensional dynamical systems, depending on the constant of integration $\frac{\delta I}{\delta u(x)} = d$; all these systems are Hamiltonian and are given by the Hamiltonian $V_d(I, J)$, which can be expressed in terms of the phase coordinates* (3), *where*

$$\frac{d}{dx} V_d(I, J) = \left(\frac{d}{dx} \frac{\delta J}{\delta u(x)}\right) \left(\frac{\delta I}{\delta u(x)} - d\right).$$

Example: let $J = \int \frac{u^2}{2}\, dx$; in this case the flow $\dot{u} = \frac{d}{dx} \frac{\delta J}{\delta u(x)} = u'$ defines the group of shifts through $x$ and commutes with $I$ if $P$ does not depend explicitly on $x$. Then we have

$$V_0(I, J) = H(p, q), \qquad \frac{dV}{dx} = u' \frac{\delta I}{\delta u(x)},$$

where $H(p, q)$ is a Hamiltonian of the stationary problem for the flow (1).

From the theorem it follows, obviously, that if two integrals $J_1$ and $J_2$ of the flow (1) commute in the non-stationary problem, $\langle J_1, J_2 \rangle = 0$, then their images $V(I, J_1)$ and $V(I, J_2)$ commute in the phase space of the stationary problem with the canonical coordinates (3).

The application of the theorem to the theory of K-dV equations is obvious, because in this theory we have many commuting flows in a function space (the higher K-dV equations according to Gardner); the images of these flows on the set of fixed points of any one, as stated in [38], give a set of commuting systems in the stationary problem. It is interesting that in [38] a complete set of integrals $J_1, \ldots, J_n$ is found for the stationary problem, which can be expressed explicitly in terms of the spectrum of the Sturm-Liouville operator (Hill, Schrödinger). In a later paper we shall show how to express these integrals of Novikov in terms of the integrals of Lax–Gel'fand–Dikii and vice versa (see [65]).

3. Very recently Krichever, developing a technique of Dubrovin and Its, and Matveev, has found a beautiful algebraic-geometric method of constructing analogues of finite-zone solutions of the periodic problem (in $x$) for "two-dimensional K-dV" equation or the equations of Kadomtsev–Petviashviki (see [37] and Ch. 1, §2 of this survey) and some other equations of Zakharov and Shabat, containing an additional coordinate $y$. It is remarkable that in the construction of solutions of "two-dimensional K-dV" equation by Krichever's scheme an arbitrary Riemann surface of genus $g$ appears. The possibility of applying this method to problems of

algebraic geometry (and of constructing explicit universal fiberings over the variety of moduli of Riemann surfaces) by the scheme of Novikov–Dubrovin (see [43] and Ch. 2, §4 of this survey and also Ch. 3, §2) is here subjected to a careful analysis.

4. Very recently a new paper of Moser has appeared (Three integrable Hamiltonian systems connected with isospectral deformations, Adv. in Math. **16** (1975), 354 ff.), in which he finds an $L-A$ pair for a very interesting class of discrete systems of classical particles on a straight line with various interaction potentials. (According to work of Calogero and Sutherland, in some of these models it is known that the corresponding quantum problem is completely soluble). An interesting group theoretical approach to the systems of Moser and Calogero has now been developed by Perelomov and Ol'shanetskii ("Complete integrable systems connected with Lie algebras"). These systems lead, in contrast to the Toda chain, to "non-local" operators $L$ and cannot be described by the methods of the present survey, even when the periodic problem is meaningful.

## References

[1] G. Borg, Eine Umkehrung der Sturm-Liouvilleschen Eigenwertaufgabe, Acta Math. **78** (1946), 1–96. MR **7**–382.

[2] V. Bargmann, Determination of a central field of force from the elastic scattering phase shifts, Phys. Rev. **75** (1949), 301–303.

[3] I. M. Gel'fand and B. M. Levitan, On the determination of a differential equation by its spectral function, Izv. Akad. Nauk SSSR Ser. Mat. **15** (1951), 309–360. MR **13**–558.

[4] V. A. Marchenko, Some problems of the theory of one-dimensional differential operators. I, Trudy Moskov. Mat. Obshch. **1** (1952), 327–420. MR **15**–315.

[5] L. D. Faddeev, Properties of the $S$-matrix of the one-dimensional Schrödinger equation, Trudy Mat. Inst. Steklov **73** (1964), 314–336. MR **31** # 2446.

[6] I. V. Stankevich, On an inverse problem of spectral analysis for the Hill equation, Dokl. Akad. Nauk SSSR **192** (1970), 34–37. MR **41** # 7195.
= Soviet Math. Dokl. **11** (1970), 582–586.

[7] E. L. Ince, Further investigations into the periodic Lamé functions, Proc. Roy. Soc. Edinburgh **60** (1940), 83–99. MR **2**–46.

[8] I. Kay and H. Moses, Reflectionless transmission through dielectrics and scattering potentials, J. Appl. Phys. **27** (1956), 1503–1508.

[9] H. Hochstadt, On the determination of a Hill's equation from its spectrum, Arch. Rational Mech. and Anal. **19** (1965), 535–562. MR **31** # 6019.

[10] N. I. Akhiezer, A continuous analogue to orthogonal polynomials on a system of intervals, Dokl. Akad. Nauk SSSR **141** (1961), 263–266. MR **25** # 4383.
= Soviet Math. Dokl. **2** (1961), 1409–1412.

[11] A. B. Shabat, On potentials with zero reflection coefficient, in the coll. *Dinamika sploshnoi sredy* (Dynamics in a dense medium), No. 5, 130–145, Novosibirsk 1970.

[12] K. M. Case and M. Kac, A discrete version of the inverse scattering problem, J. Mathematical Phys. **14** (1973), 594–603. MR **48** # 10392.

[13] I. M. Gel'fand and B. M. Levitan, On a simple identity for the eigenvalues of a Sturm-Liouville operator, Uspekhi Mat. Nauk **30**:6 (1975).

[14] L. A. Dikii, On a formula of Gel'fand–Levitan, Uspekhi Mat. Nauk **8**:2 (1953), 119–123. MR **15**–130.

[15] E. C. Titchmarsh, Eigenfunction expansions associated with second-order differential equations, Vol.2, Ch. XXI, Clarendon Press, Oxford 1958. MR **20** # 1065.
Translation: *Razlozheniya po sobstvennym funktsiyam, svyazannye s differentsial'nymi uravneniyami vtorogo poryadka*, Izdat. Inost. Lit., Moscow 1961.

[16] G. B. Whitham, Non-linear dispersive waves, Proc. Roy. Soc. Ser. **A283** (1965), 238–261. MR **31** # 996.

[17] M. D. Kruskal and N. Zabusky, Interaction of "solitons" in a collisionless plasma and the recurrence of initial states, Phys. Rev. Lett. **15** (1965), 240–243.

[18] C. S. Gardner, J. M. Green, M. D. Kruskal, and R. M. Miura, Method for solving the Korteweg-de Vries equation, Phys. Rev. Lett. **19** (1967), 1095–1098.

[19] P. D. Lax, Integrals of non-linear equations of evolution and solitary waves, Comm. Pure Appl. Math. **21** (1968), 467–490. MR **38** # 3620.
= Matematika **13**:5 (1969), 128–150.

[20] V. E. Zakharov, A kinetic equations for solitons, Zh. Eksper. Teoret. Fiz. **60** (1971), 993–1000.
= Soviet Physics JETP **33** (1971), 538–541.

[21] V. E. Zakharov and L. D. Faddeev, The Korteweg-de Vries equation is a completely integrable Hamiltonian system, Funktsional. Anal. i Prilozen. **5**:4 (1971), 18–27. MR **46** # 2270.
= Functional Anal. Appl. **5** (1971), 18–27.

[22] C. S. Gardner, Korteweg-de Vries equation and generalizations. IV, The Korteweg-de Vries equation as a Hamiltonian system, J. Mathematical Phys. **12** (1971), 1548–1551. MR **44** # 3615.

[23] V. E. Zakharov, On the problem of stochestization of one-dimensional chains of non-linear oscillators, Zh. Eksper. Teoret. Fiz. **65** (1973) 219–225.
= Soviet Physics JETP **38** (1974), 108–110.

[24] V. S. Dryuma, On an analytic solution of the two-dimensional K-dV equation, Letters to Zh. Eksper. Teoret. Fiz. **19**:12 (1974), 753–755.

[25] V. E. Zakharov and A. B. Shabat, Exact theory of two-dimensional self-focusing and one-dimensional self-modulation of waves in non-linear media, Zh. Eksper. Teoret. Fiz. **61** (1971), 118–134.
= Soviet Physics JETP **34** (1972), 62–69.

[26] V. E. Zakharov and A. B. Shabat, Interaction between solitons in a stable medium, Zh. Eksper. Teoret. Fiz. **64** (1973), 1627–1639.
= Soviet Physics JETP **37** (1973), 823–828.

[27] V. E. Zakharov and S. V. Manakov, On a resonance interaction of wave packets in non-linear media, Letters to Zh. Eksper. Teoret. Fiz. **18**:7 (1973), 413–417.

[28] L. A. Takhtadzhyan, Exact theory of propagation of ultrashort optical impulses in two-level media, Zh. Eksper. Teoret. Fiz. **66** (1974) 476–489.
= Soviet Physics JETP **39** (1974), 228–233.

[29] M. J. Ablowitz, D. J. Kaup, A. C. Newell and H. Segur, Method for solving the sine-Gordon equation, Phys. Rev. Lett. **30** (1973), 1262–1264.

[30] V. E. Zakharov, L. A. Takhtadzhyan and L. D. Faddeev, Complete description of solutions of the "sine-Gordon" equation, Dokl. Akad. Nauk SSSR **219** (1974), 1334–1337.
= Soviet Physics Dokl. **19** (1975), 824–826.

[31] M. Toda, Waves in non-linear lattice, Progr. Theoret. Phys. Suppl. **45** (1970), 174–200.
[32] M. Henon, Integrals of the Toda lattice, Phys. Rev. **B9** (1974), 1921–1923.
[33] H. Flaschka, Toda lattice. I, Phys. Rev. **B9** (1974), 1924–1925.
[34] H. Flaschka, Toda lattice. II, Progr. Theoret. Phys. **51** (1974), 703–716.
[35] S. V. Manakov, On the complete integrability and stochastization in discrete dynamical systems, Zh. Eksper. Teoret. Fiz. **67** (1974) 543–555.
[36] A. B. Shabat, On the Korteweg-de Vries equation, Dokl. Akad. Nauk SSSR **211** (1973), 1310–1313. MR **48** #9137.
= Soviet Math. Dokl. **14** (1973), 1266–1270.
[37] V. E. Zakharov and A. B. Shabat, A scheme for the integration of non-linear equations of mathematical physics by the method of inverse scattering theory. I, Funktsional. Anal. i Prilozin. **8**:3 (1974), 43–53.
[38] S. P. Novikov, The periodic Korteweg-de Vries problem I. Funktsional. Anal. i Prilozhen. **8**:3 (1974), 54–66.
[39] S. P. Novikov and B. A. Dubrovin, The periodic problem for the Korteweg-de Vries and the Sturm-Liouville equations, Petrovskii Seminar 6 March 1974, Uspekhi Mat. Nauk **29**:6 (1974), 196–197.
[40] B. A. Dubrovin, The inverse problem of scattering theory for periodic finite-zone potentials, Funktsional. Anal. i Prilozhen. **9**:1 (1975), 65–66.
[41] A. R. Its and V. B. Matveev, On Hill operators with finitely many lacunae, Funktsional. Anal. i Prilozhen. **9**:1 (1975), 69–70.
[42] B. A. Dubrovin and S. P. Novikov, Periodic and conditionally periodic analogues of multisoliton solutions of the Korteweg-de Vries equation, Zh. Eksper. Teoret. Fiz. **67** (1974), 2131–2143.
= Soviet Physics JETP **40** (1974), 1058–1063.
[43] B. A. Dubrovin and S. P. Novikov, The periodic problem for the Korteweg-de Vries and Sturm-Liouville equations. Their connection with algebraic geometry. Dokl. Akad. Nauk SSSR **219** (1974), 19–22.
[44] V. A. Marchenko, The periodic Korteweg-de Vries problem, Dokl. Akad. Nauk SSSR **217** (1974), 276–279. MR **50** #7855.
[45] V. A. Marchenko, The periodic Korteweg-de Vries problem, Mat. Sb. **95** (1974), 331–356.
[46] B. V. Dubrovin, The periodic problem for the Korteweg-de Vries equation in the class of finite-zone potentials, Funktsional. Anal. i Prilozin. **9**:3 (1975).
[47] A. R. Its and V. B. Matveev, Hill operators with finitely many lacunae and multisoliton solutions of the Korteweg-de Vries equation, Trudy Mat. Fiz. **23** (1975), 51–67.
[48] E. A. Kuznetsov and A. B. Mikhailov, The stability of stationary waves in non-linear media with weak dispersion, Zh. Eksper. Teoret. Fiz. **67** (1974), 1717–1727.
= Soviet Physics JETP **40** (1974), 855–859.
[49] I. M. Krichever, Non-reflecting potentials against the background of finite-zone potentials, Funktsional. Anal. i Prilozhen. **9**:2 (1975).
[50] P. D. Lax, Periodic solutions of the K-dV equation, Lectures in Appl. Math. **15** (1974), 85–96.
[51] B. A. Dubrovin, Finite-zone linear operators and Abelian varieties, Uspekhi Mat. Nauk **31**:4 (1976),
[52] A. R. Its, Canonical systems with a finite-zone spectrum and periodic solutions of the non-linear Schrödinger equation, Vestnik Leningrad Univ. Mat. 1976.
[53] V. B. Matveev, A new method of integrating the Korteweg-de Vries equation, Uspekhi Mat. Nauk **30**:6 (1975), 201–203.

[54] A. R. Its and V. B. Matveev, On a class of solutions of the K-dV equation, in *Problemy matematicheskoi fiziki* (Problems of mathematical physics), No 8, Leningrad University 1976.
[55] G. Springer, Introduction to Riemann surfaces, Addison–Wesley, New York 1957. Translation: *Vvedenie v teoriyu riemanovykh poverkhnostei*, Izdat. Inost. Lit., Moscow 1961.
[56] E. I. Zverovich, Boundary problems of the theory of analytic functions Uspekhi Mat. Nauk **26**:1 (1971), 113–180.
= Russian Math. Surveys **26**:1 (1971), 117–192.
[57] V. V. Golubev, *Lektsii po integrirovaniyu uravnenii dvizheniya tyazhelogo tverdogo tela s odnoi nepodvizhnoi tochkoi*, Gostekhizdat, Moscow 1953. MR **15**–904.
Translation: Lectures on the integration of the equations of motion of a heavy rigid body about a fixed point, Nat. Sci. Foundation, Israel Program Sci. Transl.
MR 22 # 7298.
[58] W. Goldberg, On the determination of a Hill's equation from its spectrum, Bull. Amer. Math. Soc. **80** (1974), 1111–1112.
[59] E. I. Dinaburg and Ya. G. Sinai, On the spectrum of the one-dimensional Schrödinger equation with quasi-periodic potential, Funktsional. Anal. i Prilozhen. **9** (1975), 8–21.
[60] V. P. Kotlyarov, The periodic problem for the non-linear Schrödinger equation, in *Matematicheskaya fizika i funktsionalnyi analiz*, (Mathematical physics and functional analysis), Trudy FTINT Akad. Nauk Ukrain. SSR **6** (1975).
[61] V. A. Kozel, On a class of solutions of the sine–Gordon equation, in *Matematicheskaya fizika i funkstionalnyu analiz* (Mathematical physics and functional analysis), Trudy FTINT Akad. Nauk Ukrain. SSR **6** (1975).
[62] I. M. Gel'fand and L. A. Dikii, Asymptotic behaviour of the resolvent of Sturm-Liouville equations and the algebra of the Korteweg-de Vries equations, Uspekhi Mat. Nauk **30**:5 (1975), 67–100.
= Russian Math. Surveys **30**:5 (1975), 77–113.
[63] P. D. Lax, Periodic solutions of the K-dV equations, Comm. Pure Appl. Math. **28** (1975), 141–188.
[64] O. I. Bogoyavlenskii and S. P. Novikov, On the connection between the Hamiltonian formalisms of stationary and non-stationary problems, Funktsional Anal. i Prilozhen. **10**:1 (1976).
[65] O. I. Bogoyavlenskii, On integrals of higher order stationary equations and the eigenvalues of Hill operators, Funktsional. Anal. i Prilozhen. **10**:2 (1976).
[66] V. E. Zakharov and S. V. Manakov, On the complete integrability of the non-linear Schrödinger equation, TMF **19** (1974).

Received by the Editors 2 June 1975

Translated by G. and R. Hudson

# METHODS OF ALGEBRAIC GEOMETRY IN THE THEORY OF NON-LINEAR EQUATIONS

I. M. Krichever

## Contents



## Introduction

The mechanism of integrating the Korteweg-de Vries equation by the method of the inverse scattering problem, which was proposed in [1] (Gardner, Green, Kruskal, Miura), was interpreted from various points of view by Lax [2], Zakharov and Faddeev [3] and Gardner [4]. Beginning with the paper by Zakharov and Shabat [5], many other physically important equations were found that can be integrated by this method over the class of rapidly decreasing functions. Among them are the following, all familiar in mathematical physics: the non-linear Schrödinger equation $\alpha u_t = u_{xx} \pm |u|^2 u$ ([5], [6]), the "sine-Gordon" equation $u_{xt} = \sin u$ ([7], [11], [12]), the Kadomtsev–Petviashvili equation

$$\frac{3}{4}\beta^2 \frac{\partial^2 v}{\partial y^2} + \frac{\partial}{\partial x}\left[\alpha \frac{\partial v}{\partial t} + h \frac{\partial v}{\partial x} + \frac{1}{4}\left(\frac{\partial^3 v}{\partial x^3} + 6v \frac{\partial v}{\partial x}\right)\right] = 0 \text{ ([8])},$$

and many others [9]–[17].

A method for finding them was developed by Zakharov and Shabat [18], [19].

The use of scattering theory restricted the method to the class of functions that decrease rapidly in the spatial variable. The periodic problem required essentially new ideas, the first of which were derived by Novikov in [20]. (Some of the results of this paper were also obtained simultaneously by Lax [21].) In subsequent papers by Dubrovin [44], Dubrovin and Novikov [54], Its and Matveev [36], and Lax [47] a theory was constructed of the so-called finite-zone periodic and conditionally periodic solutions of the K-dV equation and their profound algebraic-geometrical nature was discovered.[1]

In a series of papers Marchenko and Ostrovskii ([23], [24], [25]) obtained results on the approximation of arbitrary periodic potentials by finite zone potentials with the same period.[2] Novikov and Dubrovin were the first to introduce the general concept of a finite-zone linear differential operator, for which the Bloch eigenfunction (or the Floquet function) is defined on a Riemann surface of finite genus (an algebraic curve). A survey of these results and a full bibliography are contained in [26].

The author has proposed an algebraic-geometrical construction of a broad class of periodic and conditionally periodic solutions of the general Zakharov–Shabat equation $L_t - A_y = [A, L]$, which makes it possible to express them explicitly in terms of the Riemann $\theta$-function. In particular, the non-stationary Schrödinger equation

$$i\frac{\partial\psi}{\partial t}+\frac{\partial^2\psi}{\partial x^2}-u(x, y, t)\,\psi=0.$$

is incidentally solved by explicit formulae for the so constructed solutions $u(x, y, t)$ of the physically important Kadomtsev–Petviashvili equation [27]. In addition, this construction gives a solution of the problem of classifying commutative rings of differential operators in one variable, in the first instance when the ring contains a pair of operators of relatively prime order (see [28], [29], [30]).

A fruitful discussion of these problems took place in Gel'fand's seminar at the Moscow State University, after talks the author gave in November – December 1975. Drinfeld [31] indicated an abstract-algebraic exposition of the author's construction, which gave rise to some useful generalizations and, in particular, made advances in the problem of classifying commutative rings of differential operators on the real line, without assuming that they are prime in pairs. A complete solution of this problem was then obtained by the author (see §2 of this survey). Gel'fand and Dikii have investigated the Hamiltonian structure of equations of Lax type $L_t = [L, A]$ in which $L$

[1] Some of the results of Novikov–Dubrovin–Matveev–Its were later obtained by McKean and van Moerbeke [22].

[2] An approximation in the class of conditionally periodic potentials clearly follows from [36], [44], but to establish an approximation with the same period is difficult.

and $A$ are operators with scalar coefficients and the order of $L$ is greater than two. The corresponding analogue to the Hamiltonian formalism of Gardner–Zakharov–Faddeev proved to be rather complicated. The results are given in [32] and [33]. So far the Hamiltonian formalism even for the stationary equations of the type of the Novikov equation $[L, A] = 0$ solved by the author, which give rise to commutative algebras, has not been worked out (see Appendix 1).

The concept of a finite-zone differential operator can be generalized to the case of several independent variables. Roughly speaking, a linear differential operator in $n$ variables is said to be $k$-algebraic if it has a family of eigenfunctions, parametrized by points of a $k$-dimensional complex algebraic variety $M^k$ with "good" analytical properties, similar to the properties of Bloch functions of a finite-zone one-dimensional Schrödinger operator. The broadest is the case $k = 1$ for an arbitrary number of variables.

Dubrovin, Novikov, and the author have solved the inverse problem of the reconstruction of a 1-algebraic (weakly algebraic, in the terminology of [34]) two-dimensional Schrödinger operator. They have shown that systems of compatible 1-algebraic operators with a common variety $M^1$ in the two-dimensional case form an analogue to a commutative algebra. The commutator of any pair of operators from such an "algebra" can be factored on the right by one and the same Schrödinger operator $H$:

$$[L_i, L_j] = D_{ij}H, \quad [L_i, H] = D_iH,$$

where the $D_{ij}$ and $D_i$ are linear differential operators.

§4 contains an investigation of $k$-algebraic ($k > 1$) linear differential operators. As yet we have no solutions of the inverse problems for them. This leads to interesting new problems in algebraic geometry.

Finally, in the concluding section of the survey we give an account of the results of Moser, McKean and Airault, who have discovered a remarkable connection between the behaviour of singularities of rational and elliptic solutions of the K-dV equation and the motion of $n$ particles on a straight line [51].

## §1. The Akhiezer function and the Zakharov–Shabat equations

We consider the non-linear partial differential equations for the coefficients of the operators

$$(1.1) \qquad L_1 = \sum_{\alpha=0}^{n} u_\alpha(x, y, t) \frac{\partial^\alpha}{\partial x^\alpha}, \quad L_2 = \sum_{\beta=0}^{m} v_\beta(x, y, t) \frac{\partial^\beta}{\partial x^\beta},$$

which are equivalent to the operator equation

$$(1.2) \qquad \left[L_1 - \frac{\partial}{\partial t}, L_2 - \frac{\partial}{\partial y}\right] = 0, \quad \text{where} \quad [A, B] = AB - BA.$$

Zakharov and Shabat [19] were the first to study these equations. They developed a method of obtaining certain exact solutions that are rapidly decreasing as $|x| \to \infty$.

In this survey we limit ourselves, for the sake of definiteness, to operators with scalar coefficients. All the results carry over easily to the general case of matrix coefficients (see [29], [30]).

In the subsequent construction of exact solutions of the Zakharov–Shabat equations a central role is played by the concept of an Akhiezer function.

LEMMA 1.1. *For each regular complex curve* $\mathfrak{R}$ *of genus g with a distinguished point* $P_0$ *and a non-special effective divisor of degree g (that is, for a set of g points* $p_1, \ldots, p_g$ *in general position) there exists a unique function* $\psi(x, y, t, P)$, $P \in \mathfrak{R}$, *having the following properties.*

1°. *Except at* $P_0$ *it is meromorphic, with poles at* $p_1, \ldots, p_g$.

2°. *Near* $P_0$ *it can be represented in the form*

$$\psi(x, y, t, P) = \exp(kx + Q(k)y + R(k)t)\left(1 + \sum_{s=1}^{\infty} \xi_s(x, y, t) k^{-s}\right), \tag{1.3}$$

*where* $k^{-1} = k^{-1}(P)$ *is a fixed local parameter,* $k^{-1}(P_0) = 0$, *and* $Q(k) = q_m k^m + \ldots + q_0$ *and* $R(k) = r_n k^n + \ldots + r_0$ *are polynomials.*

Functions of this kind were first considered by Akhiezer [35] in the case of the hyperelliptic curve $w^2 = \prod_{i=1}^{2n+1} (E - E_i)$, with $P_0, p_1, \ldots, p_n$ as branch points.

Without proving the lemma, we pass on to the main theorem of this section, which was first established in [28].

THEOREM 1.1. *For each Akhiezer function there exist unique operators* $L_1$ *and* $L_2$ *of the form* (1.1) *such that*

$$L_1\psi = \frac{\partial}{\partial t}\psi, \quad L_2\psi = \frac{\partial}{\partial y}\psi.$$

PROOF. For any formal series (1.3) there is a unique operator $L_1$ such that

$$L_1\psi(x, y, t, P) \equiv \frac{\partial}{\partial t}\psi(x, y, t, P) \pmod{O(k^{-1})\, e^{kx+Q(k)y+R(k)t}}.$$

Its coefficients can be found from a system of equations equivalent to this congruence:

$$\sum_{\alpha=0}^{n} u_\alpha \sum_{l=0}^{\alpha} C_\alpha^l \frac{\partial^{\alpha-l}}{\partial x^{\alpha-l}} \xi_{s+l} = \sum_{i=0}^{n} r_i \xi_{s+i}, \qquad (\xi_j = 0,\ j < 0). \tag{1.4}$$

Then $u_n = r_n$, $u_{n-1} = r_{n-1}$, $u_{n-2} = r_{n-2} - n r_n \frac{\partial}{\partial x}\xi_1, \ldots$

The purpose of the compact curve $\mathfrak{R}$ is that an exact equation for the Akhiezer function can be derived from the congruence above. This is a characteristic feature in the solution of all the following inverse problems.

We consider the function $\left(L_1 - \frac{\partial}{\partial t}\right) \psi(x, y, t, P) = 0$. It satisfies all the requirements defining the Akhiezer function except one. The expansion of the regular factor for an exponent in $P_0$ begins with $O(k^{-1})$. From the uniqueness of $\psi(x, y, t, P)$ it follows that this function vanishes. The operator $L_2$ can be found similarly.

COROLLARY. *The operators so constructed satisfy the equation* $\left[L_1 - \frac{\partial}{\partial t}, L_2 - \frac{\partial}{\partial y}\right] = 0.$

PROOF OF THE COROLLARY. The kernel of the operator $\left[L_1 - \frac{\partial}{\partial t}, L_2 - \frac{\partial}{\partial y}\right]$ contains a one-parameter family of functions $\psi(x, y, t, P)$. Since the operator itself contains differentiation only with respect to $x$, its kernel; if it is not zero, is finite-dimensional. This contradiction proves the assertion of the corollary.

We consider an important example of the construction of solutions of the Kadomtsev–Petviashvili equation, according to this scheme.

Let $Q(k) = q_2 k^2 + q_0$, $R(k) = r_3 k^3 + r_1 k + r_0$. By what has been proved, each regular complex curve $\mathfrak{R}$ of genus $g$ with a distinguished point $P_0$ and a non-special effective divisor of degree $g$ defines the operators

$$L_1 = q_2\left(\frac{\partial^2}{\partial x^2} + v_0(x, y, t)\right), \quad L_2 = r_3\left(\frac{\partial^3}{\partial x^3} + u_1(x, y, t)\frac{\partial}{\partial x} + u_0(x, y, t)\right),$$

which satisfying (1.2). Eliminating $u_1$ and $u_2$ from the equivalent system of equations we obtain for $v(x, y, t)$ the equation

$$\frac{3}{4}\beta^2\frac{\partial^2 v}{\partial y^2} + \frac{\partial}{\partial x}\left\{\alpha\frac{\partial v}{\partial t} + h\frac{\partial v}{\partial x} + \frac{1}{4}\left(\frac{\partial^3 v}{\partial x^3} + 6v\frac{\partial v}{\partial x}\right)\right\} = 0,$$

where $\alpha = q_2^{-1}$, $\beta = d_3^{-1}$.

From (1.4) it follows that $v = q - 2\frac{\partial}{\partial x}\xi_1$. To find an explicit expression for $v(x, y, t)$, we express $\psi(x, y, t, P)$ in terms of the Riemann $\theta$-function. In passing we also prove Lemma 1.1. Its [58] first obtained corresponding expressions for the case of a hyperelliptic curve with a branch point $P_0$.

On the regular algebraic curve $\mathfrak{R}$ of genus $g$ we fix a basis of cycles $a_1, \ldots, a_g, b_1, \ldots, b_g$, with the intersection matrix

$$a_i \circ a_j = b_i \circ b_j = 0, \quad a_i \circ b_j = \delta_{ij}.$$

We now introduce a basis of holomorphic differentials $\omega_i$ on $\mathfrak{R}$, normalized by the conditions $\oint_{a_i} \omega_k = \delta_{ik}$. We denote by $B$ the matrix of $b$-periods:

$B_{ik} = \oint_{b_i} \omega_k$. This matrix is known to be symmetric and to have positive-definite imaginary part.

The integer linear combinations of vectors in $C^g$ with coordinates $\delta_{ik}$ and $B_{ik}$ form a lattice, which determines the complex torus $J(\mathfrak{R})$, the so-called Jacobian variety of the curve.

Let $\bar{P}$ be the distinguished point on $\mathfrak{R}$; then there is a well-defined mapping $\omega: \mathfrak{R} \to J(\mathfrak{R})$ the coordinates of $\omega(P)$ are $\int_{\bar{P}}^{P} \omega_k$.

From $B$ we construct the Riemann $\theta$-function, the entire function of $g$ complex variables

$$\theta(u_1, \ldots, u_g) = \sum_{m \in Z^g} \exp(\pi i (Bm, m) + 2\pi i (m, u)),$$

where $(m, u) = m_1 u_1 + \ldots + m_g u_g$.

It has the following easily verifiable properties:

$$(1.5) \quad \begin{cases} \theta(u_1, \ldots, u_j + 1, \ldots, u_g) = \theta(u_1, \ldots, u_j, \ldots, u_g), \\ \theta(u_1 + B_{1k}, \ldots, u_g + B_{gk}) = e^{-\pi i B_{kk} - 2\pi i u_k} \theta(u_1, \ldots, u_g). \end{cases}$$

In addition, for any non-special effective divisor $D = \sum_{j=1}^{g} p_j$ of degree $g$ there is a vector $W(D)$ such that the function $\theta(\omega(P) + W(D))$ defined on $\mathfrak{R}$, cut along the cycles $a_i$, $b_j$, has exactly $g$ zeros, which coincide with the $p_j$ (see [37]).

We denote by $\omega_2$, $\omega_Q$, and $\omega_R$ respectively, the normalized Abelian differentials of the second kind [38] that have a unique singularity at $P_0$ of the form $-\frac{\partial z}{z^2}$, $dQ\left(\frac{1}{z}\right)$, and $dR\left(\frac{1}{z}\right)$ in the local parameter $z(P)$. Let $2\pi i U_1$, $2\pi i U_2$, and $2\pi i U_3$ be the vectors of their $b$-periods.

From (1.5) it follows that the function

$$\exp\left\{\int_{\bar{P}}^{P} (x\omega_2 + y\omega_Q + t\omega_R)\right\} \frac{\theta(\omega(P) + U_1 x + U_2 y + U_3 t + W(D))}{\theta(\omega(P) + W(D))}$$

does not change its value in a circuit around the cycles $a_i$ and $b_j$ and is, therefore, well defined. Normalizing it at $P_0$ we obtain $\psi(x, y, t, P)$ in a form first suggested by Its.

Expanding it near $P_0$ we arrive at the following formula for the solutions of the Kadomtsev–Petviashvili equation

$$(1.6) \qquad v(x, y, t) = q + 2\frac{\partial^2}{\partial x^2} \ln \theta(U_1 x + U_2 y + U_3 t + W),$$

where $W$ is an arbitrary point of the Jacobian of the curve.

If $U_2 = 0$ or $U_3 = 0$, which means that there is on $\mathfrak{R}$ a function with a unique pole of the second or third order at $P_0$, then $v(x, y, t)$ satisfies either the K-dV equation or one of the variants of the equation of the non-linear string [9]

$$\pm\frac{\partial^2 v}{\partial y^2}\mp\frac{\partial^2 v}{\partial x^2}+\frac{1}{4}\frac{\partial^4 v}{\partial x^4}+\frac{3}{2}\frac{\partial}{\partial x}\left(v\frac{\partial v}{\partial x}\right)=0,\quad \left(h=-\frac{3}{4}\beta^2=\pm 1\right).$$

In the first case $\mathfrak{R}$ is hyperelliptic, and (1.6) reduces to the Matveev–Its formula [36].

From the expression for $\psi(x, y, t, P)$ and the fact that the $U_i$ determine rectilinear portions on the Jacobian of the curve, we derive the following important corollary.

COROLLARY. *All the so constructed solutions of the Zakharov–Shabat equations are conditionally periodic functions (the surface $\mathfrak{R}$ is regular).*

## §2. Commutative rings of differential operators

**2.1. General properties.** We consider the system of non-linear equations in the coefficients of the operators $L_1 = \sum_{\alpha=0}^{n} u_\alpha(x)\frac{d^\alpha}{dx^\alpha}$, $L_2 = \sum_{\beta=0}^{m} v_\beta(x)\frac{d^\beta}{dx^\beta}$, that is equivalent to the condition that they commute. It is assumed a priori that these are equations in the class of germs of matrix functions of a real variable $u_\alpha^{ij}(x)$, $v_\beta^{ij}(x)$, $1 \leqslant i, j \leqslant l$. It turns out that all these solutions admit a meromorphic continuation to the whole complex plane. Almost all the solutions are conditionally periodic.

Novikov and Dubrovin in [54] have integrated the equation $[L_1, L_2] = 0$ for the case of scalar operators and $n = 2$. Dubrovin [39], [40] has discussed the case of commuting matrix operators, one of which is of the first order. Recently, Manakov [41] has found an interesting new example of their application. He has shown [42] that the equations of motion of an $n$-dimensional rigid body are equivalent to the condition that the operators

$$L_1=I^{-1}\frac{d}{dt}-I^{-1}\Omega,\qquad L_2=I\frac{d}{dt}+\Omega I,$$

commute, where $I$ is the inertia tensor. The present author in [29] and [30] has completely integrated the equations for the commutativity of matrix operators of relatively prime order.

We recall that within the framework of this survey we limit ourselves to the case of scalar operators since this permits the most complete and clear presentation of the ideas involved in applying the methods of algebraic geometry in the theory of non-linear equations. The matrix version gives rise to an insignificant technical modification of all the constructions.

Let us agree that all the relevant operators have constant leading coefficients. In addition, let $u_{n-1}(x) \equiv 0$. This can always be achieved by means of a gradient transformation.

The following proposition is the basis for the applicability of methods of algebraic geometry to solve equations of the Novikov type $[L_1, L_2] = 0$.

THEOREM 2.1. *There is a polynomial in two variables $Q(w, E)$ such that $Q(L_2, L_1) = 0$.*

Apparently, Shabat was the first to obtain a theorem of this kind for the case $n = 2$.

PROOF. The operator $L_2$ defines a linear operator $L_2(E)$ on the space $\mathcal{L}(E)$ of solutions of the equation $L_1 y = Ey$. Its matrix elements in the canonical basis

$$c_j(x, E); \quad \frac{d^r}{dx^r} c_j(x_0, E) = \delta_{rj}, \quad 0 \leqslant r, \quad j \leqslant n-1,$$

are polynomials in $E$. Let $Q(w, E) = \det (w \cdot 1 - L_2^{st}(E))$ be its characteristic polynomial. The kernel of $Q(L_2, L_1)$ contains $\mathcal{L}(E)$ for all $E$, hence, it is infinite-dimensional. Therefore, the operator itself is zero.

**2.2. The case of one-dimensional fiberings. Operators of relative prime orders.** First we consider the case when for almost all $E$ the eigenvalues of $L_2(E)$ are distinct. Then to each point $P = (w, E)$ of the algebraic curve $\mathfrak{R}^2$ given by the equation $Q(w, E) = 0$ there corresponds a one-dimensional eigenspace of $L_2(E)$. This gives a one-dimensional fibering[1] over $\mathfrak{R}$. In each fibrė over $\mathfrak{R} \setminus \infty$ we select a vector with first coordinate 1 in the basis $c_j(x, E)$. The remaining coordinates are all meromorphic functions on $\mathfrak{R}$ $\lambda_j(P)$. Since the $c_j(x, E)$ are entire functions in $E$, the joint eigenfunction $\psi(x, P) = \sum_{j=1}^{n-1} \lambda_j(P) c_j(x, E)$ of $L_1$ and $L_2$ is meromorphic in the affine part of $\mathfrak{R}$. Its poles do not depend on $x$.

To find the form of $\psi(x, P)$ at infinity, we construct for each operator germ the formal Bloch function.

LEMMA 2.1. *There is a unique solution, which we denote by $\psi(x, k; x_0)$, of the equation*

$$L_1 \psi(x, k) = k^n \psi(x, k) \tag{2.1}$$

*in the space of formal series of the form*

$$\psi(x, k) = \left( \sum_{s=N}^{\infty} \xi_s(x) k^{-s} \right) e^{k(x-x_0)} \tag{2.2}$$

*(where $N$ is an integer) with the "normalization conditions" $\xi_s = 0$, $s < 0$; $\xi_0(x) \equiv 1$, $\xi_s(x_0) = 0$. Any other solution of this kind is of the form*

$$\psi(x, k) = \psi(x, k; x_0) A(k), \qquad A(k) = \sum_{s=N}^{\infty} A_s k^{-s}.$$

PROOF. Equating the coefficients of $k^{-1}$, $s \geqslant -n$ on both sides of (2.1) we obtain

[1] In his paper [31] Drinfeld took as the starting point an axiomatization of the properties of this fibering in contemporary abstract algebraic language.

$$\sum_{\alpha=0}^{n} u_\alpha \sum_{l=0}^{\alpha} C_\alpha^l \frac{d^{\alpha-l}}{dx^{\alpha-l}} \xi_{s+l} = \xi_{s+n}.$$

We find $\xi'_{s+n-1}(x)$ from the $s$-th equation, since it can easily be brought to the form $0 = n\xi'_{s+n-1}$ + (terms containing $\xi_j$, $j < s + n - 1$).

The operator $L_2$ leaves the solution space of (2.1) invariant. Hence, by the lemma just proved,

$$(2.3) \qquad (\psi^{-1}(x, k; x_0) L_2 \psi(x, k; x_0))\,|_{x=x_0} = k^m + \sum_{s=-m+1}^{\infty} A_s k^{-s}.$$

In consequence, the coefficients of the left-hand side, which are polynomials in the $u_\alpha(x_0)$ and their derivatives, and in which the $v_\beta(x_0)$ occur linearly, give first integrals of the original equations. Inverting the first $m$ integrals we obtain the following corollary.

COROLLARY 1. *The coefficients of $L_2$ are polynomials in the $u_\alpha(x)$, their derivatives, and the constants $A_s$, $-m \leqslant s \leqslant 0$.*

NOTE. To prove the corollary it is sufficient that $L_1$ and $L_2$ commute to within an operator of order $n - 2$.

From (2.3) it follows that $\psi(x, k; x_0)$ is an eigenfunction for all operators commuting with $L_1$.

COROLLARY 2. *The ring of operators that commute with a given one is commutative.*

(This was apparently first pointed out in [55].)

The functions $\psi(x, k_j; x_0)$, $k_j^n = E$ form a basis of $\mathcal{L}(E)$ consisting of eigenvectors for $L_2(E)$. Then $Q(w, E) = \prod_{j=0}^{n} (w - A(k_j))$. Hence, if $n$ is relatively prime to $m$, then for large (and hence for almost all) values of $E$ the eigenvalues of $L_2(E)$ are distinct. Therefore, the affine part of $\mathfrak{R}$ can be completed at infinity by the single point $P_0$ in the neighbourhood of which $(E(P))^{-1/n}$ is a local parameter. The expansion in this local parameter $\psi(x, P)$ has the form (2.2).

Thus, with each pair of commuting operators, and so also with the commutative ring generated by them, we can associate the complex curve $\mathfrak{R}$, the so-called spectrum of the operators, with the distinguished point $P_0$ and the joint eigenfunction $\psi(x, P)$, which is meromorphic away from $P_0$, with the divisor of the poles $p_1, \ldots, p_g$, where $g$ is the genus of the curve $\mathfrak{R}$, which has the form (2.2) in the neighbourhood of $P_0$. Hence, $\psi(x, P)$ is a function of Akhiezer type.

By Lemma 1.1, the spectral data $\mathfrak{R}$, $P_0$, $D = \sum_{j=1}^{g} p_j$ uniquely define the Akhiezer function $\psi(x, P)$. For any function $E(P)$ having a pole only at $P_0$ (we denote the ring of such functions by $A(\mathfrak{R}, P_0)$) Theorem 1.1

associates with $\psi(x, P) \exp(E(P)t)$ the operator $L$ such that $L\psi(x, P) = E(P)\psi(x, P)$. Its coefficients do not depend on $t$. And so $\mathfrak{R}$, $P_0$ and $D$ determine a homomorphism $\lambda$ form $A(\mathfrak{R}, P_0)$ into the ring of differential operators.

THEOREM 2.2. *For any commutative ring $A$ of differential operators containing a pair of operators of relatively prime order there is a complex curve $\mathfrak{R}$ of genus $g$ with a distinguished point $P_0$ and an effective divisor $D$ of degree $g$ such that $\lambda$: $A(\mathfrak{R}, P_0) \to A$ is an isomorphism.*

**2.3. Multi-dimensional fiberings. General commutative rings.** We now relax the condition that the operators are of relatively prime orders. The operator $L_2(E)$ can have multiple eigenvalues. This means that then $A(k) = \widetilde{A}(k^l)$, where $l$ is the greatest common divisor of $n$ and $m$.

To each point of $\mathfrak{R}$ given by the equation

$$Q(w, E) = \prod_{j=1}^{n_1} (w - \widetilde{A}(k_j)) = 0, \quad n_1 l = n,$$

there corresponds an $l$-dimensional subspace of eigenvectors of $L_2(E)$. This defines an $l$-dimensional fibering over $\mathfrak{R}$. In each fiber over $\mathfrak{R} \setminus \infty$ we select vectors such that $\frac{d^r}{dx^r}\varphi_i(x_0, P) = \delta_{ri}$, $0 \leqslant r, i \leqslant l-1$. As before, all the $\varphi_i(x, P)$ are meromorphic in the affine part of $\mathfrak{R}$ and the divisor of their poles $D_i$ is of degree $g$.

To find the form of $\varphi_i(x, P)$ near the "point at infinity" $P_0$, we construct the matrix and function $\psi(x, P)$ whose columns are $\varphi_i(x, P)$, $\varphi_i^{(1)}(x, P)$, and $\varphi_i^{l-1}(x, P)$. The matrix function $\Psi'(x, P)\Psi^{-1}(x, P)$ does not depend on the choice of the base $\varphi_i(x, P)$, therefore, to find it in the neighbourhood of $P_0$ we can use the functions $\psi(x, \widetilde{k}_j; x_0)$, $\widetilde{k}_j^l = k$. In the local parameter $k^{-1}(P)$ it has the form

$$(2.4) \qquad \begin{pmatrix} 0 & 1 & 0 & \dots & 0 & 0 \\ 0 & 0 & 1 & \dots & 0 & 0 \\ \cdot & \cdot & \cdot & \cdot & \cdot & \cdot \\ 0 & & & \dots & 0 & 1 \\ k+\widetilde{u}_0, & \widetilde{u}_1 & & \dots & \widetilde{u}_{l-2} & 0 \end{pmatrix} + O(k^{-1}),$$

where the $\widetilde{u}_\alpha(x)$, $0 \leqslant \alpha \leqslant l-2$, are polynomials in the coefficients of $L_1$ and their derivatives.

We introduce the operator $\widetilde{L} = \Sigma \widetilde{u}_\alpha(x) \frac{d^\alpha}{dx^\alpha}$; $\widetilde{u}_l \equiv 1$, $\widetilde{u}_{l-1} = 0$. Let $\widetilde{c}_j(x, k)$ be the canonical basis of the space of solutions of $\widetilde{L}y = ky$. From (2.4) we deduce the following result.

LEMMA 2.2. *Near $P_0$ the function $\varphi_j(x, P)$ can be represented in the form*

$$(2.5) \qquad \varphi_j(x, P) = \widetilde{c}_j(x, k)\left(1 + \sum_{s=1}^{\infty} \xi_j(x) k^{-s}\right).$$

We now consider the inverse problem of recovering commuting operators from $\mathfrak{R}$, $P_0$, $D_1, \ldots, D_l$, $\tilde{u}_0(x), \ldots, \tilde{u}_{l-2}(x)$.

LEMMA 2.3. *Given a non-singular complex curve $\mathfrak{R}$ of genus $g$, a distinguished point $P_0$, and non-special effective divisors $D_j$ of degree $g$, there exist unique functions $\varphi_j(x, P)$, meromorphic away from $P_0$, with the divisor of poles $D_j$, and having near $P_0$ the form* (2.5).

PROOF. Let $\omega$ be a normalized Abelian differential with a single pole at $P_0$ of the form $-\frac{dz}{z^2}$, $z(P) = k^{-1}(P)$. The function $c_j(x, \int_{\bar{p}}^{p} \omega)$ is defined on $\mathfrak{R}$ cut along the cycles $a_i$ and has an essential singularity at $P_0$. We denote by $G_i(x, t)$, $t \in a_i$, the ratio of its values on the two edges of the cut.

We now pose Riemann's problem of finding a function $f_j(x, P)$ that is meromorphic on $\mathfrak{R}$ cut along the $a_i$, with the divisor of poles $D_j$, and satisfying the boundary condition on $a_i$

$$f_j^+(x, t) = G_i^{-1}(x, t)\, f_i^-(x, t); \qquad f_j(x, P_0) = 1.$$

The existence, uniqueness, and explicit construction of $f_i(x, P)$ in terms of the Cauchy kernel on $\mathfrak{R}$ and the Riemann $\theta$-functions is contained in [37]. The required function is

$$\varphi_j(x, P) = c_j\left(x, \int_{\bar{p}}^{p} \omega\right) f_j(x, P).$$

THEOREM 2.3. *For each function $E(P) \in A(\mathfrak{R}, P_0)$ there is a unique operator $L$ of order $nl$ where $n$ is the multiplicity of the pole $E(P)$, such that $L\varphi_j(x, P) = E(P)\varphi_j(x, P)$.*

PROOF. We construct from $\varphi_j(x, P)$ a matrix Akhiezer function $\Psi(x, P)$. By (2.4), in the neighbourhood of $P_0$, $\left(\frac{d^{\alpha l}}{dx^{\alpha l}} \Psi(x, P)\right) \Psi^{-1}(x, P)$ has the form $k^\alpha 1 + O(k^{\alpha-1})$ where 1 is the unit matrix. As in §1, we find the coefficients of the matrix operator $\bar{L} = \sum_{\alpha=0}^{n} w_\alpha(x) \frac{d^{\alpha l}}{dx^{\alpha l}}$ from the congruence

$$(\bar{L}\Psi)\,\Psi^{-1} \equiv E(P)\cdot\bar{L}\,(\mathrm{mod}\; O(k^{-1})).$$

From the uniqueness of the matrix function $\Psi(x, P)$ it follows that $\bar{L}(\Psi(x, P) = E(P)\Psi(x, P)$. Recalling that the columns of $\Psi(x, P)$ consist of the derivatives $\varphi_j(x, P)$, we find that the action of $\bar{L}$ on the column vectors is the same as that of $L$ on $\varphi_j(x, P)$.

Thus, we have arrived at the following theorem.

THEOREM 2.4. *For any commutative ring $A$ of differential operators there exist: a curve $\mathfrak{R}$ of genus $g$ with a distinguished point $P_0$, a set of* *divisors* $D_1, \ldots, D_l$ *of degree* $g$, *a set of arbitrary functions* $\widetilde{u}_0(x), \ldots, \widetilde{u}_{l-2}(x)$ *such that the homomorphism* $\lambda\colon A(\mathfrak{R}, P_0) \to A$ *defined by them in accordance with Theorem* 2.3 *is an isomorphism.*

The curve $\mathfrak{R}$ is called the spectrum of $A$ of multiplicity $l$. The problem of selecting commuting operators with coefficients that are polynomials in $x$ is interesting. An example of such operators is constructed in [43]. Their joint spectrum is the elliptic curve $w^2 = E^3 - \alpha$. The multiplicity of the spectrum is 3.

## §3. The two-dimensional Schrödinger operator and the algebras associated with it

Here we give an account of the basic ideas in the paper by Dubrovin, Novikov, and the author [34], in which the inverse problem of recovering from "algebraic" spectral data an operator depending essentially on some spatial variables was first posed and solved. (We recall that in the operators considered in §1 derivatives with respect to $y$ occurred only to the first power.)

In this context a new problem arises naturally: to describe the subrings $A$ of the ring $\mathcal{O}$ of differential operators in two variables whose quotient rings $A$ (mod $H$) by the left principal ideal generated by the Schrödinger operator $H$ in $\mathcal{O}$ are commutative, where $H = \dfrac{\partial^2}{\partial z \partial \bar{z}} + v(z, \bar{z}) \dfrac{\partial}{\partial \bar{z}} + u(z, \bar{z})$. We call such rings "commutative modulo $H$". This means that for arbitrary operators $L_1, L_2 \in A$ there are operators $D_1$, $D_2$, and $D_3$ such that

$$[L_1, L_2] = D_1 H; \ [L_1, H] = D_2 H; \ [L_2, H] = D_3 H. \tag{3.1}$$

The latter equations are equivalent to a system of non-linear differential equations for the coefficients of $L_1$, $L_2$, $H$. As we shall see, their solutions can be expressed explicitly in terms of the Riemann $\theta$-function.

As before, we assume that the leading terms of all the operators in question are homogeneous differential operators with constant coefficients.

THEOREM 3.1. *The operators* $L_1$ *and* $L_2$ *satisfying the compatibility equations* (3.1) *are connected by an algebraic relation, that is* $Q(L_2, L_1)\varphi = 0$ *on the solution space of* $H\varphi = 0$.

The theorem follows from the fact that on the solution space $\mathcal{L}(E)$ of the equation

$$L_1 \varphi(z, \bar{z}, E) = E \varphi(z, \bar{z}, E); \quad H \varphi(z, \bar{z}, E) = 0 \tag{3.2}$$

$L_2$ defines a linear operator $L_2(E)$ whose matrix elements in a canonical basis are polynomials in $E$, (with $\dim \mathcal{L}(E) = 2n$). Then $Q(w, E) = \det(w \cdot 1 - L_2^{st}(E))$ is the characteristic polynomial of $L_2(E)$.

We assume that the eigenvalues of $L_2(E)$ are distinct for almost all $E$.

Then to each point of the algebraic curve $\mathfrak{R}$ given by the equation $Q(w, E) = 0$ there corresponds an eigenvector of $L_2(E)$ whose coordinates in the canonical base $c_j(z, \bar{z}, E)$ are all meromorphic functions $\lambda_j(P)$ on $\mathfrak{R}$.

The corresponding function $\psi(z, \bar{z}, P) = \Sigma\, \lambda_j(P)c_j(z, \bar{z}, P)$ is meromorphic on $\mathfrak{R}$ away from "infinity". Its divisor of poles is of degree $g$, the genus of $\mathfrak{R}$.

To find the behaviour of $\psi(z, \bar{z}, P)$ at "infinity", we construct, as before, the germ of the formal Bloch function. Without loss of generality, we may take the leading terms of $L_1$ and $L_2$ to be $\frac{\partial^n}{\partial z^n} + q_1 \frac{\partial^n}{\partial \bar{z}^n}$ and $\frac{\partial^m}{\partial z^m} + q_2 \frac{\partial^m}{\partial \bar{z}^m}$, respectively.

LEMMA 3.1. *There are unique formal solutions of* (3.2) *of the form*

$$(3.3)\qquad \begin{cases} \psi_1(z, \bar{z}, k_1) = e^{k_1 z}\left(1 + \sum_{s=1}^{\infty} \xi_s(z, \bar{z})\, k_1^{-s}\right), & k_1^n = E, \\ \psi_2(z, \bar{z}, k_2) = e^{k_2 \bar{z}}\left(\sum_{s=0}^{\infty} \chi_s(z, \bar{z})\, k_2^{-s}\right), & q_1 k_2^n = E \end{cases}$$

*with the "normalization" conditions* $\xi_s(0, 0) = \chi_s(0, 0) = 0$, $s \geqslant 1$, $\chi_0(0, 0) = 1$.

The series $\psi_1^{-1} L_2 \psi_1 = k_1^m + O(k_1^{m-1})$ and $\psi_2^{-1} L_2 \psi_2 = q_2 k_2^m + O(k_2^{m-1})$ are expansions of the eigenvalues of $L_2(E)$ in the neighbourhoods of the two "points at infinity" $P_1$ and $P_2$ of $\mathfrak{R}$, provided that $n$ and $m$ are relatively prime and $q_1^m \neq q_2^n$.

Local parameters in the neighbourhoods $P_1$ and $P_2$ are $(E(P))^{-\frac{1}{n}}$ and $(E(P)q_1^{-1})^{-\frac{1}{n}}$. In terms of these the expansion of $\psi(z, \bar{z}, P)$ has the form (3.3).

As in the case of the Akhiezer function the properties of $\psi(z, \bar{z}, P)$ make it possible to reconstruct it from the "algebraic" data.

LEMMA 3.2. *For any non-singular complex curve $\mathfrak{R}$ of genus $g$, with fixed local parameters $w_1 = k_1^{-1}(P)$ and $w_2 = k_2^{-1}(P)$ in the neighbourhoods of two distinguished points $P_1$ and $P_2$ and an effective non-special divisor $D$ of degree $g$, there is a unique function $\psi(z, \bar{z}, P)$ that is meromorphic away from $P_1$ and $P_2$ with the divisor of poles $D$ and whose expansion in the neighbourhood of $P_j$ in the local parameter $k_j^{-1}(P)$ has the form* (3.3).

Two local parameters $w_j(P)$ and $w_j'(P)$ in the neighbourhood of $P_j$ are said to be equivalent if $(w_j^{-1} w_j')(P_j) = 1$.

COROLLARY. *The function $\psi(z, \bar{z}, P)$ depends only on the equivalence class of $w_j(P)$.*

LEMMA 3.3. *There is a unique operator* $H = \frac{\partial^2}{\partial z \partial \bar{z}} + v(z, \bar{z}) \frac{\partial}{\partial \bar{z}} + u(z, \bar{z})$ *such that* $H\psi = 0$.

Any operator $L$ such that $L\psi = 0$ is divisible on the right by $H$, that is, $L = DH$.

PROOF. For any two series $\psi_1(z, \bar{z}, k)$ and $\psi_2(z, \bar{z}, k)$ of the form (3.3) there is an operator $H$ such that $H\psi_1 \equiv 0(\text{mod } O(k^{-1}))$ and

$H\psi_2 \equiv 0(\mathrm{mod}\ O(1))$. In a standard way within the framework of our construction, it follows from the uniqueness of $\psi(z, \bar{z}, P)$ that $H\psi(z, \bar{z}, P)$ then vanishes identically.

LEMMA 3.4. *For any function $E(P) \in A(\mathfrak{R}, P_1, P_2)$ having poles only at $P_1$ and $P_2$ there is a unique operator $L$ of the form*

$$\sum_{\alpha=0}^{n_1} u_\alpha(z, \bar{z}) \frac{\partial^\alpha}{\partial z^\alpha} + \sum_{\beta=1}^{n_2} v_\beta(z, \bar{z}) \frac{\partial^\beta}{\partial \bar{z}^\beta},$$

*where the $n_j$ are the orders of the poles of $E(P)$ at $P_j$, such that*

$$L\psi(z, \bar{z}, P) = E(P)\psi(z, \bar{z}, P).$$

The coefficients of $L$ and $H$ can be expressed in the standard way by the Riemann $\theta$-function. For example, for $H$ we have [34]

$$v(z, \bar{z}) = -\frac{\partial}{\partial z}\log\left[\frac{\theta(U_1 z + U_2 \bar{z} + V_1 + W)}{\theta(U_1 z + U_2 \bar{z} + V_2 + W)}\right],$$

$$u(z, \bar{z}) = \frac{\partial^2}{\partial z\, \partial \bar{z}} \log\theta(U_1 z + U_2 \bar{z} + W).$$

Let us summarize our results.

THEOREM 3.2. *For any ring $A$ that is "commutative modulo $H$" and contains operators of relatively prime order with the leading terms $\frac{\partial^n}{\partial z^n} + q_1 \frac{\partial^n}{\partial \bar{z}^n}$ and $\frac{\partial^m}{\partial z^m} + q_2 \frac{\partial^m}{\partial \bar{z}^m}$, $q_1^m \neq q_2^n$, there exist a curve $\mathfrak{R}$ of genus $g$ with two distinguished points $P_1$ and $P_2$, an equivalence class of local parameters in neighbourhoods of $P_1$ and $P_2$ and an effective non-special divisor of degree $g$ such that the homomorphism $\lambda$: $A(\mathfrak{R}, P_1, P_2) \to A\ (\mathrm{mod}\ H)$. defined by them is an isomorphism.*

Let us dwell on some open problems. We have constructed a class of Schrödinger operators with almost-periodic potentials for which the Bloch eigenfunctions can be found exactly at the zero energy level. It is not clear when the parameters of our construction can vary with the energy, in other words: if $H$ is defined by $\mathfrak{R}$, the points $P_1$ and $P_2$, and the divisor $D$, is there a family of eigenfunctions of $H$ with arbitrary energy ($H\psi = E\psi$, $E \neq 0$), parametricized by points of the algebraic curves $\mathfrak{R}(E)$? If there is, then how does one find $\mathfrak{R}(E)$ from the initial data? How can one find the space $M^2$ of fiberings over the complex plane $C$ with fibres $\mathfrak{R}(E)$? Is there an algebraic variety $\hat{M}^2$, a compactification of $M^2$?

This last question is closely connected with the problem of selecting among the operators we have constructed the purely potential ones, that is, the operators of the form $H = \frac{\partial^2}{\partial z \partial \bar{z}} + u(z, \bar{z})$. Only in the class of these operators does the condition of reality of coefficients, which arises naturally within the framework of our constructions, turn into the condition of being Hermitian, which must hold for the physical Schrödinger operators.

LEMMA 3.5. *If on* $\mathfrak{R}$ *there is an anti-involution* $T_1$ *leaving* $D$ *invariant and such that* $T_1(P_1) = P_2$ *and* $T_1^* w_1 = w_2$, *where* $w_1$ *and* $w_2$ *are local parameters near* $P_1$ *and* $P_2$, *then for the operator* $H$ *constructed from these data the function* $v(z, \bar{z})$ *is purely imaginary, and the potential* $u(z, \bar{z})$ *is real. Hence,* $H$ *becomes real after a gauge transformation.*

NOTE 1. The real solutions of the Zahkarov–Shabat equations can be distinguished similarly.

Apparently, $H$ is a potential if and only if the original ring $A$ is commutative, which implies the existence of the variety $\hat{M}^2$, that is, in our terminology, $H$ is 2-algebraic.

NOTE 2. A necessary condition for $H$ to be a real potential operator is the existence on $\mathfrak{R}$ of a second anti-involution such that $T_2^* w_1 = -\bar{w}_2$. As Novikov has pointed out, the presence of two anti-involutions $T_1$ and $T_2$ (under certain restrictions on the situation of $D$ on the set of fixed points of $T_1$, which are indicated in [34], Lemma 3) is sufficient for the Bloch eigenfunction $\psi(z, \bar{z}, P)$, where $T_2 P = P$, to be bounded in $z$ and $\bar{z}$. (We recall that $H\psi = 0$). The set of fixed points $T_2 P = P$ is called the "real Fermi-surface".

## §4. The problem of multi-dimensional $n$-algebraic operators

Let $A$ be a commutative ring of differential operators in $n$ variables

$$L = \sum_{|\alpha| \leq l} u_\alpha(x) \frac{\partial^\alpha}{\partial x^\alpha},$$

where, as usual, $x = (x_1, \ldots, x_n)$, $\alpha = (\alpha_1, \ldots, \alpha_n)$,

$$|\alpha| = \sum_{i=1}^{n} \alpha_i, \quad \frac{\partial^\alpha}{\partial x^\alpha} = \frac{\partial^{|\alpha|}}{\partial x_1^{\alpha_1} \ldots \partial x_n^{\alpha_n}}.$$

It is assumed that all the leading coefficients of the $u_\alpha(x)$, $|\alpha| = l$, are constant.

Suppose that the symbols of the leading terms of the $L_2 \in A$ $(i = 1, \ldots, n)$, the polynomials $P_i(k)$, $k = (k_1, \ldots, k_n)$, are algebraically independent. Then the quotient ring of the ring of polynomials $C[k_1, \ldots, k_n]$ by the ideal generated by $P_i(k) - E_i$ is finite-dimensional. We denote by $G_\alpha(x)$ $(\alpha = 1, \ldots, N)$ representatives of its generators.

LEMMA 4.1. *A basis* $\mathcal{L}(E)$ *of the solution space of the equations* $L_i y = E_i y$ *is formed by functions satisfying the "normalization" conditions*

$$G_\alpha\left(\frac{\partial}{\partial x}\right)c_\beta(x_0, E)=\delta_{\alpha\beta}, \quad E=(E_1, \ldots, E_n).$$

Each operator $L_0 \in A$ determines a linear operator $L_0(E)$ in $\mathscr{L}(E)$.

LEMMA 4.2. *The matrix elements $L_0^{st}$ of $L_0(E)$ in the base $c_\alpha(x, E)$ are polynomials in the variables $E_i$ and $E_i^{-1}$.*

THEOREM 4.1. *The operators $L_i$, $0 \leqslant i \leqslant n$, are connected by the algebraic relations $Q(L_0, \ldots, L_n) = 0$, where $Q(w, E_1, \ldots, E_n) = \det(w \cdot 1 - L_0^{st}(E))$.*

If the symbol $P_0(k)$ of the leading terms of $L_0$ assumes almost always distinct values at the roots of the equations $P_i(k) = E_i$, then for almost all $E$ the eigenvalues of $L_0(E)$ are distinct. As before, by associating with each point of the affine variety $M^n$ given in $C \times C^n$ by the equation $Q(w, E) = 0$ the eigenvectors of $L_0(E)$, we obtain the following lemma.

LEMMA 4.3. *There is a meromorphic function $\psi(x, m)$, $m \in M^n$, that is an eigenfunction for each of the operators $L_i$,*

$$L_i\psi(x, m) = E_i(m)\psi(x, m), \quad L_0(\psi(x, m)) = w(m)\psi(x, m).$$

In contrast to the case $n = 1$, the compactification of $M^n$ under which $\psi(x, m)$ has "good" properties in the neighbourhood $D^\infty$ of the divisor of infinity, is not self-evident when $n > 1$. Here $D^\infty$ denotes a divisor of the compact algebraic variety $\hat{M}^n$ for which $\hat{M}^n \setminus D^\infty$ is isomorphic to $M^n$.

THEOREM 4.2. *There exists the system of equations*

$$L_i\psi(x, k) = P_i(k)\psi(x, k) \ (i = 1, \ldots, n),$$

*has a unique solution of the form*

$$\psi(x, k) = e^{(k, x)}\left(\sum_{s=0}^{\infty} \xi_s(x, k)\right), \tag{4.1}$$

*where $(k, x) = k_1x_1 + \ldots + k_nx_n$; the $\xi_s(x, k)$ are homogeneous rational functions in $k$ of degree $-s$, and $\xi_0(0, k) = 1$; $\xi_s(0, k) = 0$, $s \geqslant 1$.*

PROOF. The functions $\xi_s(x, k)$ can be found successively from the system of equations

$$\sum_{|\alpha| \leqslant l_i} u_{\alpha, i}(x) \sum_{|r| \leqslant |\alpha|} \frac{\partial^r P_i(k)}{\partial k^r} \frac{\partial^{\alpha-r}}{\partial x^{\alpha-r}} \xi_{s+|r|} = P_i(k)\xi_s.$$

From the $s$-th equation we find $\frac{\partial}{\partial x_j}\xi_s(x, k)$. Since the $L_i$ commute, we can integrate these partial derivatives to find $\xi_s(x, k)$. Then $\xi_s(x, k)$ has the form $F(x, k)G^{-s}(k)$, where $G(k) = \det\left\|\frac{\partial P_i}{\partial k_j}\right\|$.

NOTE. If $\deg L_i = l_i$, $0 \leqslant i \leqslant n$, then by a gradient transformation the $L_i$ can be reduced to the form $P_i\left(\frac{\partial}{\partial x}\right) + \widetilde{L}_i$, $\deg \widetilde{L}_i \leqslant l_i - 2$, if and only if $\xi_0(x, k) = \xi_0(x)$ does not depend on $k$.

COROLLARY 1. *The coefficients of the series* $\psi^{-1}(x, k)L_0\psi(x, k) = A(k)$ *are polynomials in the system of first integrals of the equations equivalent to the conditions* $[L_i, L_j] = 0$.

COROLLARY 2. *The ring of operators commuting with* $L_i$, $1 \leqslant i \leqslant n$, *is commutative, and* $\psi(x, m)$ *is an eigenfunction for all the operators* $L \in A$.

COROLLARY 3. *The characteristic polynomial* $Q(w, E)$ *is* $Q(w, E) = \prod_{\alpha} (w - A(k_\alpha))$, *where the* $k_\alpha$ *are the roots of the system of equations* $P_i(k) = E_i$.

Let us introduce a grading in the ring $C[w, E_1, \ldots, E_n]$, by ascribing to these variables the degrees $l_0, \ldots, l_n$, respectively.

COROLLARY 4. *Let* $Q^0$ *be a polynomial of degree* $Nl_0$ *connecting the* $P_i(k)$, *that is,* $Q^0(P_0(k), \ldots, P_n(k)) = 0$. *Then* $Q(w, E) = Q^0(w, E) + \widetilde{Q}(w, E)$, $\deg \widetilde{Q}(w, E) \geqslant Nl_0 - 1$.

We now describe the required compactification of $M^n$. To do this we regard the "weighted" projective space $CP(w)$, $w = (l_0, \ldots, l_n)$, as quotient space of $C^{n+2} \setminus \{0\}$ under the following action of the multiplicative group of complex numbers. A point $(z_0, \ldots, z_{n+1})$ is equivalent to $(t^{l_0}z_0, \ldots, t^{l_n}z_n, tz_{n+1})$, $t \neq 0$.

Then $\hat{M}^n$ can be specified in $CP(w)$ by the equation

$$0 = z_{n+1}^{Nl_0} Q\left(\frac{z_0}{z_{n+1}^{l_0}}, \ldots, \frac{z_n}{z_{n+1}^{l_n}}\right) = Q^0(z_0, \ldots, z_n) + z_{n+1}Q_1(z_0, \ldots, z_{n+1}).$$

The open subvariety of $\hat{M}^n$: $z_{n+1} \neq 0$ is isomorphic to $M^n$.

The regular mapping $\varphi$: $CP^n \to CP(w)$ defined in homogeneous coordinates by

$$\varphi(v_1, \ldots, v_{n+1}) = (\ldots, P_i(v_1, \ldots, v_n), \ldots, v_{n+1}),$$

establishes a birational isomorphism between the hyperplane $v_{n+1} = 0$ and the divisor $D^\infty$ defined by the equations $z_{n+1} = 0$, $Q^0(z_0, \ldots, z_n) = 0$. Hence, the functions $k_i(m) = \dfrac{v_i}{v_{n+1}}$ are defined in a small neighbourhood of $D^\infty$ in $\hat{M}^n$.

THEOREM 4.3. *Near* $D^\infty$ *the function* $\psi(x, m)$ *can be expanded in the form* (4.1), *where* $k_i(m) = \dfrac{v_i}{v_{n+1}}$.

Our assumption is that the variety $\hat{M}^n$ and the divisor of poles $\psi(x, m)$ uniquely determine the commutative ring $A$. The solution of the inverse problem is complicated by the fact that $\hat{M}^n$ has singularities at the images under $\varphi$ of the points $v_{n+1} = 0$, $G(v_1, \ldots, v_n) = 0$; $G(k) = \det \left\| \frac{\partial P_i}{\partial k_j} \right\|$. The dimension of the variety of singularities $n-2$. For $n = 1$ this means that $\mathfrak{R}$ is smooth. Therefore, we can use the theory of Abelian differentials to recover $\psi(x, P)$. For $n > 1$ the theory of meromorphic differentials is not effective enough, even on smooth manifolds.

The answer to the problem we have discussed in the previous section of constructing potential Schrödinger operators must yield a solution to the inverse problem for the variety $\hat{M}^n$ whose equation $Q(w, E) = 0$ has the following form: if $Q^0(P^0(k), \ldots, P_n(k))$ is an algebraic relation between the homogeneous polynomials

$$P_0(k) = \sum_{i=1}^{n} k_i^2, \text{ then } Q(w, E) = Q^0(w, E) + \tilde{Q}(w, E), \quad \deg \tilde{Q} \leqslant \deg Q^0 - 2.$$

## APPENDIX 1

## THE HAMILTONIAN FORMALISM IN EQUATIONS OF LAX AND NOVIKOV TYPE

The K-dV equation and its higher analogues determine flows on function spaces, which according to Gardner [4] and Zakharov and Faddeev [3] are formally of the Hamiltonian form $u_t = \frac{\partial}{\partial x}\frac{\delta I_n}{\delta u}$ on any space where the $I_n = \int L_n(u, u', \ldots)\, dx$ are meaningful and commute. Here $I_n$ is the system of K-dV integrals first found in [45], and $\frac{\delta I}{\delta u} = \sum_{k=0}^{n} (-1)^k \frac{d^k}{dx^k}\frac{\partial L}{\partial u^{(k)}}$. The skew-symmetric operator $\frac{\partial}{\partial x}$ defines the Poisson bracket (the Gardner–Zakharov– Faddeev bracket) on the space of functionals, by the formula

$$\{\tilde{F}_1, \tilde{F}_2\} = \int \frac{\delta F_1}{\delta u}\frac{\partial}{\partial x}\frac{\delta F_2}{\delta u}\, dx; \quad \tilde{F}_i = \int F_i(u, u', \ldots)\, dx \qquad (i = 0, 1).$$

Zakharov and Faddeev [3] have shown that on the space of rapidly decreasing functions all the "higher K-dV analogues" are completely integrable Hamiltonian systems for which the scattering data of the Sturm–Liouville operator $-\frac{d^2}{dx^2} + u(x, t)$ are variables of "action-angle" type.

Novikov has shown that a displacement in $x$ determines on the phase-space of solutions of the stationary equations $\sum_{n=0}^{N} c_n \frac{\delta I_n}{\delta u} = h$ a completely integrable finite-dimensional Hamiltonian flow [20]. A general proposition on the connection of Hamiltonian formalisms of stationary and non-stationary equations of the form $u_c = \frac{\partial}{\partial x}\frac{\delta I}{\delta u}$ was obtained by Novikov and Bogoyavlenskii [46].

Suppose that the flows $X_i$ $(u_t = \frac{\partial}{\partial x}\frac{\delta I_i}{\delta u}), i = 1, 2$ commute. Hence, the fixed points $T_h$ of the flow $X_1 \left(\frac{\delta I_i}{\delta u} = h\right)$ form an invariant set of $X_2$. We denote the restriction of $X_2$ to $T_h$ by $\varphi_n(X_1, X_2)$. The commutativity of the flows is equivalent to the fact that $\{\tilde{I}_1, \tilde{I}_2\} = 0$ or

$$\frac{\delta I_1}{\delta u}\frac{\partial}{\partial x}\frac{\delta I_2}{\delta u} = \frac{d}{dx}\, Q(u, u', \ldots).$$

In K-dV theory such a construction of integrals in the stationary problem was proposed by Gel'fand and Dikii [48] and Lax [47].

THEOREM [46]. *The flow $\varphi_n(X_1, X_2)$ on the phase space $T_h$ is Hamiltonian with the Hamiltonian* $-Q - h\frac{\delta I_2}{\delta u}$.

In the case of higher K-dV equations there are remarkable canonically conjugate variables in the phase space $T$, which were obtained in [49].

The coefficients of the formal series $V(u, k) = \sum_{i=0}^{\infty} b_i k^i$ satisfying the equation $-V''' + 4V'(u - \frac{1}{k}) + 2Vu' = 0$ (which is equivalent to the recurrent system of equations $4b'_{n+1} = -b''_n + 4b_n u + 2b_n u'$) are uniquely determined by the initial data

$$2c(k) = V''V - \frac{(V')^2}{2} - 2V^2\left(u - \frac{1}{k}\right).$$

The higher analogues of the K-dV equation have the form $u_t = b'_{n+1}$.

Let $W = \sum_{i=0}^{\infty} w_i k^i = -\frac{1}{2}\frac{V'}{V}$; then the variables $b_i w_{n-1}$ are canonically conjugate in the phase space $T$ of solutions of the stationary equation $b_{n+1} = 0$. A shift in $x$ defines a Hamiltonian flow in $T$ with the Hamiltonian $H_{n+1}$, where $H = \sum_{i=0}^{\infty} H_i k^i = W^2 V + \frac{c(k)}{V} - V(u - \frac{1}{k})$.

In the language of the Bloch eigenfunction $\psi(x, P)$ of the finite-zone Sturm–Liouville operator: if $\chi(x, P) = -i\,(\log\psi)' = \chi_R + i\chi_I$, then $\chi_R = \frac{\sqrt{c(k)}}{\prod_i (k - \gamma_i(x))} = \frac{\sqrt{c(k)}}{V}$, $\chi_I = -\frac{1}{2}(\log\chi_R)' = W = \sum w_i k^i$. Hence, the $b_k$ are the elementary symmetric polynomials $b_k = \sigma_k(\gamma_1(x), \ldots, \gamma_n(x))$. Flaschka and MacLaughlin [50] have constructed canonically conjugate variables to the $b_k$, but the spectral meaning remains unclear.

Solutions of the Zakharov–Shabat equations, independent of the variable $y$ are described by non-linear equations for the coefficients of the operators $L_1 = \sum_{\alpha=0}^{n} u_\alpha(x, t) \frac{\partial^\alpha}{\partial x^\alpha}$ and $L_2 = \sum_{\beta=0}^{m} v_\beta(x, t) \frac{\partial^\beta}{\partial x^\beta}$, which are equivalent to the operator equation

$$\text{(A.1.1)} \qquad \left[L_1, L_2 - \frac{\partial}{\partial t}\right] = 0 \Leftrightarrow \frac{\partial L_1}{\partial t} = [L_2, L_1].$$

Since the coefficients of $L_2$, which commutes with $L_1$ by (A.1.1), can be expressed to within operators of order $n-2$ by polynomials in the derivatives $u_\alpha(x, t)$ and constants $h_s$, $-m \leqslant s \leqslant 0$ (see Corollary 1 to Lemma 2.1), these equations are equivalent to systems of equations for the functions $u_\alpha(x, t)$, known as *equations of Lax type*. In addition to what was indicated in §2, algorithms for the construction of operators $L_2$, commuting in this manner with $L_1$, are contained in [19] and [32]. In the latter paper, Gelfand and Dikii have shown that equations of Lax type can be represented in the form

$$\text{(A.1.2)} \qquad \frac{\partial u}{\partial t} = l \sum_{p=1}^{N} c_p \frac{\delta A_p}{\delta u},$$

where $u = (u_0, \ldots, u_{n-2})$, $\frac{\delta}{\delta u} = \left(\frac{\delta}{\delta u_0}, \ldots, \frac{\delta}{\delta u_{n-2}}\right)$, and $l$ is a skew-symmetric operator whose matrix elements are

$$l_{rs} = \sum_{\gamma=0}^{n-1-r-s} \binom{\gamma+r}{r} u_{r+s+\gamma+1} \left(-i\frac{\partial}{\partial x}\right)^\gamma - \binom{\gamma+r}{s} \left(i\frac{d}{dx}\right)^\gamma u_{r+s+\gamma+1}.$$

The construction of the integrals $A_p$ of equations of Lax type uses the expansion in fractional powers of the resolvent of $L_1$.

The operator $l$ determines the Poisson bracket (Gelfand–Dikii bracket) on the space of functionals, by the formula

$$\{\tilde{F}_1, \tilde{F}_2\} = -\int \left[\sum_{r,s} \left(l_{rs}\frac{\delta F_2}{\delta u_s}\right) \frac{\delta F_1}{\delta u_r}\right] dx.$$

The proof of the Jacobi identity for this bracket is non-trivial. Gelfand and Dikii have told the author that a complete proof, not only for scalar but also for matrix operators, is in [33]. As in the case of the "higher K-dV analogues", all the flows $u_t = l\frac{\delta A_p}{\delta u}$ commute among each other.

The Lagrangian nature of the equations for stationary solutions of equations of Lax type does not follow directly from (A.1.2), since it is necessarily connected with the inversion of the operator $l$. The latter equations, which indicate that for $L_1$ there is a commuting operator $L_2$ are called equations of Novikov type, $[L_1, L_2] = 0$. (The construction of polynomial integrals for these, and also the complete integration of equations of Novikov type due to the present author, were quoted in §2 of this survey.)

The Lagrangian part of the Novikov equations $\sum_{p=1}^{N} c_p \frac{\delta A_p}{\delta u}$ was considered explicitly by Gelfand and Dikii only under the additional assumption that the Lagrangian $\sum_{p=1}^{N} c_p A_p$ is non-degenerate. (This assumption seems to be equivalent to our requirement that the orders of the operators $L_1$ and $L_2$ be relatively prime. For these equations there is an algorithm for the construction integrals in involution. A count of the number of independent integrals must yield the complete integrability of the corresponding Hamiltonian system.[1]

As the solutions of the equations $[L_1, L_2] = 0$ show, when the orders of the operators are not relatively prime, an interesting variant of the Hamiltonian formalism with constraints must hold for the corresponding system.

## APPENDIX 2

## ELLIPTIC AND RATIONAL SOLUTIONS OF THE K-dV EQUATIONS AND SYSTEMS OF MANY PARTICLES

In October 1976 I received a preprint of the paper by Airault, McKean, and Moser [51] in which a remarkable connection is discovered between the evolution of poles of rational and elliptic solutions of the K-dV equation and the motion of a discrete system of interacting particles on a line.[2]

It is easy to show that all elliptic solutions of the K-dV equation are of the form $u(x, t) = \sum_{j=1}^{n} 2\wp(x - x_j(t))$, where $\wp$ is the Weierstrass function. The K-dV equation for them is equivalent to the system

$$\dot{x}_j = 6 \prod_{k \neq j} \wp(x_j - x_k), \tag{A.2.1}$$

$$\sum_{k \neq j} \wp'(x_j - x_k) = 0, \tag{A.2.2}$$

where $x_j \neq x_k$ $(j = 1, \ldots, n)$.

[1] See the concluding remarks.

[2] In January 1977 Olshanitskii and Kolodzhevo pointed out to the author that in the paper [52] of G.V. and D.V. Chudnovskii the evolution of the poles of elliptic solutions of the K-dV and Burgers–Hopf equations and certain others is interpreted in terms of the motion of a Hamiltonian system of particles on a line. Some of their results on the K-dV equation overlap with the results of [51] reported here.

In this way the question of describing elliptic solutions of the K-dV equation rests upon that of describing $L^n$, given in $C^n$ by the equations (A.2.2). Apart from the case $n = 3$, practically nothing is known about it. We do not even know the dimension of $L^n$. Apart from the degenerate cases of "travelling" waves $f(x - ct)$, elliptic solutions of the K-dV equation with three poles reduce to the two-zone solutions $u(x, t)$ (first found by Novikov and Dubrovin [54]) for which

$$x_1+x_2+x_3=0, \quad t=\int_0^{x_1-x_2} \frac{dz}{\sqrt{12\,(g_2-3\wp^2(z))}}$$

$$x_2-x_3=\frac{1}{2}\wp^{-1}\left[-\wp(x_1-x_2)+\sqrt{g_2-3\wp^2(x_1-x_2)}\right].$$

As is well known, the function $x^{-2}$ is a degenerate form of the Weierstrass function.

Thus, if we let both periods of $\wp(x)$ tend to infinity, we obtain rational solutions of the K-dV equation of the form $2 \sum_{j=1}^{3} (x - x_j(t))^{-2}$.

However, we can obtain more complete results. It is easy to prove that rational solutions of the K-dV equation must be of the form $u(x, t) = 2 \sum_{j=1}^{n} (x - x_j(t))^{-2}$. The equations (A.2.1) and (A.2.2) for the rational case reduce to the system $\dot{x}_j = \sum_{k\neq j} 6(x_j - x_k)^{-2}$, $\sum_{k\neq j} (x_j - x_k)^{-3} = 0$ $(j = 1, \ldots, n)$.

What is remarkable is the fact that the variety of rational solutions of the K-dV equation is invariant under the flows $X_i$ determined by the "higher K-dV analogues". We denote by $X_i$ the images of these flows on the variety $L^n$. Since $\dim L^n \leqslant n$, there is a flow $\widetilde{X}_k$ that vanishes on $L^n$. Consequently, all rational solutions of the K-dV equation are stationary for one of the higher K-dV equations, that is, they form a separatrix family of finite-zone potentials of the Sturm–Liouville operator.

The expansion at infinity of the function $X_i u(x, t)$ has the form

$$c_i\left[\prod_{d=0}^{i}\left(n-\frac{d(d+1)}{2}\right)\right]x^{-2i+1}+O(x^{-2i}), \quad c_i\neq 0.$$

When we introduce the functions $\pi_k = \sigma_k(x_1, \ldots, x_n)$, then $\widetilde{X}_i\pi_k = 0$ if $k < 2i - 1$. Since the $\pi_k$, $k \leqslant n$ form a coordinate system on $L^n$, the flow $\widetilde{X}_i$ must vanish on $L^n$ for $2i - 1 > n$. Now $X_i u(x, t) = 0$ only when $n$ takes one of the values $d(d + 1)/2$. Otherwise, $L^n$ is empty.

If $n = d(d + 1)/2$, then the closure of $L^n$ is isomorphic to $C^d$. This isomorphism is determined by the mapping under which to $t_1, \ldots, t_d$ there correspond the poles of the function $u(x, 1)$, where $u(x, t)$ is the solution of the Cauchy problem with the initial data $u(x, 0) = d(d + 1)x^{-2}$ for the flow $t_1X_1 + \ldots + t_dX_d$.

In [53] Moser has established the complete integrability of a system of particles on a line with the pair potential $2x^{-2}$. The Hamiltonian of this system is $H = \frac{1}{2}\sum_{i=1}^{n} p_i^2 + \sum_{i<k} 2(x_j - x_k)^{-2}$. He found a representation of Lax type: $B_t = [A, B]$ for the equations of motion of this system, where the matrix elements of $A$ and $B$ are

$$A_{jj} = p_j, \quad A_{jk} = i(x_j - x_k)^{-1}, \quad j \neq k,$$

$$B_{jj} = -i\sum_{k\neq j}(x_j - x_k)^{-2}, \quad B_{jk} = i(x_j - x_k)^{-2}, \quad j \neq k.$$

From this representation, obviously, $F_k = \operatorname{tr} B^k$ and $F_2 = H$ are integrals in involution. Hence, the flows defined in the phase space by the $F_k$ commute. The set of fixed points of the initial system, that is, grad $F_2 = 0$ or $p_j = 0$, $\sum_{k\neq j}(x_j - x_k)^{-3}$ $(j = 1, \ldots, n)$, is $L^n$. A direct comparison of the formulae shows that the flows on $L^n$ corresponding to the motion of the poles of the solutions of the K-dV equation and the restriction of the flow grad $F_3$ to $L^n$ are the same. Apparently, there is a hitherto unproven proposition that the flows $\widetilde{X}_i$ and $(\operatorname{grad} F_i|_{L^n})$ coincide on $L^n$.

## CONCLUDING REMARKS

1. After the main text of this survey had been sent to the printers, the author learned that Veselov has answered a number of the questions mentioned in Appendix 1. He proved that the kernel of the operator $l$ is formed by linear combinations $\sum_{p=-n+2}^{0} c_p \frac{\delta A_p}{\delta u}$. Hence, the stationary equations $l\frac{\delta \mathcal{L}}{\delta u} = 0$, $\mathcal{L} = \sum_{p=1}^{N} c_pA_p$ are Lagrangian, $\frac{\delta}{\delta u}(\mathcal{L} - \sum_{p=-n+2}^{0} c_pA_p) = 0$. He proved that when $N$ and $n$ are relatively prime, the Lagrangian is non-degenerate and the equation $l\frac{\delta\mathcal{L}}{\delta u} = 0$ is an $(n - 1)$-parameter family of completely integrable Hamiltonian systems.

2. In [56] Petviashvili, using numerical computations stated a proposition on the existence of solutions for the Kadomtsev–Petviashvili equations.

Their explicit form was found by Matveev,

$$u(x, y, t) = \mp \frac{1 \pm 4y^2 - 4(x - 12t)^2}{2(x \mp 12t)^2 + y^2 \mp \frac{1}{4}}.$$

All $N$ soliton solutions of this equation were found in the paper [57], by Bordag, Its, Matveev, Manakov, and Zakharov.

$$u(x, y, t) = 2\frac{\partial^2}{\partial x^2} \log \det A,$$

where $A_{nn} = (x - i\nu_n y - (\xi_n + 3\nu_n^2 t))$, $A_{nm} = \dfrac{2}{\nu_n - \nu_m}$, $n \neq m$. For the solution to have no singularities the constants must be determined by $N = 2k$, $\operatorname{Re} \nu_n > 0$, $\nu_{n+k} = -\bar{\nu}_n$, $\xi_{n+k} = \bar{\xi}_n$.

It is interesting that the Kadomtsev–Petviashvili equation turns out to have no interaction of solitons even of phase shift type.

3. Very recently the author has discovered an algorithm for the construction of a broad class of rational and elliptic solutions of the Zakharov–Shabat equations. The evolution of the poles of these solutions, as in the case of rational solutions of the K-dV equation, is closely connected with the motion of systems of particles on a line.

4. Recently, the author proved that a function $u(x, y, t)$ is a rational solution of the Kadomtsev–Petviashvili equation, decreasing as $x \to \infty$, if and only if $u(x, y, t) = -2 \sum_{j=1}^{N} (x - x_j(y, t))^{-2}$ (where $N$ is arbitrary), and that the dynamics of the poles $x_j(y, t)$ in the variable $y$ coincides with the motion of the Moser system of particles with the Hamiltonian $H$ (see Appendix 2), while in the variable $t$ it coincides with the flow given by the Hamiltonian $F_3$. Explicit forms can be found for $u$. Thus, the theory of discrete integrable systems is covered by the theory of algebraic-geometrical solutions of the Zakharov–Shabat equations as a theory of special solutions.

NOTE IN PROOF. Very recently the author learned of three remarkable long forgotten papers: J. L. Burchnall and T. W. Chaundy, Proc. London Math. Soc. **21** (1922), 420–440; Proc. Royal Soc. London Ser. A **118** (1928), 557–573; Ser. A **134** (1931), 471; H. E. Baker, Proc. Royal Soc. London Ser. A **118** (1928), 573–580.

In these papers the problem of the classification of commutative algebras of ordinary scalar differential operators containing a pair of operators of relatively prime orders is posed and solved. For algebras of general type the problem is reduced to Abelian integrals, although finite formulae for the coefficients of the operators are not obtained. This result was rediscovered by the author and forms part of the results of [30]. Some degenerate cases are considered in the 1931 paper. In the 1922 paper commutative algebras are found containing a Sturm-Liouville operator; an algorithm is indicated for the reduction of a potential to hyperelliptic integrals. Formulae in terms of the $\theta$-function discovered in the 70's were not known in [26], [36], [30]. It is natural to compare these results with the theory of exact periodic solutions of the K-dV equation ([26]) and its subsequent development, which is reflected in this survey.

1. The K-dV equation and its higher analogues are of Lax form. An important consequence of the results of Gardner–Zakharov–Faddeev in K-dV theory consists in the fact that all these systems commute, and as a result of this, the K-dV equation and higher K-dV equations define a simultaneous deformation of all commutative algebras containing a Sturm–Liouville operator. This fact was the starting point of the modern theory of periodic solutions of the K-dV equation [20].

2. The works of the 20's and 30's we have mentioned are entirely of local character in $x$. The periodicity (quasiperiodicity) of the coefficients of the operators is not obtained. Hence, the connection between commutative algebras and the Floquet theory of linear equations with periodic coefficients is not noted, where the eigenfunction of the operators is determined non-locally in terms of the translation operator through a period. The key observation of the modern theory of the K-dV equation consists in the fact that Hill operators with finitely many of lacunae can automatically be embedded in a commutative algebra. The converse is also true [20], [44], [21], [22], [36]. The omission of this connection probably accounts for the fact that the remarkable results of the 20's were unknown in operator spectral theory and had no influence on the solution of direct and inverse problems. For example, these papers are not quoted in articles by Ince (1939–40) and Hochstadt (1965), which study special examples of periodic operators with finitely many lacunae.

3. In these old papers there is no discussion of all the problems concerning the construction of polynomial integrals, of the commutativity equations, of the theory of completely integrable Hamiltonian systems, of the temporal dynamics by virtue of the K-dV equation [26], nor of the algebraic-geometrical method of constructing exact solutions of the Zakharov–Shabat equations [28], [30]. A classification of commutative rings of matrix operators or of commuting scalar operators whose orders are not relatively prime is not achieved; nor are rings of multi-dimensional operators discussed (see §§3 and 4 of this survey).

## References

[1] C. Gardner, J. Green, M. Kruskal, and R. Miura, A method for solving the Korteweg-de Vries equation, Phys. Rev. Lett. **19** (1967), 1095–1098.

[2] P. D. Lax, Integrals of non-linear equations of evolution and solitary waves, Comm. Pure Appl. Math. **21** (1968), 467–490. MR **38** #3620.
= Matematika **13**:5 (1969), 128–50.

[3] V. E. Zakharov and L. D. Faddeev, The Korteweg-de Vries equation is a completely integrable Hamiltonian system, Funktsional. Anal. i Prilozhen. **5**:4 (1971), 18–27. MR **46** #2270.
= Functional Anal. Appl. **5** (1971), 280–287.

[4] C. Gardner, Korteweg-de Vries equation and generalization. IV, The Korteweg-de Vries equation as a Hamiltonian system, J. Mathematical Phys. **12** (1971), 1548–1551. MR **44** #3615.

[5] V. E. Zakharov and A. B. Shabat, An exact theory of two-dimensional self-focussing and of one-dimensional self-modulation of waves in non-linear media, Zh. Eksper. Teoret. Fiz. **61** (1971). MR **53** #9966.
= Soviet Physics JETP **34** (1972), 62–69.

[6] V. E. Zakharov and A. B. Shabat, On the interaction of solitons in a stable medium, Zh. Eksper. Teoret. Fiz. **64** (1973), 1627–1639.
= Soviet Physics JETP **37** (1973), 823–828.

[7] M. A. Ablowitz, D. J. Kaup, A. C. Newell, and H. Segur, Method for solving the sine-Gordon equation, Phys. Rev. Lett. **30** (1973), 1262–1264. MR **53** #9967.

[8] V. S. Dryuma, On the analytic solution of the two-dimensional Korteweg-de Vries equation, Zh. Eksper. Teoret. Fiz. Lett. **19** (1973), 219–225.

[9] V. E. Zakharov, On the problem of the stochastization of one-dimensional chains of non-linear oscillators, Zh. Eksper. Teoret. Fiz. **65** (1973), 219–225.
= Soviet Phys. JETP **38** (1974), 108–110.

[10] V. E. Zakharov and S. V. Manakov, The theory of resonance interaction of wave packets in non-linear media, Zh. Eksper. Teoret. Fiz. **69** (1975), 1654–1673. MR **54** #14617.
= Soviet Physics JETP **42** (1975), 842–850.

[11] L. A. Takhtadzhyan, An exact theory of the propagation of ultra-short optical impulses in two-level media, Zh. Eksper. Teoret. Fiz. **66** (1974), 476–489.
= Soviet Phys. JETP **39** (1975), 228–233.

[12] V. E. Zakharov, L. A. Takhtadzhyan, and L. D. Faddeev, A complete description of the solutions of the "sine-Gordon" equation, Dokl. Akad. Nauk SSSR **219** (1974), 1334–1337. MR **52** #9881.
= Soviet Physics Dokl. **19** (1974), 824–826.

[13] M. Toda, Waves in non-linear lattice, Progr. Theor. Phys. Suppl. **45** (1970), 174–200.

[14] M. Henon, Integrals of the Toda lattice, Phys. Rev. B(3) **9** (1974), 1921–1923. MR **53** #12410.

[15] H. Flaschka, The Toda lattice. I, Phys. Rev. B(3) **9** (1974), 1924–1925. MR **53** #12411.

[16] H. Flaschka, The Toda lattice. II, Progr. Theoret. Phys. **52** (1974), 703–716. MR **53** #12412.

[17] S. V. Manakov, The complete integrability and stochastization in discrete dynamic systems, Zh. Eksper. Teoret. Fiz. **67** (1974), 543–555. MR **52** #9938.
= = Soviet Physics JETP **40** (1974), 269–274.

[18] A. B. Shabat, On the Korteweg-de Vries equation, Dokl. Akad. Nauk SSSR **211** (1973), 1310–1313. MR **48** # 9137.
= Soviet Math. Dokl. **14** (1973), 1266–1270.

[19] V. E. Zakharov and A. B. Shabat, A scheme for the integration of non-linear equations of mathematical physics by the method of the inverse scattering problem. I, Funktsional. Anal. i Prilozhen. **8**:3 (1974), 43–53.
= Functional Anal. Appl. **8** (1974), 226–235.

[20] S. P. Novikov, The periodic Korteweg-de Vries problem. I, Funktsional. Anal. i Prilozhen. **8**:3 (1974), 54–56. MR **52** #3760.
= Functional Anal. Appl. **8** (1974), 236–246.

[21] P. D. Lax, Periodic solutions of the K-dV equations, Lect. in Appl. Math. **15** (1974), 85–96. MR **49** #384.

[22] H. McKean and Van Moerbeke, The spectrum of Hill's equation, Invent. Math. **30** (1975), 217–274. MR **53** #936.

[23] V. A. Marchenko, The periodic Korteweg-de Vries problem, Dokl. Akad. Nauk SSSR **217** (1974), 276–279. MR **50** #7855.
= Soviet Math. Dokl. **15** (1974), 1052–1056.

[24] V. A. Marchenko and I. V. Ostrovskii, The periodic Korteweg-de Vries problem, Mat. Sb. **95** (1974), 331–356.
= Math. USSR–Sb. **24** (1974), 319–344.

[25] V. A. Marchenko, *Operator Sturm–Liuvillya i ego prilozheniya*, (The Sturm–Liouville operator and its applications), Naukova Dumka, Kiev 1977.

[26] V. A. Dubrovin, V. B. Matveev, and S. P. Novikov, Non-linear equations of Korteweg-de Vries type, finite-zone linear operators, and Abelian manifolds, Uspekhi Mat. Nauk **31**:1 (1976) 55–136.
= Russian Math. Surveys **31**:1 (1976), 59–146.

[27] B. B. Kadomtsev and V. I. Petviashvili, On the stability of solitary waves in weakly dispersing media, Dokl. Akad. Nauk SSSR **192** (1970), 753–756.
= Soviet Phys. Dokl. **15** (1971), 539–541.

[28] I. M. Krichever, An algebraic-geometrical construction of the Zakharov–Shabat equation and their periodic solutions, Dokl. Akad. Nauk SSSR **227** (1976), 291–294.
= Soviet Math. Dokl. **17** (1976), 394–397.

[29] I. M. Krichever, Algebraic curves and commuting matrix differential operators, Funktsional. Analiz i Prilozhen. **10**:2 (1976), 75–76.
= Functional Anal. Appl. **10** (1976), 144–146.

[30] I. M. Krichever, The integration of non-linear equations by methods of algebraic geometry, Funktsional. Analiz i Prilozhen. **11**:1 (1977), 15–31.
= Functional Anal. Appl. **11** (1977), 12–26.

[31] V. E. Drinfeld, On commutative subrings of some non-commutative rings, Funktsional. Analiz i Prilozhen. **11**:1 (1977), 11–14.
= Functional Anal. Appl. **11** (1977), 9–12.

[32] I. M. Gel'fand and L. A. Dikii, Fractional powers of operators and Hamiltonian systems, Funktsional. Analiz i Prilozhen. **10**:4 (1976), 13–29.
= Functional Anal. Appl. **10** (1976), 259–273.

[33] I. M. Gel'fand and L. A. Dikii, Resolvents and Hamiltonian systems, Funktsional. Analiz i Prilozhen. **11**:2 (1977). 11–27.
= Functional Anal. Appl. **11** (1977), 93–105.

[34] V. A. Dubrovin, I. M. Krichever, and S. P. Novikov, The Schrödinger equation in a periodic magnetic field, and Riemann surfaces, Dokl. Akad. Nauk SSSR **229** (1976), 15–18.
= Soviet Math. Dokl. **17** (1976), 947–951.

[35] N. I. Akhiezer, A continuous analogue of orthogonal polynomials on a system of intervals, Dokl. Akad. Nauk SSSR **141** (1961), 263–266. MR **25** #4383.
= Soviet Math. Dokl. **2** (1961), 1409–1412.

[36] A. R. Its and V. B. Matveev, Schrödinger operators with a finite-zone spectrum and $N$-soliton solutions of the Korteweg-de Vries equation, Teoret. Mat. Fiz. **23**:1 (1975), 51–67.

[37] E. I. Zverovich, Boundary value problems of the theory of analytic functions, Uspekhi Mat. Nauk **26**:1 (1971), 113–181.
= Russian Math. Surveys **26**:1 (1971), 117–192.

[38] G. Springer, Introduction to Riemann surfaces, Addison–Wesley, Reading, Mass., 1957. MR **19**–1169.
Translation: *Vvedenie v teoriyu riemanovykh poverkhnostei*, Izdat. Inostr. Lit., Moscow 1961.

[39] B. A. Dubrovin, Finite-zone linear operators and Abelian manifolds, Uspekhi Mat. Nauk **31**:4 (1976), 259–260.

[40] B. A. Dubrovin, Completely integrable systems, associated with matrix operators, Funktsional. Anal. i Prilozhen. **11**:4 (1977), 28–41.
= Functional Anal. Appl. **11** (1977), 265–277.

[41] S. V. Manakov, Note on the integration of the Euler equations for the dynamics of an $n$-dimensional rigid body, Funktsional. Anal. i Prilozhen. **10**:4 (1976), 93–94.
= Functional Anal. Appl. **10** (1976), 328–329.

[42] V. I. Arnold, *Matematicheskie metody klassicheskoi mekhaniki* (Mathematical methods of classical mechanics), Nauka, Moscow 1974.

[43] J. Dixmier, Sur les algebres de Weyl, Bull. Soc. Math. France **96** (1968), 209–242. MR **39** # 4224.
= Matematika **13**:4 (1969), 27–40.

[44] B. A. Dubrovin, The inverse problem of scattering theory for periodic finite-zone potentials, Funktsional. Anal. i Prilozhen. **9**:1 (1975), 65–66. MR **52** # 3755.
= Functional Anal. Appl. **9** (1975), 61–62.

[45] M. Kruskal and N. Zabusky, Interaction of 2 solitons in a collisionless plasma and the recurrence of initial states, Phys. Rev. Lett. **19** (1967), 1095–1098.

[46] O. I. Bogoyavlenskii and S. P. Novikov, On the connection of Hamiltonian formalisms of stationary and non-stationary problems, Funktsional. Anal. i Prilozhen. **10**:1 (1976), 9–13.
= Functional Anal. Appl. **10** (1976), 8–11.

[47] P. D. Lax, Periodic solution of the K-dV equation, Comm. Pure Appl. Math. **38**:1 (1975), 141–188. MR **51** # 6192.

[48] I. M. Gelfand and L. A. Dikii, The asymptotic behaviour of the resolvent of Sturm–Liouville operators and an algebra of Korteweg-de Vries equations, Uspekhi Mat. Nauk **30**:5 (1975), 67–100.
= Russian Math. Surveys **30**:5 (1975), 77–113.

[49] S. I. Alber, The study of equations of the Korteweg-de Vries type by the method of recurrence relations, preprint, Inst. Khim. Fiz. Akad. Nauk SSSR, Chernogolovka 1976.

[50] H. Flaschka and D. McLaughlin, Canonically conjugate variables for the Korteweg-de Vries equation and the Toda lattice with periodic boundary conditions, Progr. Theoret. Phys. **53**:2 (1976), 438–451.

[51] H. Airault, H. McKean, and J. Moser, Rational and elliptic solutions of the Korteweg-de Vries equation and a related many-body problem, preprint, Courant Inst. **8** (1976).

[52] D. V. Chudnovskii and G. V. Chudnovskii, Pole expansions of non-linear partial differential equations, preprint, Inst. Marconi 1977.

[53] J. Moser, Three integrable Hamiltonian systems connected with isospectral deformations, Adv. in Math. **16** (1976), 197–220. MR **51** # 12058.

[54] B. A. Dubrovin and S. P. Novikov, Periodic and conditionally periodic analogues of multisoliton solutions of the Korteweg-de Vries equation, Zh. Eksper. Teoret. Fiz. **67** (1974), 2131–2143. MR **52** # 3759.
= Soviet Physics JETP **40** (1974), 1058–1063.

[55] S. A. Amitsur, Commutative linear differential operators, Pacific J. Math. **8** (1958), 1–10. MR **20** # 1808.

[56] V. I. Petviashvili, The formation of three-dimensional Langmuir solitons under the action of a powerful radio-wave in the ionosphere, Fizika plazmy **2**:3 (1976), 650–655.

[57] L. A. Bordag, A. R. Its, V. B. Matveev, S. V. Manakov, and V. E. Zakharov, Two-dimension solitons of Kadomtsev–Petviashvili equations, Phys. Rev. Lett. **29** (1977).

[58] A. R. Its and V. B. Matveev, On a class of solutions of the K-dV equation, in *"Problemy matematicheskoi fiziki"* (Problems in mathematical physics) No 8, Leningrad University (1976)

Received by the Editors 24 March 1977

Translated by R. L. and G. Hudson

## ALGEBRAIC CURVES AND NON-LINEAR DIFFERENCE EQUATIONS

I.M. Krichever

In [1] we have given an account of a scheme for the integration of certain non-linear differential equations by methods of algebraic geometry. After a slight modification, the main ideas and results of the scheme can be carried over to difference equations.

**1.** Let

$$L_1^{ij}=\sum_{\alpha=-n_1}^{n_2} u_\alpha(s)\,\delta_{i,\,j-\alpha},\qquad L_2^{ij}=\sum_{\beta=-m_1}^{m_2} v_\beta(s)\,\delta_{i,\,j-\beta}$$

be difference operators whose coefficients are $(l\times l)$-matrices. We stipulate that their highest and lowest coefficients are non-singular diagonal matrices with distinct diagonal elements.

We consider equations in the coefficients of these operators that are equivalent to the equality $[L_1, L_2]=0$.

The operator $L_2$ induces on the solution space of the equation $L_1y=Ey$ a finite-dimensional linear operator $L_2(E)$. Its characteristic polynomial $Q(w,E)$ defines a complex curve $\mathfrak{R}$, and the projection $(w,E)=P\to E$ defines a meromorphic function on it.

THEOREM 1. *For any pair of commuting difference operators we can find a polynomial in two variables such that* $Q(L_2, L_1)=0$.

If all the eigenvalues of $L_2(E)$ are distinct, as in the case of pairwise coprime numbers $n_2$, $m_2$ and $n_1$, $m_1$, then to each point $(w,E)$ of $\mathfrak{R}$ there corresponds an eigenvector of $L_2(E)$ that is unique up to a proportionality factor.

THEOREM 2. *If* $(n_2,m_2)=1$ *and* $(n_1,m_1)=1$, *then* $E(P)$ *has* $l$ *poles* $(P_1^+,\dots,P_l^+)$ *of order* $n_2$ *and* $l$ *poles* $(P_1^-,\dots,P_l^-)$ *of order* $n_1$. *The coordinates* $\psi_j(i,P)$ *of the eigenvector-functions of* $L_1$ *and* $L_2$ *belong to the space associated with the divisor* $\Delta=D+(i-1)D_\infty+P_j^+-P_j^-$, *where* $D$ *is an effective divisor whose degree* $g$ *is equal to the genus of the curve for almost all solutions of the original equations, and* $D_\infty=(P_1^++\dots+P_l^+)-(P_1^-+\dots+P_l^-)$.

We consider the inverse problem of recovering the operators from a curve with distinguished points $P_j^\pm$ and a divisor $D$ of degree $g$.

Since $\deg\Delta=g$, by the Riemann–Roch theorem the $\psi_j(i,P)$ are uniquely determined by the conditions of Theorem 2 up to a normalization. Having fixed one, we have the following theorem.

THEOREM 3. *For any function* $E(P)$ *with poles on* $\mathfrak{R}$ *only at the points* $P_j^\pm$, *there exists a unique operator* $L$ *such that* $L\psi(i,P)=E(P)\psi(i,P)$.

**2.** In this section we construct exact solutions for certain non-linear differential-difference equations.

Suppose that we are given a set of polynomials $Q_j^\pm(k)$ and $R_j^\pm(k)$.

THEOREM 4. *For every effective divisor* $D$ *on a curve* $\mathfrak{R}$ *of genus* $g$ ($\deg D=g$) *with fixed local coordinates* $k_{j\pm}^{-1}(P)$ *in neighbourhoods of the* $P_j^\pm$ *there exists one and (apart from a proportionality factor) only one function* $\varphi_{j_1}(i,y,t,P)$ *that is meromorphic outside* $P_j^\pm$, *and for which* $D$ *is the divisor of the poles. In a neighbourhood of* $P_j^\pm$ *the function*

$$\varphi_{j_1}(i,\ y,\ t,\ P)\exp\{Q_j^\pm(k_{j\pm}(P))\,y+R_j^\pm(k_{j\pm}(P))\,t\}$$

*has a pole (zero) of order* $i$ *if* $j=j_1$, *and of order* $i-1$ *if* $j\neq j_1$.

By defining the normalization of $\varphi_j(i,y,t,P)$ arbitrarily we obtain the vector-valued function $\psi(i,y,t,P)$.

THEOREM 5. *There exist unique difference operators whose coefficients depend on* $y$ *and* $t$, *such that*

$$\left(L_1-\frac{\partial}{\partial y}\right)\psi(s,\ y,\ t,\ P)=0\quad\text{and}\quad\left(L_2-\frac{\partial}{\partial t}\right)\psi(s,\ y,\ t,\ P)=0.$$

COROLLARY. *These operators satisfy the equation*

$$(1)\qquad [L_1,\ L_2]=\frac{\partial L_2}{\partial y}-\frac{\partial L_1}{\partial t}.$$

3. EXAMPLE. We consider the equations of a Toda chain:

$$\dot v_n=c_{n+1}-c_n,\qquad \dot c_n=c_n\,(v_n-v_{n-1}).$$

By Theorem 4, there is a unique function $\psi(n, t, P)$ with poles at the points $d_1, \ldots, d_g$ of $\mathfrak{R}$ defined by $w^2 = \prod_{i=1}^{2g+2} (E-E_i)$, and with the following asymptotic expansion at the inverse images of $E=\infty\ (P^{\pm})$:

$$\psi^{\pm}(n,\ t,\ E)=i^n\lambda_n^{\pm 1}E^{\pm n}(1+\xi_1^{\pm}(n,\ t)\,E^{-1}+\ldots)\exp\left(\mp\frac{1}{2}tE\right).$$

By Theorem 5, the operators

$$L^{nm}=i\sqrt{c_n}\,\delta_{n,\ m+1}+v_n\delta_{n,\ n}-i\sqrt{c_{n+1}}\,\delta_{n,\ m-1},$$
$$A^{nm}=\frac{i}{2}\sqrt{c_n}\,\delta_{n,\ m+1}+w_n\delta_{n,\ n}+\frac{i}{2}\sqrt{c_{n+1}}\,\delta_{n,\ m-1}$$

satisfy the equations $L\psi = E\psi$ and $A\psi = \partial\psi/\partial t$. Here $\sqrt{c_n} = \lambda_{n-1}/\lambda_n$, $v_n = \xi_1^{+}(n+1, t) - \xi_1^{+}(n, t)$, and $w_n = v_n/2 + \lambda_n/\lambda_n$.

The equations (1) are equivalent to the system $\dot v_n = c_{n+1} - c_n$,

$$\frac{\dot c_n}{c_n}=(v_n-v_{n-1})-(w_n-w_{n-1})=\frac{1}{2}(v_n-v_{n-1})-\frac{1}{2}\frac{\dot c_n}{c_n},$$

which is the same as the equations of a Toda chain.

We must remark that this representation of equations is different from the commutation representation, used in earlier work (for a bibliography, see [2]).

By expressing $\psi(n, t, P)$ in terms of Riemann's theta-function as in the formula of Its [3] and also § 3 of [1], we obtain the following formulae in which we have used the notation of [1]:

$$\log c_n=\frac{d}{dn}\log\frac{\theta(\omega^{+}+\mathbf{W})\,\theta((n-1)\mathbf{U}+t\mathbf{V}+\mathbf{W}+\omega^{-})}{\theta(\omega^{-}+\mathbf{W})\,\theta((n-1)\mathbf{U}+t\mathbf{V}+\mathbf{W}+\omega^{+})}+\text{const},$$
$$v_n=\frac{d}{dn}\frac{d}{dt}\log\frac{\theta(n\mathbf{U}+t\mathbf{V}+\omega^{+}+\mathbf{W})}{\theta(t\mathbf{V}+\omega^{+}+\mathbf{W})}+\text{const},$$

where the vectors $\boldsymbol{\omega}^{+}$, $\mathbf{V}$, and $\mathbf{U}$ and the constants depend only on the curve $\mathfrak{R}$, and $d/dn$ denotes the difference derivative. A formula for the variables $v_n$ analogous to ours was first derived by Novikov [2]. In 1977 the author became aware of a similar paper of Mumford.

## References


[1] I. M. Krichever, Integration of non-linear equations by methods of algebraic geometry, Funktsional. Analiz i Prilozhen. **11** (1977), 15–31.
= Functional Anal. Appl. **11** (1977), 12–26.

[2] B. A. Dubrovin, V. B. Matveev, and S. P. Novikov, Non-linear equations of Korteweg-de Vries type, finite zone linear operators, and Abelian varieties, Uspekhi Mat. Nauk **31**:1 (1976), 55–136. MR **55** # 899.
= Russian Math. Surveys **31**:1 (1976), 59–146.

[3] A. R. Its and V. B. Matveev, On a class of solutions of the Korteweg-de Vries equation, in: *Problemy matematicheskoi fiziki* (Problems of mathematical physics), No. 8, Leningrad State Univ., Leningrad 1976.


Received by the Editors 12 October 1976.

# THE STRUCTURES OF HAMILTONIAN MECHANICS

A. M. Vinogradov and B. A. Kupershmidt

## Contents



## Introduction

The account we present of the foundations of Hamiltonian mechanics and its mathematical language – the Hamiltonian formalism – represents the contents of a course of lectures given by A. M. Vinogradov in the Faculty of Applied Mathematics at the Moscow Institute of Electronic Machine Construction in the Faculty of Mechanics and Mathematics at the Moscow State University, and at the eighth Voronezh winter school. What these lectures aimed at was a logical account of mechanics, for example, the question when basic facts can be derived from basic concepts by "rules of grammar". Since the language of Hamiltonian mechanics is the calculus of differential forms and vector fields on smooth manifolds, the basic formulae of this calculus play the role of such grammatical rules. A pleasant consequence of this approach is the possibility of avoiding the manner of arguing by calculations which is usual in analytical mechanics.

Lack of space prevents us from considering more special questions (non-holonomic systems, the theory of impact and refraction, friction, optimal control).[1] So we only remark that they too admit a very simple invariant treatment starting out from the point of view we have taken. We have also not described the natural connection of mechanics with the general theory of differential equations, which is a problem of paramount importance. The algebraic–logical side of this connection is indicated in [16]; for the geometrical basis see [10] and [13].

We analyse very briefly the connection between the Hamiltonian and Langrangian approaches to mechanics. We remark, therefore, that we do not share the widespread view that the Langrangian approach is more fundamental than the Hamiltonian. In this context we draw the reader's attention to §10, where we make an attempt to show how the Hamiltonian point of view can be placed directly (that is, without the Lagrangian) at the basis of mechanics as a physical theory.

So that the reader can get an idea of areas where the methods of Hamiltonian mechanics can be applied and developed fruitfully, we have included in the list of references works on Hamiltonian field theory [54]–[61] and non-linear differential equations [62]–[74], which are of interest from this point of view. From the works [31]–[53] one can see what topics connected with the Hamiltonian formalism are of interest to modern mathematicians, and also to specialists in mathematical and theoretical physics. The list of references does not pretend to be complete, and as a rule we restrict ourselves to indicating works of a survey and general character, in which the reader can find further references.

It is a pleasure for the authors to state that this paper was written thanks to the initiative of S. G. Krein, and also thanks to the support it found from S. V. Fomin.

## §0. Definitions, notation, and basic facts

We present the necessary facts from analysis on manifolds (for more details, see [1] or [2]).

If $M$ is a smooth ($=C^\infty$) manifold, then $\mathscr{F}(M)$ denotes the ring of smooth functions on $M$, $\mathscr{D}(M)$ the $\mathscr{F}(M)$-module of vector fields on

1 See Appendix II.

$M$, $\Lambda^k(M)$ $(0 \leqslant k \leqslant n = \dim M)$ the $\mathscr{F}(M)$-module of differential $k$-forms on $M$ $(\Lambda^0(M) = \mathscr{F}(M))$, and $\Lambda(M) = \bigoplus_{k=0}^{n} \Lambda^k(M)$.

The homomorphism of rings of functions $\mathscr{F}(N) \to \mathscr{F}(M)$ induced by a smooth map $F$: $M \to N$ is denoted by $F^*$: $F^*(f)(x) = f(F(x))$, $f \in \mathscr{F}(N)$.

**The tangent space.** By a tangent vector at a point $a \in M$ we mean an $\mathbf{R}$-linear map $\xi$: $\mathscr{F}(M) \to \mathbf{R}$ satisfying Leibniz' rule: $\xi(fg) = g(a)\xi(f) + f(a)\xi(g)$, $f, g \in \mathscr{F}(M)$. The totality of tangent vectors at a point $a \in M$ forms an $n$-dimensional linear space $T_a(M)$ over $\mathbf{R}$, the so-called tangent space at $a$. $\bigcup_{a \in M} T_a(M)$, the tangent space $T(M)$ (to $M$), is furnished with a smooth structure in a natural way: if $(x_1, \ldots, x_n)$ are local coordinates on $M$, then $(\bar{\pi}^*(x_1), \ldots, \bar{\pi}^*(x_n), v_1, \ldots, v_n)$ are special local coordinates on $T(M)$; here $\bar{\pi}$: $T(M) \to M$, $\bar{\pi}(T_a(M)) = a$, is the natural projection, and the point $\tilde{\xi} \in T(M)$ with the above local coordinates gives rise to the tangent vector $\xi = \sum_{i=1}^{n} v_i \frac{\partial}{\partial x_i}$ at the point $a \in M$ with local coordinates $(x_1, \ldots, x_n)$. If $F$: $M \to N$ is a smooth map, $\xi \in T_a(M)$, then the tangent vector $F_a(\xi) = dF_a(\xi)$ in $T_{F(a)}(N)$ is defined by the equation $F_a(\xi) = \xi \circ F_a^*$.

**Vector fields.** A vector field $X$ on $M$ is an $\mathbf{R}$-linear map $X$: $\mathscr{F}(M) \to \mathscr{F}(M)$ satisfying Leibniz' formula: $X(fg) = gX(f) + fX(g)$. A vector field $X$ gives rise at every point $a \in M$ to a tangent vector $X_a$: $X_a(f) = X(f)(a)$. Conversely, a smooth field of tangent vectors $X_a \in T_a(M)$ determines a vector field $X$: $X(f)(a) = X_a(f)$. If $F$: $M \to N$ is a smooth map and there is a vector field $Y \in \mathscr{D}(N)$ for which $F^* \circ Y = X \circ F^*$, then $X$ is said to be compatible with $F$, and $Y = F(X)$ is the image of $X$. When $F$ is a diffeomorphism, any field $X$ has an image $F(X) = F^{*-1} \circ X \circ F^*$.

**Trajectories and shift operators.** Let $\frac{d}{dt} \in \mathscr{D}(\mathbf{R})$, $X \in \mathscr{D}(M)$. A local trajectory of $X$ is a map $F$: $I \to M$ of the interval $I \subset \mathbf{R}$ such that $F_\tau(\frac{d}{dt}|_\tau) = X_{F(\tau)}$, $\forall \tau \in I$. $F$ is a local trajectory if and only if $\frac{d}{dt} \circ F^* = F^* \circ X$. If $(x_1, \ldots, x_n)$ are local coordinates in the chart $U \ni a = F(\tau)$, $X|_U = \sum_{i=1}^{n} \alpha_i(x) \frac{\partial}{\partial x_i}$, then $F$ is a local trajectory if and only if $\frac{dx_i}{dt} = \alpha_i(x)$ $(i = 1, \ldots, n)$. If $F_i$: $I_i \to M$ $(i = 1, 2)$ are two local trajectories passing through the point $a = F_1(\tau_1) = F_2(\tau_2)$, then $F_1(t) = F_2(t + \tau_2 - \tau_1)$ for all $t$ sufficiently close to $\tau_1$. In the case $I = \mathbf{R}$ the trajectory is called global, and if there is a global trajectory through every point $a \in M$, then $X$ is called full. The maps $A_t: a \to F(F^{-1}(a) + t)$, which are defined, generally speaking, for sufficiently small $|t|$ (depending on $a \in M$), determine a smooth (local) one-parameter group $\{A_t\}$ of diffeomorphisms of $M$; this is denoted by $\{A_t\} \Leftrightarrow X$. Here $X = \lim_{t\to 0} (A_t^* - E)/t$, $X \circ A_t^* = A_t^* \circ X = \frac{d}{dt} \circ A_t^*$. If $f \in \mathscr{F}(M)$, and $X(f) = \alpha = \text{const}$, then $A_t^*(f) = f + \alpha t$. Let $F: M \to N$ be a smooth map and $F(X) = Y \in \mathscr{D}(N)$, $\{B_t\} \Leftrightarrow Y$; then $F \circ A_t = B_t \circ F$. If $X, Y \in \mathscr{D}(M)$, $\{A_t\} \Leftrightarrow X$, $\{B_t\} \Leftrightarrow Y$, then $[X, Y] = 0$ if and only if $A_t \circ B_s = B_s \circ A_t$ (locally).

**A vector field dependent on time.** A field $X \in \mathscr{D}(M \times \mathbf{R})$ is said to be time dependent on $M$ if it is compatible with $\pi_2: M \times \mathbf{R} \to \mathbf{R}$ and $\pi_2(X) = \frac{d}{dt}$. A family of fields $X_t \in \mathscr{D}(M)$ is defined by the formula $X_t = g_t^* \circ X \circ \pi_1^*$, where $\pi_1: M \times \mathbf{R} \to M$, and $g_t: M \to M \times \mathbf{R}$, $g_t(x) = (x, t)$. If $\{A_t\} \Leftrightarrow X$, then the family of diffeomorphisms $G_t: M \to M$, $G_t = g_t^{-1} \circ A_t \circ g_0$, is a shift operator on $t$ along the family of fields $X_t$. Here $X_t = (G_t^*)^{-1}(\frac{d}{dt} G_t^*)$. Conversely, for a smooth family of diffeomorphisms $G_t$ this formula defines a time-dependent field $X_t$, for which the shift operator along its trajectories is $G_t$.

***p*-covectors, the cotangent space, differential forms.** A $p$-covector $\omega$ at a point $a \in M$ is a $p$-polylinear function on $T_a(M)$. The totality of 1-covectors, or simply covectors, at $a \in M$, forms an $n$-dimensional cotangent space $T_a^*(M)$. By analogy with $T(M)$, the cotangent, or phase, space $T^*(M) = \Phi = \bigcup_{a\in M} T_a^*(M)$ is a smooth manifold with special local coordinates $(\pi^*(x_1), \ldots, \pi^*(x_n), p_1, \ldots, p_n)$; here $\pi: \Phi \to M$, $\pi(T_a^*(M)) = a$, is the natural projection, and the point $\tilde{\omega}$ with the above local coordinates corresponds to the covector $\omega = \sum_{i=1}^{n} p_i dx_i$ at the point $a \in M$ with the local coordinates $(x_1, \ldots, x_n)$.

The totality of $p$-covectors at a point $a$ forms a $\binom{n}{p}$-dimensional linear space $\Lambda_a^p(M)$. If $F: M \to N$ is a smooth map, $\omega \in \Lambda_{F(a)}^p(N)$, then the $p$-covector $F_a^*(\omega) \in \Lambda_a^p(M)$ is defined by the equation $F_a^*(\omega)(\xi_1, \ldots, \xi_p) = \omega(F_a(\xi_1), \ldots, F_a(\xi_p))$, $\xi_1, \ldots, \xi_p \in T_a(M)$.

A differential $p$-form, or $p$-covector field $\omega$, is a smooth function on $\mathscr{D}(M)$ that is skew $p$-polylinear over $\mathscr{F}(M)$. It properly determines a smooth field of $p$-covectors $\{\omega_a\}$, $a \in M$, by the equation $\omega_a(\xi_1, \ldots, \xi_p) = (\omega(X_1, \ldots, X_p))(a)$, where $X_i|_a = \xi_i \in T_a(M)$. Conversely, a smooth field of $p$-covectors $\omega_a$, $a \in M$, defines a differential $p$-form $\omega$ by the same formula. If $F: M \to N$ is a smooth map, $\omega \in \Lambda^p(N)$, then the $p$-form $F^*(\omega) \in \Lambda^p(M)$ is determined by the equation

$F^*(\omega)\,(X_1, \ldots, X_p)\,(a) = \omega_{F(a)}(F_a(X_1), \ldots, F_a(X_p))$, $X_i \in \mathscr{D}(M)$. The most important example of a 1-form is the differential of a function: if $f \in \mathscr{F}(M)$, then $df \in \Lambda^1(M)$: $(df)\,(X) = X(f)$, $X \in \mathscr{D}(M)$.

**The operator of outer differentiation (the differential) $d: \Lambda(M) \to \Lambda(M)$.** This operator is uniquely determined by the properties: 1) **R**-linearity; 2) $d^2 = 0$; 3) $d(\omega_1 \wedge \omega_2) = d\omega_1 \wedge \omega_2 + (-1)^p\, \omega_1 \wedge d\omega_2, \omega_1 \in \Lambda^p(M)$; 4) if $f \in \mathscr{F}(M) = \Lambda^0(M)$, then $(df)\,(X) = X(f)$. If $\omega \in \Lambda^p(M)$, $p > 0$, and $X_1, \ldots, X_{p+1} \in \mathscr{D}(M)$, then

$$(0.1)\quad (d\omega)\,(X_1, \ldots, X_{p+1}) = \sum_{i=1}^{p+1} (-1)^{i+1} X_i\,(\omega\,(X_1, \ldots, \hat{X}_i, \ldots, X_{p+1})) + \sum_{i<j} (-1)^{i+j}\,\omega\,([X_i, X_j], X_1, \ldots, \hat{X}_i, \ldots, \hat{X}_j, \ldots, X_{p+1}),$$

where the symbol "$\wedge$" signifies the omission of the corresponding argument. The operator $d$ is natural: if $F: M \to N$ is a smooth map, then $d_M \circ F^* = F^* \circ d_N$.

A form $\omega \in \Lambda^p(M)$ is said to be closed if $d\omega = 0$ and exact if $\omega = dv$, $v \in \Lambda^{p-1}(M)$. Locally, every closed form is exact (Poincaré's lemma).

**Lie derivatives.** The shift operators $\{A_t\}$ allow us to define the Lie derivative of objects of a different nature. In particular, if $Y \in \mathscr{D}(M)$, then

$$(0.2)\qquad X\,(Y) = \lim_{t\to 0} \frac{A_t\,(Y) - Y}{t} = \frac{d}{dt} A_t\,(Y)\,|_{t=0} = [Y, X];$$

and if $\omega \in \Lambda^p(M)$, then

$$(0.3)\quad X\,(\omega) = \lim_{t\to 0} \frac{A_t^*\,(\omega) - \omega}{t} = \frac{d}{dt} A_t^*\,(\omega)\,|_{t=0} = X \,\lrcorner\, d\omega + d\,(X \,\lrcorner\, \omega),$$

where the symbol "$\lrcorner$" denotes the inner product: $(X \,\lrcorner\, \omega)\,(X_2, \ldots, X_p) = \omega(X_1, X_2, \ldots, X_p)$. We note that $X \circ d = d \circ X$ and $X(\omega_1 \wedge \omega_2) = X(\omega_1) \wedge \omega_2 + \omega_1 \wedge X(\omega_2)$.

LEMMA 0.1. $X(Y) = 0$ *or* $X(\omega) = 0$ *if and only if* $A_t(Y) = Y$ *or* $A_t^*(\omega) = \omega$.

**Submanifolds. The implicit function theorem.** Suppose that $x_0 \in M$, and that the functions $f_1, \ldots, f_k$ in $\mathscr{F}(M)$, regarded as maps $f_i$: $M \to \mathbf{R}$, are such that $df_1|_{x_0}, \ldots, df_k|_{x_k}$ are linearly independent. Then in some neighbourhood $U$ of $x_0$, the subset $N$ defined by $N = \{x \in U | f_i(x) = f_i(x_0),\ i = 1, \ldots, k\}$ is a closed submanifold of $U$. Conversely, any closed submanifold of $M$ can be obtained in this way. Let $F$: $M \to N$ be a smooth map, $\dim M \leqslant \dim N$, $\operatorname{Ker} F_x = 0 \quad \forall x \in M$ and $F(x_1) \neq F(x_2)$ if $x_1 \neq x_2$. If $F(M)$ is a closed or open submanifold of $N$, then $F$ is called an embedding; in general, $F(M)$ is called an embedded submanifold.

## §1. The universal 1-form. Lagrangian manifolds

1.1. So as to geometrize the concept of a 1-form, we define its graph. By the graph of a 1-form $\omega$ on $M$ we mean the map $S_\omega\colon M \to \Phi$, $S_\omega(a) = \omega_a \in T_a^*(M)$. Here $S_\omega$ is a smooth section of the vector foliation $\pi\colon \Phi \to M$. (We recall that $\Phi = T^*(M)$ is the phase space.) Sometimes the submanifold $S_\omega(M) \subset \Phi$ is called the graph of $\omega$.

1.2. From the technical point of view, the basic structural unit of Hamiltonian mechanics is a 1-form $\rho \in \Lambda^1(\Phi)$, uniquely characterized by the following universal property:

$$S_\omega^*(\rho) = \omega \ \forall \omega \in \Lambda^1(M). \tag{1.1}$$

Thus, the study of the set of forms $\omega \in \Lambda^1(M)$ reduces to that of one universal form $\rho$. The form $\rho$ defined by the equation

$$\rho_\nu = (d\pi_\nu)^*(\nu) \tag{1.2}$$

satisfies (1.1). In (1.2) the point $\nu \in \Phi$ is at the same time interpreted as a covector at the point $\pi(\nu) \in M$. We claim that $\rho$ is unique. Let $\rho'$ be another form satisfying (1.1), $\tilde{\rho} = \rho - \rho'$. Then $S_\omega^*(\tilde{\rho}) = 0 \quad \forall \omega \in \Lambda^1(M)$, or $\tilde{\rho}_\nu(\xi) = 0 \quad \forall \xi \in T_\nu(S_\omega(M)) \subset T_\nu(\Phi)$. But the vectors $\xi \in T_a(\Phi)$ tangent to all the graphs $S_\omega(M)$ passing through $\nu$ clearly generate $T_\nu(\Phi)$. Therefore, $\tilde{\rho}_\nu = 0 \quad \forall \nu \in \Phi$, that is, $\rho = \rho'$.

In special local coordinates $(p, x)$ in $\Phi$, (1.2) gives

$$\rho_{(p,x)} = d\pi^*_{(p,x)}\Big(\sum_{i=1}^n p_i dx_i\Big) = \sum_{i=1}^n p_i dx_i, \text{ that is,}$$

$$\rho = p\, dx = \sum_{i=1}^n p_i\, dx_i. \tag{1.3}$$

The reader acquainted with the usual mechanics course will recognize in our universal form $\rho$ the differential of the classical action.

1.3. Let us see what geometrical property distinguishes graphs of closed 1-forms among others. Since the operator $d$ is natural, $d\omega = dS_\omega^*(\rho) = S_\omega^*(d\rho)$. Therefore, if $\omega$ is closed, that is, $d\omega = 0$, then $S_\omega^*(d\rho) = 0$. Denoting the embedding $S_\omega(M) \subset \Phi$ by $i$, we have $i^*(d\rho) = d\rho|_{S_\omega(M)} = 0$. Manifolds of dimension $n$ in the phase space $\Phi$ of dimension $2n$ on which the form $d\rho$ $(= dp \wedge dx)$ vanishes are called *Lagrangian*. They play an important role in mechanics. Thus, we have proved that a manifold $S_\omega(M) \subset \Phi$ is Lagrangian if and only if $d\omega = 0$. Lagrangian manifolds are not exhausted by the graphs of closed forms. For example, the manifold $T_a^*(M) \subset \Phi$ is Lagrangian, because $\rho|_{T_a^*(M)} = 0$ by (1.3). Since $(d\rho)|_L = d(\rho|_L)$, $L$ is Lagrangian if and only if the form $\rho|_L$ is closed.

How do we recover $f$ when $L = S_{df}(M)$ is known? Let $\gamma\colon I \to M$ be a curve, $\bar{\gamma} = S_{df} \circ \gamma$. Since $\bar{\gamma}(I) \subset L$, we have

$$\int_{\gamma(I)} df = \int_{\gamma(I)} S^*_{df}(\rho) = \int_I \gamma^*(S^*_{df}(\rho)) = \int_I \bar{\gamma}^*(\rho) = \int_{\bar{\gamma}(I)} \rho = \int_{\bar{\gamma}(I)} \rho|_L.$$

As $\int_{\gamma(I)} df$ does not depend on the path of integration, nor does $\int_{\bar{\gamma}(I)} \rho|_L$.

Therefore, we can write $\int_{y_0}^{y} \rho|_L$ in place of $\int_{\bar{\gamma}(I)} \rho|_L$ and

$f(x) - f(x_0) = \int_{x_0}^{x} df = \int_{y_0}^{y} \rho|_L$, where $S_{df}(x_0) = y_0$, $S_{df}(x) = y$, or

$\pi(y_0) = x_0$, $\pi(y) = x$. Thus,

$$f(x) = \int_{y_0}^{y} \rho|_L + \text{const}, \quad x = \pi(y). \tag{1.4}$$

If $L$ is an arbitrary Lagrangian manifold, then $\rho|_L$ is closed and $\int_{y_0}^{y} \rho|_L$ is, generally speaking, a many-valued function on $L$. In this case (1.4) determines a many-valued function on $M$, the "graph of the differential" (it makes sense to take this to be $L$). We note that this function is many-valued for two reasons: because $\int_{y_0}^{y} \rho|_L$ is many-valued and because the inverse image $(\pi|_L)^{-1}(x)$ is.

1.4. As the coordinate notation shows, the form $d\rho_\nu$ is a skew-bilinear non-degenerate form on $T_\nu(\Phi)$, $\nu \in \Phi$. We therefore examine the geometry of a skew-bilinear non-degenerate form $\Omega$ on an even-dimensional vector space $V$, $\dim V = 2n$. We remark that every bilinear form $\Omega$ on $V$ defines a homomorphism $\gamma: V \to V^*$ by the formula $[\gamma(\xi)](\eta) = \Omega(\xi, \eta)$ and, conversely, the same formula associates a bilinear form $\Omega$ with a given $\gamma$. The non-degeneracy of $\Omega$ is equivalent to $\gamma$ being an isomorphism. We call two vectors $\xi$ and $\eta$ *skew-orthogonal* (relative to $\Omega$) if $\Omega(\xi, \eta) = 0$. The non-degeneracy of $\Omega$ means that the only vector skew-orthogonal to the whole space is $\{0\}$.

A basis $(\xi_1, \eta_1, \dots, \xi_n, \eta_n)$ of $V$ is said to be *canonical* for $\Omega$ if $\Omega(\xi_i, \xi_j) = \Omega(\eta_i, \eta_j) = 0$, $\Omega(\xi_i, \eta_j) = \delta_{ij}$. As is well known, there always is a canonical basis. Its existence allows us to construct an isomorphism between any two $2n$-dimensional linear spaces $V_1$ and $V_2$ with skew non-degenerate forms $\Omega_1$ and $\Omega_2$ given on them that carries one form into the other. This isomorphism is an operator carrying some canonical basis for $\Omega_1$ into some canonical basis for $\Omega_2$.

We now consider *Lagrangian planes* – maximal subspaces on which $\Omega$

vanishes.

LEMMA 1.1. *If $\Lambda$ is a Lagrangian plane, then* $\dim \Lambda = n$.

Let $\dim \Lambda = k$, $(\xi_1, \ldots, \xi_k)$ be a basis of $\Lambda$, $\Lambda' = \bigcap_{i=1}^{k} \operatorname{Ker} \gamma(\xi_i)$. Then $\Lambda' \supset \Lambda$, for if $\eta \in \Lambda$, then $0 = \Omega(\xi_i, \eta) = \gamma(\xi_i)(\eta)$ $(i = 1, \ldots, k)$. Therefore, $\dim \Lambda' = 2n - k \geqslant k = \dim \Lambda$, hence $k \leqslant n$. If $k < n$, then $\Lambda' \neq \Lambda$ and, taking $\xi \in \Lambda' \setminus \Lambda$, we see that $\Omega|_{l(\Lambda, \xi)} = 0$, where $l(\Delta, \xi)$ is the linear span of the set $\Delta \cup \xi$. But this contradicts the maximality of $\Lambda$. Thus, $k = n$.■

Let $V^{2n-1}$ be a hyperplane in $V$. We consider functionals $\Phi \in V^*$ for which $\operatorname{Ker} \varphi = V^{2n-1}$. The totality of such functionals, together with zero, forms a one-dimensional subspace (that is, a line) Ann $V^{2n-1}$ in $V^*$. Let $A(V^{2n-1}) = \gamma^{-1}(\operatorname{Ann} V^{2n-1})$. If $0 \neq \xi \in A(V^{2n-1})$, then $0 = \Omega(\xi, \xi) = \gamma(\xi)(\xi)$. Hence it follows that $\xi$, and therefore also $A(V^{2n-1})$, belongs to $V^{2n-1}$. We call the line $A(V^{2n-1})$ (or a vector $\xi \in A(V^{2n-1})$) the *direction line* (*vector*) of $V^{2n-1}$. Thus, $A(V^{2n-1})$ is the one-dimensional subspace consisting of all vectors $\xi$ such that $\Omega(\xi, \eta) = 0$ for all $\eta \in V^{2n-1}$. An important fact is the *linear absorption principle*: a Lagrangian plane $\Lambda$ lying in a hyperplane $V^{2n-1}$ contains the direction line of that hyperplane. For otherwise $\Omega$ would vanish on the $(n+1)$-dimensional subspace $l(\Lambda, A(V^{2n-1}))$.

1.5. We shall need the following lemma.

LEMMA 1.2. *Let $L$ be a Lagrangian plane. Then any basis $(\xi_1, \ldots, \xi_n)$ of $L$ can be extended to a canonical basis $(\xi_1, \eta_1, \ldots, \xi_n, \eta_n)$.*

The proof is by induction. For $n = 1$ the assertion is obvious. Let $n > 1$. We consider $V_1 = \bigcap_{i>1} \operatorname{Ker} \gamma(\xi_i)$. Now $V_1 \supset L$ and $\dim V_1 = n + 1$. Therefore, $\Omega(\eta_1, \xi_1) \neq 0$ for any $\eta_1 \in V \setminus L$, otherwise, $l(L, \eta_1)$ would be an $(n+1)$-dimensional Lagrangian plane. We normalize $\eta_1$ so that $\Omega(\xi_1, \eta_1) = 1$. It is not hard to see that $l(\xi_1, \eta_1)$ and $\bar{V} = \operatorname{Ker} \gamma(\xi_1) \cap \operatorname{Ker} \gamma(\eta_1)$ give a decomposition of $V$ into a skew-orthogonal sum: $V = l(\xi_1, \eta_1) \oplus \bar{V}$, and the restriction of $\Omega$ to $\bar{V}$ is non-degenerate. Now $\bar{V}$ contains (with respect to $\Omega|_{\bar{V}}$) the Lagrangian plane $L = l(\xi_2, \ldots, \xi_n)$ with the basis $(\xi_2, \ldots, \xi_n)$, and by the inductive hypothesis it can be extended to a canonical one (with respect to $\Omega|_{V}$) by vectors $\eta_2, \ldots, \eta_n \in \bar{V}$. Clearly, $(\eta_1, \ldots, \eta_n)$ are the required vectors.■

We now consider the question in how many ways a basis $(\xi_1, \ldots, \xi_n)$ of a Lagrangian plane $L$ can be complemented to a canonical basis $(\xi_1, \ldots, \xi_n, \eta_1, \ldots, \eta_n)$ of $V$. Every such complementation gives rise to a decomposition of $V$ into the direct sum of Lagrangian planes $L$ and $L' = l(\eta_1, \ldots, \eta_n)$, $V = L \oplus L'$. We claim that the converse is also true.

LEMMA 1.3. *For given $L$, $\xi_1, \ldots, \xi_n$ and $L'$ satisfying the above conditions, the complementary vectors $\eta_1, \ldots, \eta_n$ can be uniquely recovered.*

The required vector $\eta_j$ must belong to $V_j = \bigcap_{i \neq j} \operatorname{Ker} \gamma(\xi_i)$ and $V_j \cap L'$; now $V_j \supset L$ and $\dim V_j = n + 1$. Since $L$ and $L'$ form a direct sum, $\dim (V_j \cap L') = 1$. If $0 \neq \eta_j' \in V_j \cap L'$, then $\Omega(\xi_j, \eta_j') \neq 0$, otherwise

$\Omega|_{V_j} = 0$, contradicting the fact that dim $V_j = n + 1$. Setting $\eta_j = \eta_j'/\Omega(\xi_j, \eta_j')$, we obtain the required basis. Its uniqueness follows from the fact that $\Omega(\xi_j, \eta) = 1$ for only one vector $\eta$ on the line $V_j \cap L'$.■

LEMMA 1.4. *The space of Lagrangian planes complementary to a given Lagrangian plane L is homotopically trivial (= contractible).*

(For facts about homotopy theory, see [5]).

Let $L'$ be a Lagrangian plane complementing $L$, $(\xi_1, \ldots, \xi_n)$ a basis of $L$, $(\eta_1, \ldots, \eta_n)$ a basis of $L'$ such that $\Omega(\xi_i, \eta_j) = \delta_{ij}$ (Lemma 1.3), and $(\alpha_1, \ldots, \alpha_n, \beta_1, \ldots, \beta_n)$ coordinates with respect to the basis $(\xi_1, \ldots, \xi_n, \eta_1, \ldots, \eta_n)$. Then every complementary plane $L''$ is given by a system of equations $\alpha_i = \sum_{k=1}^{n} a_i^k \beta_k$ $(i = 1, \ldots, n)$, where the matrix $\| a_i^k \|$ is uniquely determined by $L''$. A straightforward verification shows that $L''$ is Lagrangian if and only if $\| a_i^k \|$ is symmetric. Thus, the space of complementing Lagrangian planes is homeomorphic to the space of symmetric matrices of order $n$, which is contractible.■

## §2. Fields and 1-forms on $\Phi$

2.1. Having studied the geometry of the localization of $d\rho$ at some point $\nu \in \Phi$, we now consider the situation globally. A global analogue to $\gamma$ is the map $\Gamma: \mathscr{D}(\Phi) \to \Lambda^1(\Phi)$ defined by the formula $\Gamma(X) = d\rho(\ , X) = -X \lrcorner\, d\rho$. This $\Gamma$ is clearly $\mathscr{F}(\Phi)$-linear, and its localization is $(-\gamma)$, since $[\Gamma(X)(Y)](\nu) = d\rho(Y, X)(\nu) = d\rho_\nu(Y_\nu, X_\nu) = -\gamma_\nu(X_\nu)(Y_\nu)$, $\nu \in \Phi$, that is, $\Gamma(X)_\nu = -\gamma_\nu(X_\nu)$. Since $\gamma$ is an isomorphism at every point $\nu \in \Phi$, $\Gamma$ is also an isomorphism.

We remark that if $U$ is a domain in $\Phi$, then in the same way we can define an isomorphism $\Gamma_U: \mathscr{D}(U) \to \Lambda^1(U)$.

In special coordinates $(p, q)$ on $\Phi$ (earlier written $(p, x)$ we have $d\rho = dp \wedge dq$, therefore,

$$\Gamma\left(\alpha \frac{\partial}{\partial q} + \beta \frac{\partial}{\partial p}\right) = \alpha\, dp - \beta\, dq. \tag{2.1}$$

**2.2. Canonical manifolds and canonical coordinates.** We have seen that there is a *canonical structure* on the manifold $\Phi = T^*(M)$, that is, a non-degenerate closed 2-form $d\rho$. Similarly, there is such a structure on any domain $U \subset \Phi$. However, there also occur other situations where it is necessary to study a closed non-degenerate 2-form $\Omega$ on a manifold $K$ (necessarily even-dimensional owing to the skew-symmetry and non-degeneracy of $\Omega$) that need not be the phase space of any manifold. In this context we introduce the concept of a canonical manifold, so that some of our subsequent constructions and results can be stated in reasonable generality, and also so that it is clear what part of the principles of Hamiltonian mechanics derives solely from the presence of a canonical structure on $\Phi$, and what from more subtle circumstances.

By a *canonical structure* on a $2n$-dimensional manifold $K$ we mean a non-degenerate closed 2-form $\Omega$ on $K$, and by a *canonical manifold* the pair $(K, \Omega)$.

By *canonical coordinates* on $K$ we mean (local) coordinates $(p,q)$ in which $\Omega$ can be written in the form $\Omega = dp \wedge dq$. As we shall prove in §7, such coordinates exist in a neighbourhood of any point $a \in K$. An example of canonical coordinates are special coordinates on $\Phi$.

Just as in the case of a phase space, we can introduce the concept of a Lagrangian manifold $L$ on an arbitrary canonical manifold $K$. Namely, $L$ is called Lagrangian if $\Omega|_L = 0$ and $L$ has maximal dimension, namely, $n$. We can also define a map $\Gamma$ by the formula $\Gamma(X) = \Omega(\ ,X)$. From the non-degeneracy of $\Omega$ it follows, as before, that $\Gamma$ is an isomorphism. We note that $\Gamma$ is also given in canonical coordinates by (2.1).

We shall make use of the isomorphism $\Gamma$ and call a vector field $X$ *Hamiltonian* if the 1-form $\Gamma(X)$ is the differential of some function $H \in \mathscr{F}(K)$. In classical mechanics, every function on the phase space is called Hamiltonian. We extend this term to apply also in the case of arbitrary canonical manifolds. We write the fact that $H$ is a Hamiltonian of a field $X$ in this way: $\Gamma(X) = dH$, or $H \Leftrightarrow X$, or $X = X_H$.

If a field $X$ is Hamiltonian and in the canonical coordinate system $(p, q)$ we have $X = \alpha \dfrac{\partial}{\partial q} + \beta \dfrac{\partial}{\partial p}$, then, as follows from (2.1), $\beta_i = -H_{q_i}$, $\alpha_i = H_{p_i}$. The differential equations corresponding to the Hamiltonian field $X$ are

$$\dot{q} = H_p, \quad \dot{p} = -H_q. \tag{2.2}$$

These then are the famous Hamilton equations corresponding to a Hamiltonian function $H$.

2.3. By a *canonical transformation* in $K$ we mean a diffeomorphism $A: K \to K$ preserving $\Omega$, that is, $A^*(\Omega) = \Omega$. This also means that $\Omega(X, Y) = (A^*(\Omega))(X, Y) = \Omega(A(X), A(Y))$, $\forall X, Y \in \mathscr{D}(K)$. If $L \subset K$, then $A^*(\omega)|_L = A^*\{\omega|_{A(L)}\}$ for any diffeomorphism $A: K \to K$ and any form $\omega \in \Lambda^p(K)$. If $A^*(\omega) = \omega$, then $\omega|_L = A^*\{\omega|_{A(L)}\}$. Hence a canonical transformation carries Lagrangian manifolds into Lagrangian manifolds. If $\{A_t\}$ is a one-parameter group of canonical transformations, $\{A_t\} \Leftrightarrow X$, then $X(\Omega) = \lim\limits_{t\to 0} \dfrac{A_t^*(\Omega)-\Omega}{t} = 0$. In this connection, a vector field $X$ for which $X(\Omega) = 0$ is called an *infinitesimal canonical transformation.* The one-parameter group $\{A_t\}$ corresponding to an infinitesimal canonical transformation consists of canonical transformations (Lemma 0.1).

2.4. If the form $\omega \in \Lambda^1(K)$ is closed, but not exact, then it is only locally the differential of some function $f \in \mathscr{F}(K)$. Therefore, it is natural to call vector fields corresponding to closed forms under the isomorphism $\Gamma$ *locally Hamiltonian.* For such fields the Hamiltonian exists only locally. We remark that if $K$ is simply-connected (in the case $\Phi = T^*(M)$ this is equivalent to $M$ being simply-connected), then locally Hamiltonian fields are Hamiltonian. The connection between locally Hamiltonian fields and infinitesimal canonical transformations is described by the so-called fundamental theorem of mechanics.

FUNDAMENTAL THEOREM OF MECHANICS. *Locally Hamiltonian vector fields are infinitesimal canonical transformations, and vice versa.*

We have $X(\Omega) = d(X \lrcorner\, \Omega) + X \lrcorner\, d\Omega = -d\Gamma(X)$. Therefore, if $X$ is locally Hamiltonian, that is, $d\Gamma(X) = 0$, then $X(\Omega) = 0$; and if $X$ is an infinitesimal canonical transformation, that is, $X(\Omega) = 0$, then $d\Gamma(X) = 0$.■

Since the form $\Omega_a$ on $T_a(K)$ is non-degenerate, $\underbrace{\Omega_a \wedge \ldots \wedge \Omega_a}_{n \text{ times}} \neq 0$ (this can be checked by writing $\Omega_a$ in some canonical basis). Therefore, $\bar{\Omega} = \underbrace{\Omega \wedge \ldots \wedge \Omega}_{n \text{ times}}$ is a volume form on $K$. If $X$ is an infinitesimal canonical transformation, $\{A_t\} \Leftrightarrow X$, then $X(\bar{\Omega}) = 0$, so that $A_t^*(\bar{\Omega}) = \bar{\Omega}$. In particular, by virtue of the fundamental theorem of mechanics we have the following result.

LIOUVILLE'S THEOREM. *Shift operators along trajectories of a Hamiltonian vector field preserve the ("phase") volume, that is, the form* $\bar{\Omega}$.

**2.5. The form $\rho$ is natural.** Let $A: M_1 \to M_2$ be a diffeomorphism. We construct a map $\bar{A}: T^*(M_1) \to T^*(M_2)$ by the rule $\bar{A}_{|T_x^*(M_1)} = (A_x^*)^{-1}$. Now $\bar{A}$ is a foliated smooth map, with a smooth inverse $\overline{A^{-1}}$, therefore, $\bar{A}$ is a foliated diffeomorphism: $\pi_2 \circ \bar{A} = A \circ \pi_1$. Let $\omega_2 \in \Lambda^1(M_2)$, $\omega_1 = A^*(\omega_2)$, and $S_{\omega_i}$ be sections corresponding to the forms $\omega_i$, $i = 1, 2$. Then $\forall\, x \in M_1$, $\bar{A}(S_{\omega_i}(x)) =$ $= (A_x^*)^{-1}(\omega_1|_x) = (A_x^*)^{-1}(A^*(\omega_2|_{A(x)})) = \omega_2|_{A(x)} = S_{\omega_2}(A(x))$, that is, $\bar{A} \circ S_{\omega_1} = S_{\omega_2} \circ A$.

COROLLARY 2.1. *Let $\rho_i$ be the universal form on $\Phi_i$ $(i = 1, 2)$. Then $\bar{A}^*(\rho_2) = \rho_1$, that is, the universal form $\rho$ is natural.*

Let $\omega_1 \in \Lambda^1(M_1)$, $\omega_2 = (A^*)^{-1}(\omega_1)$; then $S^*_{\omega_1}(\bar{A}^*(\rho_2)) = (\bar{A} \circ S_{\omega_1})^*(\rho_2) = (S_{\omega_2} \circ A)^*(\rho_2) =$ $= A^*(S_{\omega_2}(\rho_2)) = A^*(\omega_2) = \omega_1$, that is, $S^*_{\omega_1}(A^*(\rho_2)) = \omega_1$ $\forall\, \omega_1 \in \Lambda^1(M_1)$. Since the universal form $\rho_1$ is unique, $\bar{A}^*(\rho_2) = \rho_1$.■

COROLLARY 2.2. *If $A: M \to M$ is a diffeomorphism, then $\bar{A}^*(\rho) = \rho$.*

We now turn to the infinitesimal version of the preceding arguments. Let $X \in \mathscr{D}(M)$, $\{A_t\} \Leftrightarrow X$, $\{\bar{A}_t\} \Leftrightarrow \hat{X}$. Since $A_t^*(\rho) = \rho$, we have $\hat{X}(\rho) = 0$, and from $\pi \circ \bar{A}_t = A_t \circ \pi$ it follows that $\pi(X) = X$: $\hat{X} \circ \pi^* =$ $= (d/dt)\, \bar{A}_t^*|_{t=0} \circ \pi^* = (d/dt)\, (\bar{A}_t^* \circ \pi^*)|_{t=0} = (d/dt)(\pi^* \circ A_t^*)|_{t=0} = \pi^*(d/dt)\, A_t^*|_{t=0} = \pi^* \circ X$. Since $\Gamma(\hat{X}) = d(\rho(\hat{X})) - \hat{X}(\rho) = d(\rho(\hat{X}))$, $\hat{X}$ is a Hamiltonian field with the Hamiltonian $H = \rho(\hat{X})$. Let us compute $\rho(X)$ at $\nu \in \Phi$. Since $\rho_\nu = \pi_\nu^*(\nu)$, we see that $\rho(\hat{X})(\nu) = \rho_\nu(\hat{X}_\nu) = \pi_\nu^*(\nu)(\hat{X}_\nu) = \nu(\pi_\nu(\hat{X}_\nu)) = \nu(X_{\pi(\nu)})$. Thus

$$\rho(\hat{X})(\nu) = \nu(X_{\pi(\nu)}). \tag{2.3}$$

We mention without proof a fact that we do not use later: let $B: \Phi \to \Phi$ $(Y \in \mathscr{D}(\Phi))$ be a diffeomorphism such that $B^*(\rho) = \rho$ $(Y(\rho) = 0$; then $B(Y)$ has the form $\bar{A}(\hat{X})$ for some $A(X)$.

**2.6. Transformation properties of canonical manifolds.** Let $(K_i, \Omega_i)$, $i = 1, 2$, be canonical manifolds, and $A: K_1 \to K_2$ a canonical diffeomorphism, that is, $A^*(\Omega_2) = \Omega_1$.

LEMMA 2.1. $\Gamma_2 \circ A = (A^{-1})^* \Gamma_1$.

$\forall X, Y \in \mathscr{D}(K_1)$, $\Gamma_1(X)(Y) = \Omega_1(Y, X) = A^*(\Omega_2(A(Y), A(X))) = A^*[\Gamma_2(A(X)), (A(Y))] = [A^*(\Gamma_2(A(X)))]\ (Y)$, that is, $\Gamma_1(X) = A^*[\Gamma_2(A(X))]$.■

COROLLARY 2.3. *Let $h \in \mathscr{F}(K_2)$ and $g = A^*(h)$. Then $A(X_g) = X_h$.*

Applying Lemma 2.1 to the field $X = X_g$, we find that $\Gamma_2(A(X_g)) = (A^{-1})^*\,[\Gamma_1(X_g)] = (A^{-1})^*\,(dg) = dh$, hence, $A(X_g) = \Gamma_2^{-1}(dh) = X_h$.■

COROLLARY 2.4. *Let $A$: $K \to K$ be a canonical transformation. Then $\Gamma(A(X)) = (A^{-1})^*(\Gamma(X))$ $\forall X \in \mathscr{D}(K)$.*

Let $Z \in \mathscr{D}(K)$. We denote by $Z_{\mathscr{D}}$: $\mathscr{D}(K) \to \mathscr{D}(K)$ and $Z_\Lambda$: $\Lambda^1(K) \to \Lambda^1(K)$ the corresponding Lie derivatives.

LEMMA 2.2. *Let $Z$ be an infinitesimal canonical transformation. Then $\Gamma \circ Z_{\mathscr{D}} + Z_\Lambda \circ \Gamma = 0$.*

The transformations $\{G_t\}$, $\{G_t\} \iff Z$ are canonical, and by Corollary 2.4, $\Gamma(G_t(X)) = (G_g^{-1})^*\,(\Gamma(X))$, $\forall X \in \mathscr{D}(K)$. Hence $[\Gamma(G_t(X)) - \Gamma(X)]/t = \Gamma[(G_t - E)/t(X)] = [G^*_{-t}(\Gamma(X)) - \Gamma(X)]/t = (G^*_{-t} - E)/t(\Gamma(X))$. As $t \to 0$, we obtain $\Gamma(Z(X)) = -Z(\Gamma(X))$.■

## §3. Poisson brackets

3.1. Let $X_1$ and $X_2$ be two Hamiltonian fields on $K$ with Hamiltonians $H_1$ and $H_2$, respectively. Applying Lemma 2.2 for $Z = X_1$ to $X_2$, we obtain $0 = \Gamma((X_1)(X_2)) + X_1(\Gamma(X_2)) = \Gamma([X_2, X_1]) + X_1(dH_2)$, that is, $\Gamma([X_1, X_2]) = d(X_1(H_2))$. Thus, the commutator of two Hamiltonian fields $X_1$ and $X_2$ is another Hamiltonian field, with Hamiltonian $X_1(H_2) = \{H_1, H_2\}$, which is called the *Poisson bracket* of $H_1$ and $H_2$. By definition, $\{H_1, H_2\} = X_1(H_2)$ is the derivative of $H_2$ along the field $X_1 = X_{H_1}$. We note that $\{H_1, H_2\} = dH_2(X_1) = \Gamma(X_2)(X_1) = \Omega(X_1, X_2)$. Hence, the Poisson bracket is bilinear over $\mathbf{R}$ and skew-symmetric. Furthermore, the Jacobi identity is satisfied: $\forall H_1, H_2, H_3 \in \mathscr{F}(K)$,

$$\{H_1, \{H_2, H_3\}\} + \{H_2, \{H_3, H_1\}\} + \{H_3, \{H_1, H_2\}\} = 0, \tag{3.1}$$

because the left-hand side of (3.1) is $\{H_1, X_2(H_3)\} - \{H_2, X_1(H_3)\} - [X_1, X_2](H_3) = X_1(X_2(H_3)) - X_2(X_1(H_3)) - [X_1, X_2](H_3) = 0$. Thus, the Poisson bracket makes $\mathscr{F}(K)$ into a Lie algebra, which is denoted by $\mathscr{D}_{\mathscr{P}}$. If $\mathscr{D}_{\mathscr{H}}$ is the Lie algebra of Hamiltonian fields on $K$, we come to a homomorphism of Lie algebras $\mathscr{L} = \Gamma^{-1} \circ d$: $\mathscr{D}_{\mathscr{P}} \to \mathscr{D}_{\mathscr{H}}$.

The following formulae are obtained straight from the definition:

$$\{H_1, H_2H_3\} = \{H_1, H_2\}H_3 + \{H_1, H_3\}\,H_2. \tag{3.2}$$

If $\{H, H_i\} = 0$ $(i = 1, \ldots, k)$, $\varphi \in \mathscr{F}(\mathbf{R}^k)$, then

$$\{H, \varphi(H_1, \ldots, H_k)\} = 0. \tag{3.3}$$

If $(p, q)$ are canonical coordinates, then

$$(3.4)\qquad \{H_1, H_2\} = \sum_{i=1}^{n} \left( \frac{\partial H_1}{\partial p_i} \frac{\partial H_2}{\partial q_i} - \frac{\partial H_1}{\partial q_i} \frac{\partial H_2}{\partial p_i} \right) = \frac{\partial H_1}{\partial p} \frac{\partial H_2}{\partial q} - \frac{\partial H_1}{\partial q} \frac{\partial H_2}{\partial p}.$$

**3.2.** Let $A\colon K_1 \to K_2$ be a canonical diffeomorphism, and let $\{,\}_i$ denote the Poisson bracket with respect to the form $\Omega_i$ $(i = 1, 2)$.

LEMMA 3.1. *If $f, g \in \mathcal{F}(K_2)$, then $A^*(\{f, g\}_2) = \{A^*(f), A^*(g)\}_1$.*

Let $\bar{f} = A^*(f)$, $\bar{g} = A^*(g)$. Then $A(X_{\bar{f}}) = X_f$, $A(X_{\bar{g}}) = X_g$ (Corollary 2.3). Therefore $A^*(\ f, g\ _2) = A^*(\Omega_2(X_f, X_g)) = \Omega_1(A^{-1}(X_f), A^{-1}(X_g)) = \Omega_1(X_{\bar{f}}, X_{\bar{g}}) = \{\bar{f}, \bar{g}\}_1$. ■

LEMMA 3.2. *Let $\{H, f\} = \alpha = \mathrm{const}$, $\{A_t\} \Leftrightarrow X_H$. Then $A_t^*(f) = f + \alpha t$.*

$\{H, f\} = X_H(f) = \alpha = \mathrm{const}$. For details, see §0.■

**3.3. Vertical fields on $\Phi$.** The following useful facts follow from (2.1).

LEMMA 3.3. *Let $\omega \in \Lambda^1(M)$, $X = \Gamma^{-1}(\pi^*(\omega))$. Then $X$ is vertical, that is, is tangent to the fibres of the projection $\pi\colon \Phi \to M$, and is constant along these fibres. In particular, if $f \in F(M)$, then the field $X_{\pi^*(f)}$ is vertical.*

COROLLARY 3.1. *If $f, g \in F(M)$, then $\{\pi^*(f), \pi^*(g)\} = 0$.*

COROLLARY 3.2. *Let $\omega, \nu \in \Lambda^1(M)$, $\{A_t\} \Leftrightarrow \Gamma^{-1}(\pi^*(\omega))$. Then $A_t \circ S_\nu = S_{\nu - t\omega}$.*

## §4. Symmetries

**4.1.** Let $X_H$ be a Hamiltonian field on $K$. It is natural to call a canonical transformation $A\colon K \to K$ a *symmetry of the Hamiltonian system* with Hamiltonian $H$ if it preserves the field $X_H$: $A(X_H) = X_H$. By Corollary 2.3, $A(X_H) \Leftrightarrow (A^{-1})^*(H)$. Since the Hamiltonian of a Hamiltonian system is uniquely determined up to a constant, $H = (A^{-1})^*(H) + \mathrm{const}$, or $A^*(H) = H + \mathrm{const}$. Conversely, if $A^*(H) = H + \mathrm{const}$, then $A(X_H) = X_H$.

It is natural to define a symmetry of a Hamiltonian $H$ as a canonical transformation $A$ such that $A^*(H) = H$. Clearly a symmetry of the Hamiltonian $H$ is also a symmetry of the system $X_H$, but the converse is false.

EXAMPLE 4.1. A particle in a gravitational field. $K = T^*(\mathbf{R})$, $\Omega = dp$, $H = \frac{p^2}{2} + q$, where $q$ is a coordinate on $\mathbf{R}$. If $A(q, p) = (q + a, p)$, $a \in \mathbf{R}$, then $A^*(q) = q + a$, $A^*(p) = p$, $A^*(\Omega) = \Omega$, $A^*(H) = H + a$. For $a \neq 0$, $A$ is a symmetry of the system $X_H$, but not a symmetry of the Hamiltonian $H$.

Thus, there may be fewer symmetries of the Hamiltonian $H$ than of the system $X_H$. However, for "the majority" of Hamiltonians the totality of their symmetries is the same as that of symmetries of the corresponding Hamiltonian system. For if $A^*(H) = H + a$, $0 \neq a \in \mathbf{R}$, then the level surfaces $\{x \mid H(x) = c\}$ and $\{x \mid H(x) = c + a\}$ must be "identical", since they are carried into each other by $A$, that is, the family of level surfaces of the function $H$ is periodic with period $a$. Consequently, if this family is not periodic for any $a \in \mathbf{R}$, then the symmetries of $H$ and $X_H$ are the same.

EXAMPLE 4.2. The harmonic oscillator. Let $K = T^*(\mathbf{R})$, $H = (p^2 + q^2)/2$. That the level surfaces of $H$ are non-periodic follows from the fact that the surfaces $\{H = c\}$ are empty sets if $c < 0$ and non-empty if $c > 0$.

In connection with this example we remark that if the domain of definition of $H$ is not **R**, then the family of level surfaces of this function is necessarily non-periodic.

In the physical literature it is usual to consider symmetries of $H$, although from the point of view of the theory of systems of canonical equations it is more natural to consider symmetries of $X_H$. Therefore, henceforth we call symmetries of $X_H$ *non-standard* symmetries of $H$, and the usual symmetries of $H$ *standard*.

**4.2.** We now consider the infinitesimal version of the preceding arguments. Let $\{A_t\}$ be a one-parameter group of symmetries of $X_H$ and $\{A_t\} \Leftrightarrow Y$. Then 1) $Y(\Omega) = 0$; 2) $Y(X_H) = [Y, X_H] = 0$. It is natural to call a field $Y$ satisfying 1) and 2) an *infinitesimal symmetry* of $X_H$. In other words, an infinitesimal symmetry is an infinitesimal canonical transformation that commutes with $X_H$. There is a natural connection between infinitesimal and finite symmetries.

LEMMA 4.1. *Let $Y$ be an infinitesimal symmetry of $X_H$ and $\{A_t\} \Leftrightarrow Y$. Then $A_t$ is a non-standard symmetry of $H$.*

Since $Y(\Omega) = 0$ and $Y(X_H) = 0$, we have $A_t^*(\Omega) = \Omega$ and $A_t(X_H) = X_H$. By the fundamental theorem of mechanics, the field $Y$ in question is locally Hamiltonian. Let $h$ be its local Hamiltonian; then $0 = \Gamma([X_H, Y]) = d\{H, h\}$, hence $\{h, H\} = \alpha = \text{const}$. Let us explain the meaning of this constant. Since $A_t$ is a non-standard symmetry, $A_t^*(H) - H = c(t)$ is also constant. Since $\{A_t\}$ is a group, $c(t+s) + H = A_{t+s}^*(H) = A_t^*(A_s^*(H)) = A_t^*(c(s) + H) = c(t) + c(s) + H$, that is, $c(t+s) = c(t) + c(s)$. Since $c(t)$ is smooth, $c(t) = c_0 t$. Hence the symmetries occurring in $\{A_t\}$ for $t \neq 0$ are simultaneously either all standard (if $c_0 = 0$) or all non-standard (if $c_0 \neq 0$). Further, $\{h, H\} = Y(H) = = (d/dt)\,(A_t^*(H))|_{t=0} = c_0$. Thus, $\alpha = c_0$. Therefore, if $\{h, H\} = 0$ (for any local Hamiltonian $h$ of $Y$), then $Y$ should be called a standard infinitesimal symmetry of $H$, and non-standard if $\{h, H\} = \text{const} \neq 0$.

REMARK 4.1. If there is a point $x \in K$ where $dH_x = 0$, then $H$ has no non-standard symmetries, since $\{h, H\}\,(x) = dH_x(Y_h|_x) = 0 \;\; \forall h \in \mathcal{F}(K)$.

We consider the infinitesimal analogue to the example with a particle in a gravitational field. Let $Y = \dfrac{\partial}{\partial q}$, then $A_t(q, p) = (q + t, p)$, $h = p$, $\{h, H\} = 1$, $A_t^*(H) = H + t$, that is, $Y$ is a non-standard infinitesimal symmetry.

We remark that if there is a non-standard infinitesimal symmetry $Y$ and $Y$ is full, then all the level surfaces of $H$ are diffeomorphic, so that $A_t(\{H = a\}) = \{H = a + ct\}$, $c \neq 0$. But if $Y$ is a standard infinitesimal symmetry, then $Y$ is tangent to the level surfaces of $H$, so that $Y(H) = 0$.

In what follows we shall often omit the word "infinitesimal" in front of "symmetry", and we call a function $f \in \mathcal{F}(K)$ with $\{f, H\} = \text{const}$ a symmetry (standard or not) of the Hamiltonian $H$.

4.3. Having discussed symmetries of Hamiltonian systems, we now consider the general case of a canonical manifold. Here it is appropriate to introduce one particular type of symmetry that arises on the phase space $T^*(M)$. Let $H \in \mathscr{F}(\Phi)$, $X \in \mathscr{D}(M)$, $\hat{X} \in \mathscr{D}(\Phi)$. Since $\rho(\hat{X})$ is the Hamiltonian of $X$, we have the following lemma.

LEMMA 4.2. *If* $\hat{X}(H) = 0$, *then* $\{\rho(\hat{X}), H\} = 0$.

It is precisely symmetries of this special type that are, as a rule, utilized by physicists. Their popularity is explained by the fact that they are "visible", since they are directly related to the configuration space $M$ (see §10.7). The other, "hidden", symmetries, are not immediately obvious, and sometimes it is not at all easy to guess whether they exist (see §10.6).

4.4. Every domain $U$ of a canonical manifold $K$ can itself be regarded as a canonical manifold on which the canonical structure is given by the form $\Omega_{|U}$. Furthermore, with every Hamiltonian $H \in \mathscr{F}(K)$ we can associate its $U$-part $H_U = H_{|U}$, which can be regarded as a Hamiltonian on $U$. Here the restriction of $X_H$ to $U$ is, clearly, a Hamiltonian field on $U$ with Hamiltonian $H_U$. Thus, we can speak of the $U$-part of $H$ or $X_H$ and, consequently, of symmetries of all the above types for these parts. We call such symmetries partial. It is clear that the restriction to $U$ of every symmetry of one type or another is a partial symmetry. The converse is false, because the whole may be less symmetrical than a part. In particular, the Hamiltonian may fail to possess non-standard (infinitesimal) symmetries, while some $U$-part of it does have them.

EXAMPLE 4.3. The harmonic oscillator. As we have seen above, the harmonic oscillator has no non-standard symmetries. With the same notation, we take for $U$ the phase plane without the origin of coordinates, that is, $U = \mathbf{R}^2 \setminus \{0\}$. On $U$ there is defined the many-valued function Arctan $(p/q)$ whose differential is single-valued, because the separate branches of Arctan $(p/q)$ differ from one another by a constant $\pi k$: $d(\text{Arctan}\,(p/q)) = \dfrac{p\,dq - q\,dp}{p^2 + q^2} = \lambda$. The form $\lambda$ is closed, but not exact on $U$, so that the field

$X = \Gamma^{-1}(\lambda) = \dfrac{q}{q^2+q^2}\dfrac{\partial}{\partial q} + \dfrac{p}{q^2+q^2}\dfrac{\partial}{\partial p}$ is locally Hamiltonian. Since $X(\dfrac{p^2+q^2}{2}) = 1$, $X$ is an infinitesimal non-standard symmetry of the Hamiltonian $\dfrac{p^2+q^2}{2}\Big|_U$.

Apart from partial symmetries, it is sometimes useful to consider local symmetries of a Hamiltonian near some fixed point $x \in K$, that is, germs of partial symmetries at $x$. The study of local symmetries is significantly easier than that of global ones. Besides, in the most interesting cases, they can be completely computed. In §7.4 it will be shown that near non-singular points all Hamiltonians have identical structures. The Lie algebras of local symmetries close to the simplest singular points can also be described, but to do this is already rather complicated, and we omit it (we treat a particular case in §6.5).

## §5. The structure of Hamiltonian systems with a given degree of symmetry

We now study some structural features of a Hamiltonian system that arise as a consequence of its symmetry properties. More precisely, we consider an invariant that characterizes the symmetry of such a system, which we call the "degree of symmetry", and we clarify what structural properties a system with a given degree of symmetry has. We shall show, in particular, that the symmetries generate a double fibering of $K$.

Unfortunately, a theory that could be used to test the presence of one symmetry or another of a given Hamiltonian and, in particular, to calculate its degree of symmetry, is still waiting to grow out of "alchemy" into "chemistry". The classical many-body problem is a well-known illustration of our helpnessness with similar questions. Therefore, eschewing alchemy, we shall pass this problem by in silence.

5.1. We say that a system of functions $f_1, \ldots, f_k \in \mathcal{F}(K)$ is *essentially independent* if the set $\{x \in K \mid \operatorname{rank}(df_1|_x, \ldots, df_k|_x) < k\}$ has no interior points. By the *degree of symmetry* $s(H)$ of a Hamiltonian $H$ we mean the largest number $k$ such that there is an essentially independent system of functions $f_1, \ldots, f_k \in \mathcal{F}(K)$ for which $\{f_i, f_j\} = \{f_i, H\} = 0$ $(1 \leqslant i, j \leqslant k)$.

LEMMA 5.1. *If* $\{f_i, f_j\} = 0$, $1 \leqslant i, j \leqslant k$, *then the fields* $X_i = X_{f_i}$ *are tangent to the manifolds* $P_c = P(c_1, \ldots, c_k) = \{x \in K \mid f_i(x) = c_i, 1 \leqslant i \leqslant k\}$ *(provided, of course, that* $P_c$ *is a manifold).*

$X_i(f_j) = \{f_i, f_j\} = 0$, therefore $X_i$ is tangent to the surfaces $\{f_j = c_j\}$, $1 \leqslant j \leqslant k$, hence also to their intersection $P_c$.■

COROLLARY 5.1. $1 \leqslant s(H) \leqslant n$.

Let $(f_1, \ldots, f_k)$ be the system of functions figuring in the definition of $s(H)$, and $x \in K$ a point such that $df_1|_x, \ldots, df_k|_x$ are linearly independent. Let $P_{F(x)} = \{y \in K \mid f_i(y) = f_i(x), 1 \leqslant i \leqslant k\}$. Then near $x$, by the implicit function theorem, $P_{F(x)}$ is a $(2n-k)$-dimensional submanifold of $K$. By Lemma 5.1, $X_i$ is tangent to $P_{F(x)}$; in particular, $X_i|_x \in T_x(P_{F(x)})$. Since the $X_i$ are linearly independent, $\dim T_x(P_{F(x)}) = 2n - k \geqslant k$, hence $k \leqslant n$, that is, $s(H) \leqslant n$. On the other hand, if the system consisting of the function $f_1 = H$ is essentially independent, that is, the set $B = \{x \in K \mid dH_x = 0\}$ has no interior points ($\operatorname{int} B = \emptyset$), then $s(H) = 1$. If $\operatorname{int} B \neq \emptyset$ we can take for $f_1$ a function that coincides with $H$ on $K \setminus \operatorname{int} B$ and for which $\operatorname{int}\,[\{x \in K \mid df_1|_x = 0\} \cap \operatorname{int} B] = \emptyset$.

We omit the proof that such a function exists, both because it is banal and because Hamiltonians of this sort do not occur "in nature".■

5.2. We start by studying Hamiltonian systems of degree of symmetry $\geqslant k$. Let $f_1, \ldots, f_k$ be some collection of essentially independent symmetries of a Hamiltonian $H$, $X = X_H$, $X_i = X_{f_i}$ and $\{A_t^i\} \Longleftrightarrow X_i$. Since $\{f_i, f_j\} = 0$, we have $[X_i, X_j] = 0$ and $A_t^i \circ A_s^j = A_s^j \circ A_t^i$. The first important structural feature of our system consists in the possibility of fibering $K$ into a family of disjoint submanifolds (with singularities) of dimension $2n-k$, "woven" out of the trajectories of $X$. To construct these manifolds, we consider the level surfaces of the functions $f_i$: $P_i(\lambda) = \{x \mid f_i(x) = \lambda\}$ and their intersection

$$P_c = \bigcap_{i=1}^{k} P_i(c_i), \quad c = (c_1, \ldots, c_k).$$ By Lemma 5.1, the trajectories of $X$ lie on the manifolds $P_c$. In what follows it is convenient for us to assume that $f_1 = H$ (for otherwise we could add $H$ to the functions $f_1, \ldots, f_k$ and argue as in the proof of Corollary 5.1). Under this assumption, the sets $P_c$ lie on the level surfaces of $H$. If $x \in P_c$ and $df_1|_x, \ldots, df_k|_x$

are linearly independent, then near $x$ the set $P_c$ is a submanifold of dimension $2n-k$, that is, $P_c \cap U$ is a submanifold, where $U$ is some neighbourhood of $x$ in $K$. By the essential independence of the system of functions $f_1, \ldots, f_k$, the points where the differentials $df_1|_x, \ldots, df_k|_x$ are linearly dependent form the "inessential" part of $K$. Therefore, loosely speaking, $P_c$ is a submanifold near almost all points $x \in P_c$. Henceforth we call the sets $P_c$ manifolds, regardless of the fact that they may have singularities, and we bear in mind that all our arguments for which it in fact matters that the $P_c$ are genuine manifolds (without singularities) are valid for those points $x \in K$ near which $P_c$ is a manifold. Clearly, $P_c \cap P_{c'}$ is non-empty only if $c = c'$, and $\cup P_c = K$. In this sense we say that $K$ is fibred by the $k$-parameter family $\{P_c\}$, $c \in \mathbf{R}^k$, of manifolds of dimension $2n-k$.

Let $\gamma$: $I \to K$ be a trajectory of $X_H$, $I \subset \mathbf{R}$. Since $\gamma(I) \subset P_c$ for some $c \in \mathbf{R}^k$, and since $P_c$ is closed, the closure of $\gamma(I)$ also lies in $P_c$. This shows that "the more symmetrical the Hamiltonian system, the less intricate are its trajectories in $K$". Furthermore, we obtain the following necessary geometrical condition for a Hamiltonian system to have degree of symmetry $\geqslant k$: there must be room for the closures of all the trajectories of the Hamiltonian system on $(2n-k)$-dimensional submanifolds of $K$. This circumstance enables us to give fix an upper estimate for the degree of symmetry of a Hamiltonian system if we know some of its trajectories. We must, however, remark that it is very difficult to realize this possibility in practice.

5.3. It is pertinent to ask how far the functions $f_1, \ldots, f_k$ that form a maximal collection of pairwise commuting symmetries of a given Hamiltonian $H$ are uniquely determined. If we form new functions $\bar{f}_i = \varphi_i(f_1, \ldots, f_k)$ $(i = 1, \ldots, k)$, so that they are essentially independent, for which it is enough to insist that the determinant $\left|\dfrac{\partial \varphi_i}{\partial f_j}\right|$ vanishes on a set $S \subset \mathbf{R}^k = \{(f_1, \ldots, f_k)\}$ without interior points, then, by (3.3), $\{\bar{f}_i, H\} = \{\bar{f}_i, \bar{f}_j\} = 0$ $(1 \leqslant i, j \leqslant k)$, that is, $\bar{f}_1, \ldots, \bar{f}_k$ again form a maximal collection of commuting symmetries of $H$. If we form the manifolds $\bar{P}_{\bar{c}} = \{x \in K | \bar{f}_i(x) = \bar{c}_i\}$, $\bar{c} = (\bar{c}_1, \ldots, \bar{c}_k)$, then $P_c \subset \bar{P}_{\bar{c}}$, where $\bar{c}_i = \varphi_i(c_1, \ldots, c_k)$. And if, moreover, $c \notin S$, then $c_i$ can be expressed in terms of $\bar{c}_i$, $\bar{P}_{\bar{c}} \subset P_c$, hence, $P_c = \bar{P}_{\bar{c}}$. Thus, the collection of manifolds $P_c$ coincides with $\bar{P}_{\bar{c}}$ if we disregard the hypersurface $\bar{S} = \left\{x \in K \,\middle|\, \left|\dfrac{\partial \varphi_i}{\partial f_i}\right|(x) = 0\right\}$. Clearly, the converse also holds, so that we have the following result.

PROPOSITION 5.1. *The symmetries $f_1, \ldots, f_k$ of the Hamiltonian $H$ are uniquely determined up to a transformation $\bar{f}_i = \varphi_i(f_1, \ldots, f_k)$, $i = 1, \ldots, k$,* *if and only if the collection of manifolds* $P_c$ *is uniquely determined by* $H$.

An example of a non-unique collection is given by the Hamiltonian $H = q_1 \in \mathcal{F}(T^*(\mathbf{R}^2))$, where $(q_i)$ $(i = 1, 2)$ are coordinates in $\mathbf{R}^2$, and $(q_i, p_i)$ is the corresponding special system of coordinates. The collection $f_1 = q_1, f_2 = q_2$, is clearly not functionally related to the collection $q_1, p_2$.

There are, however, Hamiltonians for which $P_c$ is uniquely determined. A relevant example will be given in §6.5. We remark that the reason for uniqueness in that example lies in the fact that the trajectories of $X_H$ fill out everywhere densely "almost all" the manifolds $P_c$, which are, therefore, invariantly defined as the closures of the trajectories of $X_H$. A similar circumstance apparently occurs for "almost all" Hamiltonians whose level surfaces are compact. A priori one can suppose that there are similar but more complicated mechanisms for the behaviour of the trajectories of $X_H$ ensuring the uniqueness of the family $\{P_c\}$ in general.

To sum up, we can say that the degree of symmetry characterizes the degree of intricacy of the trajectories of $X_H$, while the character of intricacy of these trajectories is connected with the problem of the uniqueness of the family $\{P_c\}$. Arguments to be introduced in §6 below for the case $k = n$ show that the ratio $s(H)/n$, $0 < s(H)/n \leqslant 1$, can be interpreted as the degree of integrability of the corresponding Hamiltonian system, so that integrability and symmetry are practically one and the same. The interrelations between properties of symmetry and solubility (integrability) that arise in Hamiltonian mechanics are essentially the same as in the theory of algebraic equations, where they are described by Galois theory. Unfortunately, the lack of methods of finding the algebra Sym $H$ is a serious obstacle to the practical use of a "Hamiltonian Galois theory", even if it were constructed.

**5.4.** For further investigation of the structure of a Hamiltonian system in a neighbourhood of an arbitrary point $x \in K$ we make use of the map $\alpha_x\colon (t_1, \ldots, t_k) = t \to A_t(x) = A^1_{t_1} \circ \ldots \circ A^k_{t_k}(x)$. Generally speaking, $\alpha_x$ is not defined for all $t \in \mathbf{R}^k$, because the fields $X_1, \ldots, X_k$ need not be full. On the other hand, it is not hard to see that there is in $\mathbf{R}^k$ a neighbourhood $U_x$ of zero such that $\alpha_x$ is defined on $U_x$. Decreasing $U_x$ if necessary, we can choose a neighbourhood $V_x$, $0 \in V_x \subset U_x$, such that $A^i_{t_i} \circ A^j_{t_j}(x) = A^j_{t_j} \circ A^i_{t_i}(x)$ for any $t \in V_x$. Such neighbourhoods $V_x$ we call admissible. We denote by $N_x$ the set $\alpha_x(V_x) = \{y \in K \mid y = A_t(x), t \in V_x\}$. Our next aim is to describe the structure of $N_x$.

LEMMA 5.2. *Suppose that* $\operatorname{rank}(x_1|_x, \ldots, X_k|_x) = s$. *Then* $\operatorname{rank}(X_1|_y, \ldots, X_k|_y) = s$ *for any point* $y \in N_x$.

Let $y = \alpha_x(t) = A_t(x)$, $t \in V_x$. Then $X_t|_y = (dA_t|_x)(X_i|_x)$ in view of the fact that $A_t$ maps $X_i$ to itself. Furthermore, $dA_t|_x$ is an isomorphism, since $A_t$ is a diffeomorphism. Therefore, $\operatorname{rank}(X_1|_y, \ldots, X_k|_y) = \operatorname{rank}(X_1|_x, \ldots, X_k|_x) = s$. ∎

LEMMA 5.3. *If* rank $(X_1|_x, \ldots, X_k|_x) = s$, *then* $N_x$ *is an s-dimensional embedded manifold.*

1) Let $s = k$. By Lemma 5.2, rank $d\alpha_x|_t = k$, $t \in V_x$. Hence $N_x$ is an embedded manifold of dimension $k$.

2) Let $s < k$. We choose $s$ linearly independent vectors from among $X_1|_x, \ldots, X_k|_x$. Let these be, for example, $X_1|_x, \ldots, X_s|_x$. We denote by $\widetilde{V}_x$ the intersection of $V_x$ with the subspace $\{\tau\} = \{(t_1, \ldots, t_s, 0, \ldots, 0)\}$ in $\mathbf{R}^k$, and let $\widetilde{\alpha}_x = \alpha_x|_{\widetilde{V}_x}$. By Lemma 5.2, rank $d\widetilde{\alpha}_x = s$, and from the part of Lemma 5.3 already proved we conclude that $\widetilde{N}_x = \mathrm{Im}\widetilde{\alpha}_x$ is an embedded manifold of dimension $s$. Further, $\widetilde{N}_x = N_x$, since $X_{s+1}, \ldots, X_k$ are tangent to $\widetilde{N}_x$: at $x$ this is so because $X_{s+1}|_x, \ldots, X_k|_x$ can be expressed linearly in terms of $X_1|_x, \ldots, X_s|_x$, and since $(dA\tau|_x)(X_i|_x) = X_i|_{\alpha_x(\tau)}$, the vectors $X_{s+1}|_{\alpha_x(\tau)}, \ldots, X_k|_{\alpha_x(\tau)}$ can in the same way be expressed linearly in terms of $X_1|_{\alpha_x(\tau)}, \ldots, X_s|_{\alpha_x(\tau)}$.■

REMARK 5.1. The map $\alpha_x$ and the manifold $V_x$ can, clearly, be formed by means of the construction we have described, for an arbitrary system of commuting fields $X_1, \ldots, X_k$ on some manifold $M$. In particular, the fields $X_i = \Gamma^{-1}(df_i)$, where $\{f_i, f_j\} = \text{const}$, are such a collection.

REMARK 5.2. Let rank $(X_1|_x, \ldots, X_k|_x) = k$. Then $A_t(\alpha_x(t)) = \alpha_x(t + t') = \alpha_x(r_t(t'))$, where $t, t' \in V_x$, $r_t(t') = t + t'$. Since $\frac{d}{d\tau}(r_{\tau e_i}) = \frac{\partial}{\partial t_i}$, where $e_i$ is the $i$-th basis vector in $\mathbf{R}^k$, $(t_j(e_i) = \delta_{ij})$, we have $\alpha_x\left(\frac{\partial}{\partial t_i}\right) = X_i|_{N_x}$. Thus, $\alpha_x$ identifies $N_x$ with $V_x$ and the system of fields $X_i|_{N_x}$ with the system of coordinate fields, that is, $\alpha_x$ is a multidimensional analogue to the concept of a trajectory (see §0).

Let $y \in N_x$, $y = A_t(x)$, and $z \in N_y$, $z = A_{t'}(y)$. As we remarked above, the point $A_{t'+t}(x)$ may be undefined (because the $X_i$ are not full), whereas the point $A_{t'}(A_t(x)) = A_{t'}(y) = z$ may be defined. Therefore, it is natural to consider the "transitive completion" of $N_x$, that is, the set $\bar{N}_x = \{y \in K \mid \exists t_1, \ldots, t_l \in \mathbf{R}^k : y = A_{t_1} \circ \ldots \circ A_{t_l}(x)\}$. It is clear that if $\bar{N}_x \cap \bar{N}_y$ is non-empty, then $\bar{N}_x = \bar{N}_y$. By Lemmas 5.2 and 5.3, the set $\bar{N}_x$, which consists of the $\bar{N}_y$, $y \in \bar{N}_x$, is an embedded submanifold. Thus, we see that $K$ is fibred into non-disjoint fibres formed by the embedded submanifolds $\bar{N}_x$. Also, owing to the essential independence of the functions $f_1, \ldots, f_k$, the open everywhere dense set $U_0$ in $K$ where the differentials $df_1, \ldots, df_k$ are linearly independent is fibred into embedded submanifolds of dimension $k$. In the complement $V_1 = K \setminus U_0$ we can choose a set $U_1$, open in $V_1$, on which rank $(df_1, \ldots, df_k) = k - 1$, and in the same way it is fibred into embedded submanifolds of dimension $k - 1$, and so on. Thus, the collection of symmetries of a Hamiltonian system determines a "hierarchic" division of the canonical manifold $K$ into submanifolds $U_s$( $s = 0, \ldots, k$) at whose points the rank of the system $df_1, \ldots, df_k$ is equal to $k - s$ and each of the submanifolds $U_s$ is "fibred" by submanifolds of the form $N_x$ of dimension $k - s$.

Thus, the system of symmetries determines two fiberings of $K$: into the manifolds $P_c$ and into the manifolds $\bar{N}_x$. Lemma 5.1 shows that every $\bar{N}_x$ lies in $P_c$, $c = F(x)$. Thus, the $\bar{N}_x$, $x \in P_c$, in their turn fibre the manifolds $P_c$, so that we get a "double" fibering. It is useful to note that the $\bar{N}_x$ are pre-Lagrangian, that is, $\Omega|_{\bar{N}_x} = 0$. For $\Omega_x(X_i|_x, X_j|_x) = \{f_i, f_j\}\ (x) = 0$, and the vectors $X_i|_x (i = 1, \ldots, k)$ generate $T_x(\bar{N}_x)$. Therefore, $\Omega_x|_{T_x(\bar{N}_x)} = 0$.

5.5. We turn now to the case when the fields $X_i|_{\bar{N}_x}$ are full. Here we can completely describe the structure of the manifolds $\bar{N}_x$.

THEOREM 5.1. *If the fields* $X_i|_{\bar{N}_x}$ *are full and* rank $(X_1|_x, \ldots, X_k|_x) = s$, *then* $N_x$ *is diffeomorphic to a manifold of the form* $T^m \times \mathbf{R}^{s-m}$, *where* $T^m = \underbrace{S^1 \times \ldots \times S^1}_{m \text{ times}}$ *is an m-dimensional torus.*

Just as in the proof of Lemma 5.3, we can reduce the proof to the case $s = k$. Owing to the fullness of the fields $X_i|_{\bar{N}_x}$ the map $\alpha_x$ is defined on the whole of $\mathbf{R}^k$, and by Lemma 5.2 it is a local diffeomorphism. We consider Ker $\alpha_x = \{t \in \mathbf{R}^k | \alpha_x(t) = x\}$. Now Ker $\alpha_x$ is a subgroup of $\mathbf{R}^k$. For if $t_1, t_2 \in$ Ker $\alpha_x$, then $\alpha_x(t_1 + t_2) = \alpha_{\alpha_x(t_1)}(t_2) = \alpha_x(t_2) = x$ and $x = \alpha_x(0) = \alpha_x(t_1 + (-t_1)) = \alpha_{\alpha_x(t_1)}(-t_1) = \alpha_x(-t_1)$, that is, $(-t_1) \in$ Ker $\alpha_x$ together with $t_1$. It is not hard to see that Ker $\alpha_x$ is discrete. This means that every bounded set in $\mathbf{R}^k$ contains only finitely many points of Ker $\alpha_x$ or, equivalently, that there is a neighbourhood $W$ of the origin in $\mathbf{R}^k$ such that $W \cap$ Ker $\alpha_x = \{0\}$. This last is true in view of the fact that $\alpha_x$ is a local diffeomorphism.

Let $y \in \bar{N}_x$ and $\alpha_x(t_1) = \alpha_x(t_2) = y$. Then $\alpha_x(t_1 - t_2) = \alpha_{\alpha_x(t_1)}(-t_2) = \alpha_y(-t_2) = A_{-(t_2)}(y) = x$, that is, $t_1 - t_2 \in$ Ker $\alpha_x$. Conversely, if $t' \in$ Ker $\alpha_x$, then $\alpha_x(t_1 + t') = \alpha_{\alpha_x(t')}(t_1) = \alpha_x(t_1) = y$. Thus, $\alpha_x(t_1) = \alpha_x(t_2)$ if and only if $t_1 - t_2 \in$ Ker $\alpha_x$, that is, if and only if $t_1$ and $t_2$ lie in one and the same coset of Ker $\alpha_x$. For this reason, the formula $\bar{\alpha}_x([t]) = \alpha_x(t)$, where $[t]$ is the coset of $t \in \mathbf{R}^k$ under Ker $\alpha_x$, gives a well-defined map $\bar{\alpha}_x$: $\mathbf{R}^k/$Ker $\alpha_x \to \bar{N}_x$. This is clearly a one-to-one correspondence and coincides locally with $\alpha_x$ because $\alpha_x$ is discrete. Therefore, $\bar{\alpha}_x$ is a diffeomorphism. By the theorem on discrete subgroups of $\mathbf{R}^k$ (see [25]), $\mathbf{R}^k/$Ker $\alpha_x \approx T^m \times \mathbf{R}^{k-m}$, where $m$ is the dimension of the linear space spanned by Ker $\alpha_x$, which is equal to the number of integral generators of this subgroup.■

## §6. Completely integrable systems

We now examine in greater detail for $s(H) = n$ the situation described in the preceding section. In view of Corollary 5.1, we should then call the corresponding Hamiltonian system *completely symmetric.* But for reasons that will become clear later, such systems are called *completely integrable* (see also §5.3).

**6.1.** An important property of completely integrable systems is the following: the manifold $\bar{N}_x$ is Lagrangian if $x$ is a point in general position, that is, rank $(X_1|_x, \ldots, X_k|_x) = n$. For in this case dim $\bar{N}_x = n$ and, as we saw above, $\Omega|_{\bar{N}_x} = 0$. In the case in question, the manifolds $P_c$, "as a rule", also are of dimension $n$, so that if $\bar{N}_x \subset P_c$, then $\bar{N}_x$ coincides with one of the non-singular components of $P_c$. (By a non-singular component we mean a connected component of $P_c \setminus P_c^{\text{sing}}$, where $P_c^{\text{sing}}$ is the set of singular points of $P_c$.)

We now assume in addition that the fields $X_i$ $(i = 1, \ldots, n)$ are full (or full in the relevant domain of $K$). In this case all the fibres $\bar{N}_x = N_x$ of general form are of dimension $n$ and, by Theorem 5.1, $N_x = T^m \times \mathbf{R}^{n-m}$. The integer $m$ in this equation depends, of course, on $N_x$, that is, on $x$. Thus, we arrive at a function $m = m(x)$ which takes integer values and is defined for all points $x \in K$ where dim $N_x = n$. It is clear that $m(x) = m(A_t(x))$ and that $m(x)$ is constant along the fibres $N_x$. We investigate its structure in more detail.

Suppose that an $n$-dimensional submanifold $W \subset K$ satisfies the conditions (T): $x \in W$, $W$ is Lagrangian, $W \cap N_w = w \quad \forall w \in W, T_w(W) \cap T_w(N_w) = 0$, and $m$ is defined on $W$. If $W'$ is a Lagrangian manifold intersecting $N_x$ transversally at $x$, then a sufficiently small neighbourhood of $x$ in $W'$ satisfies (T) (since $N_x$ is a submanifold). This shows that there are manifolds satisfying (T). We consider the map

$\alpha = \alpha_W \colon W \times \mathbf{R}^n \to K$, $\alpha(w, t) = \alpha_w(t) = A_t(w)$, $w \in W$, $t \in \mathbf{R}^n$.

LEMMA 6.1. *$\alpha$ is a local diffeomorphism.*

We must show that $d\alpha|_{(w,t)} \colon T_{(w,t)}(W \times \mathbf{R}^n) \to T_{\alpha(w,t)}(K)$ is an isomorphism for $\forall (w, t) \in W \times \mathbf{R}^n$. But it follows from (T) that this is so for the points $(w, 0)$. We now introduce $r_\tau \colon W \times \mathbf{R}^n \to W \times \mathbf{R}^n$, $r_\tau(w, t) = (w, t + \tau)$. Then $\alpha = A_\tau \circ \alpha \circ r_{-\tau}$, so that $d\alpha|_{(w,t)} = dA_t|_w \circ d\alpha|_{(w,0)} \circ dr_{-t}|_{(w,t)}$ for $\tau = t$, from which it is clear that $d\alpha|_{(w,t)}$ is also an isomorphism.■

It is obvious that the set $\alpha(W \times R^n) = \bigcup_{w \in W} N_w$ is open, contains $N_x$, and is invariant under all $A_t$. If $y \in \alpha(W \times \mathbf{R}^n)$, then there is some $w \in W$ such that $y \in N_w$. Hence $m(y) = m(w)$ and in studying the function $m$ near $x$ it is enough to study it only on $W$.

LEMMA 6.2. *There is a neighbourhood $V$ of $x$ in $W$ such that $m(x) \leqslant m(w) \quad \forall w \in V$.*

We note that $\alpha^{-1}(W) \cap (w \times \mathbf{R}^n) = \operatorname{Ker} \alpha_w$. Since $\alpha$ is a local diffeomorphism, $\alpha^{-1}(W)$ is an $n$-dimensional submanifold of $W \times \mathbf{R}^n$ transversally intersecting the fibres $\alpha^{-1}(N_w) = w \times \mathbf{R}^n$. We denote by $W_t$ the connected component of $\alpha^{-1}(W)$ containing $t \in \operatorname{Ker} \alpha_x$, and we set $W_i = W_{t_i}$ $(i = 1, \ldots, m(x))$, where $t_1, \ldots, t_{m(x)}$ is a basis of the discrete subgroup $\operatorname{Ker} \alpha_x$. Let $V_i = \{w \in W \mid W_i \cap (w \times \mathbf{R}^n) \neq \emptyset\}$. Then $\hat{V} = \bigcap_{i=1}^{m(x)} V_i$ is a non-empty open neighbourhood of $x$ in $V$. The intersection $W_t \cap (w \times \mathbf{R}^n)$ if it is not empty, consists of a single point, which we denote by $l_t(w)$, and we set $l_i = l_{t_i}$. Since $l_t(w)$ depends smoothly on $w$, the vectors $l_i(w)$ $(i = 1, \ldots, m(x))$ are independent in some neighbourhood $V \subset \hat{V}$ of $x$ in $W$. But $l_i(w) \in \operatorname{Ker} \alpha_w$, so that rank $(\operatorname{Ker} \alpha_w) \geqslant m(x)$ if $w \in V$, that is, $m(w) \geqslant m(x)$. ■

COROLLARY 6.1. *The function $m(x)$ is locally constant, that is, the union of the open sets on which $m(x)$ is constant is everywhere dense in its domain of definition.*

This follows from Lemma 6.2 and the fact that $m(x)$ is integral-valued.■

LEMMA 6.3. *Suppose that $m(x)$ is constant near $x$ and that the neighbourhood $V$ constructed in the proof of the preceding lemma is connected. Then $\alpha^{-1}(V) = \bigcup_{t \in \operatorname{Ker} \alpha_x} W_t$, where $W_t = \widetilde{W}_t \cap \alpha^{-1}(V)$.*

Let $E$ be some connected component of $\alpha^{-1}(V)$, distinct from all the $\widetilde{W}_t$, $t \in \operatorname{Ker} \alpha_x$, and $W_E = \{w \in V \mid (w \times \mathbf{R}^n) \cap E \neq \emptyset\}$. Clearly, $W_E$ is open in $V$ and $x \notin W_E$. Let $e(w)$ be the unique point of intersection of $w \times \mathbf{R}^n$ with $E$, $w \in W_E$. Since $e(w) \in \operatorname{Ker} \alpha_w$, rank $(\operatorname{Ker} \alpha_w) = m(x)$, and the vectors $l_1(w), \ldots, l_{m(x)}(w)$ are linearly independent for $w \in V$, we see that $e(w) = \sum_{i=1}^{m(x)} \lambda_i(w) l_i(w)$, where the $l_i(w)$ are rational numbers. The functions $\lambda_i(w)$ are clearly continuous, hence are constants $\lambda_i$, so that the domain $W_E$, which is diffeomorphic to $E$, is connected. If $w_0 \in \bar{W}_E \setminus W_E$, then the point $e(w_0) = \sum_{i=1}^{m(x)} \lambda_i l_i(w_0)$ lies in $\operatorname{Ker} \alpha_{w_0}$, being a limit point for $E = \bigcup_{w \in W_E} e(w)$. For the same reason its connected component $E'$ in $\alpha^{-1}(E)$ contains $E$, and $E' \setminus E \ni e(w_0)$. This contradicts the fact that $E$ is a connected component.■

COROLLARY 6.2. *$l_1(w), \ldots, l_{m(x)}(w)$ form a basis of* $\operatorname{Ker} \alpha_w$ *for* $w \in V$.

6.2. Suppose that $W$ satisfies (T), $v_i = \alpha^*(f_i)$, $\tau_i = \pi^*(t_i)$, where $\pi(w, t) = t = (t_1, \ldots, t_n)$, $(w, t) \in W \times \mathbf{R}^n$. Since $\alpha$ is a local diffeomorphism, the form $\hat{\Omega} = \alpha^*(\Omega)$ defines a canonical structure on $W \times \mathbf{R}^n$. It is clear that $\alpha\left(\frac{\partial}{\partial \tau_i}\right) = X_i$. Hence the field $\frac{\partial}{\partial \tau_i}$ is Hamiltonian (with respect to $\hat{\Omega}$), $\alpha^*(f_i) = v_i$ is its Hamiltonian (Corollary 2.2), and $\frac{\partial}{\partial \tau_i}(\hat{\Omega}) = 0$. We calculate $\hat{\Omega}$ in the coordinates $(v, \tau)$. If

$\hat{\Omega} = \sum_{1 \leq i,j \leq n} a_{ij} dv_i \quad d\tau_j + \sum_{1 \leq k < l \leq n} (b_{kl} d\tau_k \wedge d\tau_l + c_{kl} dv_k \wedge dv_l)$, then

it follows from $\frac{\partial}{\partial \tau_s}(\hat{\Omega}) = 0$ that $\frac{\partial a_{ij}}{\partial \tau_s} = \frac{\partial b_{kl}}{\partial \tau_s} = \frac{\partial c_{kl}}{\partial \tau_s} = 0$, that is,

$a_{ij}$, $b_{kl}$, and $c_{kl}$ are functions of the variables $v$ only. Next, $N_w$, and hence also $w \times \mathbf{R}^n = \alpha^{-1}(N_w)$ are Lagrangian manifolds. Therefore, $0 = \hat{\Omega}|_{W \times \mathbf{R}^n} = \sum_{k,l} b_{kl}(v(w))\, d\tau_k \wedge d\tau_l$ from which it follows that $b_{kl} = 0$. Because $W \times \{0\} \subset W \times \mathbf{R}^n$ is Lagrangian, we have $0 = \hat{\Omega}|_{W \times \{0\}} = \sum_{k,l} c_{kl}(v) dv_k \wedge dv_l$, that is, $c_{kl} = 0$. Finally, since

$dv_j = -\Omega\partial/\partial\tau_i = \sum_{i=1}^{n} a_{ij}(v) dv_i$, we have $a_{ij}(v) = \delta_{ij}$, so that we have proved the following result.

LEMMA 6.4. $\hat{\Omega} = \sum_{i=1}^{n} dv_i \wedge d\tau_i$.

COROLLARY 6.3. (complement theorem). *Suppose that in a domain $U \subset K$ the differentials of the functions $\mathcal{P}_1, \ldots, \mathcal{P}_n$ are independent and that $\{\mathcal{P}_i, \mathcal{P}_j\} = 0$. Then for $\forall u \in U$ there is a neighbourhood $U' \subset U$ and functions $Q_1, \ldots, Q_n \in \mathcal{F}(U')$ such that the family $(\mathcal{P}, Q)$ forms a canonical system of coordinates in $U'$.*

Let $x = u$, $f_i = \mathcal{P}_i$ and $(x \times 0) \subset V' \subset W \times \mathbf{R}^n$ be a domain such that $\alpha(V') \subset U$ and $\alpha|_{V'}$ is a diffeomorphism. Then we can set $\alpha(V') = U'$ and $Q_i = (\alpha|_{V'}^{-1})^*(\tau_i)$.■

**6.3. Action-angle variables.** We consider the coordinates $q_i = f_i|_W$ on $W$, and let $(p_i, q_i)$ be special coordinates on $T^*(W)$. The map $\beta$: $T^*(W) \to W \times \mathbf{R}^n$ given by $\beta^*(v_i) = q_i$, $\beta^*(\tau_i) = p_i$ is, by Lemma 6.4, an anticanonical diffeomorphism, that is, $\beta^*(\hat{\Omega}) = -\Omega_W$. We assume henceforth that $V = W$ satisfies the conditions of Lemma 6.3 and is, in addition, simply-connected. Then $\bar{l}_i = \beta^{-1} \circ l_i$: $W \to T^*(W)$ is a section of the fibering $T^*(W) \to W$ and $\bar{l}_i(W) \subset (\alpha \circ \beta)^{-1}(W)$ is Lagrangian, since $W$ is. Therefore, $\bar{l}_i = s_{dw_i}$, where $w_i(q_1, \ldots, q_n) \in \mathcal{F}(W)$. It follows from the independence of the vectors $\bar{l}_i(w)$ that the differentials $dw_i$ are independent therefore, the system of functions $w_1, \ldots, w_{m(x)}$ can be complemented by functions $w_j(q_1, \ldots, q_n)$ $(j = m(x) + 1, \ldots, n)$ to a coordinate system in $W$. We consider the special coordinate system $(w_i, \pi_i)$ in $T^*(W)$ corresponding to the coordinate system $w_i$ in $W$. Then $(\alpha \circ \beta)^{-1}(W) = \bigcup_{n \in \mathbf{Z}^{m(x)}} \{\pi_i = n_i | i = 1, \ldots, m(x)\}$, where we have set $n = (n_1, \ldots, n_{m(x)})$, $n_i \in \mathbf{Z}$. We define a free action of the group $\mathbf{Z}^{m(x)}$

on $T^*(W)$ by the rule: $k \cdot (w_1, \ldots, w_n, \pi_1, \ldots, \pi_n) =$
$= (w_1, \ldots, w_n, \pi_1 + k_1, \ldots, \pi_{m(x)} + k_{m(x)}, \pi_{m(x)+1}, \ldots, \pi_n)$, where $k = (k_1, \ldots, k_{m(x)}) \in \mathbf{Z}^{m(x)}$. By construction $l_i(w)$ is the point of the fibre $T_w^*(W)$ with coordinates $\pi_j = \delta_{ij}$, $i = 1, \ldots, m(x)$. Hence, if $\alpha(z) = y$, $y \in N_w$, then $(\alpha \circ \beta)^{-1}(y) = \beta^{-1}(\alpha_w(y)) = \beta^{-1}(z + \operatorname{Ker} \alpha_w) =$ $= \beta^{-1}(z) + \beta^{-1}(\operatorname{Ker} \alpha_w)$ is the orbit of $\beta^{-1}(z)$ under the given action of $\mathbf{Z}^{m(x)}$. Thus, the map $\alpha \circ \beta$ induces a diffeomorphism $\gamma\colon T^*(W)/\mathbf{Z}^{m(x)} \to \alpha(W \times R^n)$, $\gamma(\mathbf{Z}^{m(x)} \cdot u) = \alpha(\beta(u))$, $u \in T^*(W)$. If $g\colon T^*(W) \to T^*(W)/\mathbf{Z}^{m(x)}$ is the natural projection, then there are many-valued functions $\varphi_i = \pi_i \circ g^{-1}$, $\mu_i = w_i \circ g^{-1}$ defined on $T^*(W)/\mathbf{Z}^{m(x)}$. In actual fact they are clearly single-valued, with the exception of $\varphi_1, \ldots, \varphi_{m(x)}$, which are defined mod 1. In particular, their differentials are always single-valued, so that the form $\widetilde{\Omega} = \sum_i d\mu_i \wedge d\varphi_i$ is well-defined and $\gamma^*(\Omega) = \widetilde{\Omega}$. Setting $\mathcal{P}_i = (\gamma^{-1})^*(\mu_i)$, $\mathcal{Q}_i = (\gamma^{-1})^*(\varphi_i)$, and noting that $\mathcal{P}_i = w_i(f_1, \ldots, f_n)$ (see above), we arrive all in all at the following theorem.

THEOREM 6.1. *Let $H$ be a completely integrable Hamiltonian, $f_1, \ldots, f_n$ be a complete collection of symmetries of it, and let the fields $X_{f_i}$ be full. Then for every point $x \in K$ near which $m(x)$ is constant, there is some neighbourhood $U$, invariant with respect to the operators $A_t$, and canonical coordinates $\mathcal{Q}_i$, $\mathcal{P}_i$, of which $\mathcal{Q}_1, \ldots, \mathcal{Q}_{m(x)}$ are many-valued functions defined* mod 1, *such that $H = \psi(\mathcal{P}_1, \ldots, \mathcal{P}_n)$.*

REMARK 6.1. When $m(x) = n$, all the coordinates $\mathcal{Q}_i$ are defined mod 1, that is, they are cyclic (or "angles"). In view of the formula below, the variables $\mathcal{P}_i$ are called "actions", and all together action-angle variables.

**6.4.** In the coordinate system $(\mathcal{Q}, \mathcal{P})$ we have constructed the Hamiltonian system in question has the form

$$\dot{\mathcal{Q}}_i = H_{\mathcal{P}_i}, \quad \dot{\mathcal{P}}_i = 0 \quad (i = 1, \ldots, n). \tag{6.1}$$

Therefore, $\mathcal{P}_i = \alpha_i =$ const, and $\mathcal{Q}_i = \beta_i t + \beta_i^0$, where $\beta_i = H_{\mathcal{P}_i}(\alpha_1, \ldots, \alpha_n)$, so that the system has been completely integrated, which explains its name. The numbers $\beta_i = \beta_i(x)$ $(i = 1, \ldots, m(x))$ can be understood as the frequencies with which a point moving along the trajectory of $X_H$ passing through $x$ runs round the torus $T^{m(x)}$ (by virtue of the diffeomorphism $N_x = T^{m(x)} \times \mathbf{R}^{n-m(x)}$ constructed above). The frequency vector $\beta = (\beta_1, \ldots, \beta_{m(x)})$ is not uniquely determined by $H$ and $x$, since it depends on the choice of the representation of $N_x$ in the form $T^m \times \mathbf{R}^{n-m}$. But in the important special case $m = n$ the frequency vector $(\beta_1, \ldots, \beta_n)$ is uniquely determined by the choice of cyclic coordinates on the torus $T^n \approx N_w$. The tori $N_w$ are, generally speaking, in this case completely determined by $H$ itself. For if all the frequencies

$\beta_1, \ldots, \beta_n$ on $N_w \approx T^n$ are independent over **Z**, then the trajectories of $X_H$ are everywhere dense on $N_w$, so that $N_w$ is in this case the closure of one of the trajectories. If the "frequency map" $N_w \to \beta(w) = (\beta_1(w), \ldots, \beta_n(w))$ is an epimorphism, then the trajectories of $X_H$ are everywhere dense on almost all tori $N_w$. Therefore, these tori and, as is not hard to see from continuity arguments, all the others are in this case uniquely determined by $H$. This is certainly so if $\left|\frac{\partial^2 H}{\partial \mathcal{P}_i \partial \mathcal{P}_j}\right| \neq 0$ on $U$. In this case the Hamiltonian system is called non-degenerate. The cyclic coordinates $\mathcal{Q}_i$ on $N_w \approx T^n$ are uniquely determined by the choice of basis in Ker $\alpha_w$, that is, up to a integral unimodular transformation. Since such a transformation must depend continuously on $w$, but is integral, it does not, in fact, depend on $w$. So we come to the following result.

PROPOSITION 6.1. *Suppose that $m = n$ and that the Hamiltonian system under discussion is non-degenerate on the domain $U$ of Theorem 6.1. Then the tori $N_w$, $w \in U$, are uniquely determined by the Hamiltonian, and the frequency function $\beta(w) = (\beta_1(w), \ldots, \beta_n(w))$ is uniquely determined up to a integral unimodular transformation of the image.*

**6.5. Cyclic "actions".** We now indicate a method of finding the coordinates $\mathcal{P}_1, \ldots, \mathcal{P}_{m(x)}$ introduced above directly. Let $\gamma_i(a)$, $a = (a_1, \ldots, a_n)$, $a_i = f_i(w)$, be the cycle on $N_w \approx T^{m(x)} \times \mathbf{R}^{n-m(x)}$ defined by the equations $\mathcal{Q}_j = 0$, $j \neq i$. If $\rho' = \sum_i \mathcal{P}_i\, d\mathcal{Q}_i$, then $\Omega|_U = d\rho'$ and $\int_{\gamma_i(a)} \rho' = \int_{\gamma_i(a)} \mathcal{P}_i\, d\mathcal{Q}_i = \mathcal{P}_i \int_{\gamma_i(a)} d\mathcal{Q}_i = \mathcal{P}_i$, since $\mathcal{P}_i|_{N_w}$ is constant. Let $\rho''$ be another form such that $d\rho'' = \Omega|_U$. Then $d(\rho' - \rho'') = 0$ and $\int_{\gamma_i(a)} (\rho' - \rho'')$ does not depend on $a$, that is, it is a constant $\sigma_i$. This follows from Stokes' formula owing to the fact that the cycles $\gamma_i(a)$ and $\gamma_i(a')$ are, obviously, homologous. Therefore, $\int_{\gamma_i(a)} \rho' = \int_{\gamma_i(a)} \rho'' + \sigma_i$. And so

$$\mathcal{P}_i(a) = \int_{\gamma_i(a)} \rho \qquad (i = 1, \ldots, m(x)) \qquad (d\rho = \Omega). \tag{6.2}$$

Complementing the system obtained in (6.2) by functions $\mathcal{P}_i = \mathcal{P}_i(f_1, \ldots, f_n)$ $(i = m + 1, \ldots, n)$ in such a way that the system $\mathcal{P}_1, \ldots, \mathcal{P}_n$ is independent, we obtain half of the coordinates introduced above, with $H = H(\mathcal{P}_1, \ldots, \mathcal{P}_n)$. This is enough to study the frequency map and the frequencies of the system in question. To find the corresponding coordinates $\mathcal{Q}_i$ we consider any Lagrangian manifold $V'$ intersecting all the fibres $N_w$, $w \in U$, transversally. Then, if $Y_i = X_{\mathcal{P}_i}$

and $\{A^i_t\} \Leftrightarrow Y_i$, the manifolds $V_{t_1,\ldots,t_n} = (A^1_{t_1} \circ \ldots \circ A^n_{t_n})(V')$ are the surfaces $\{\mathcal{Q}_i = t_i \mid i = 1, \ldots, n\}$. Therefore, the question of constructing the functions $\mathcal{Q}_i$ is reduced to the integration of the Hamiltonian fields $Y_1, \ldots, Y_n$.

If $m = n$, then (6.2) defines all the functions $\mathcal{P}_i$. In the case $K = T^*(M)$ we can take for $\rho$ the universal form $\rho = \Sigma\, p_i dq_i$, whose integrals along one path or another are customarily called the action. Therefore, the $(\mathcal{Q}_i, \mathcal{P}_i)$ are in this case called "action-angle variables".

6.6. To conclude this section, we apply the theory we have set out to the problem of the local classification of Hamiltonians at a singular point.

Let $n = 1$, that is, $\dim \Phi = 2$, $H(x) = h(x) = 0$, $dH_x = dH_x = 0$. We ask, is there a canonical transformation in a sufficiently small neighbourhood of $x$ that takes $H$ to $h$ and leaves $x$ fixed? For this, first of all, it is necessary that there is an ordinary (not canonical) transformation satisfying the stipulated conditions. By Morse's lemma (see [9]), the equivalence of the Hessians of $H$ and $h$ at $x$, provided that they are non-degenerate, is necessary and sufficient for this. We recall that the Hessian $\mathrm{Hess}(\varphi)$ of a function $\varphi$ at a point $x$ where $d(\varphi)x = 0$ is the following bilinear form on $T_x(K)$: $\mathrm{Hess}(\varphi)(\xi, \eta) = X(Y(\varphi))(x)$, $\xi, \eta \in T_x(K)$, $X_x = \xi$, $Y_x = \eta$. The formula $d\omega(X, Y) = X(\omega(Y)) - Y(\omega(X)) - \omega([X, Y])$, applied to $\omega = d\varphi$, shows that $X(d\varphi(Y))(x) = Y(d\varphi(X))(x)$, from which it follows that $\mathrm{Hess}(\varphi)$ is symmetric and well-defined, that is, does not depend on the extensions $X$ and $Y$ of the vectors $\xi$ and $\eta$. Next, we suppose that Hess $H$ and Hess $h$ are positive-definite, hence equivalent. Then the level surfaces of $H$ and $h$ near $x$ are closed curves.

We construct the action-angle variables $\mathcal{P}$, $\mathcal{Q}$ for $H$ in a neighbourhood of $x$. By §§6.4 and 6.5, the function $\mathcal{P}$ is uniquely determined up to a constant by the level lines of $H$, which we normalize by setting $\mathcal{P}(x) = 0$, and the cyclic coordinate $\mathcal{Q}$ is uniquely determined up to a constant. In the same way, we construct action-angle coordinates $\mathcal{P}'$, $\mathcal{Q}'$ for the Hamiltonian $h$. The required canonical transformation (if it exists) must take the level lines of $H$ to level lines of $h$, consequently, the normalized action $\mathcal{P}$ to the normalized action $\mathcal{P}'$; the cyclic coordinates $\mathcal{Q}$ and $\mathcal{Q}'$ must go into each other, up to a constant. Therefore, there is a canonical transformation carrying $H$ to $h$ if and only if $H$ and $h$, as functions of their normalized actions, coincide. What we have said shows that knowledge of $H$ together with all its derivatives at a singular point does not allow us to recover the Hamiltonian itself up to a canonical transformation; for example, we can add to $H$ any function of $\mathcal{P}$ that is flat at $x$.

We should mention that under the above hypotheses ($m = n = 1$, positive-definiteness of the Hessians), the algebras of local symmetries are nonetheless unique, that is, are carried into each other by a canonical transformation. For these algebras, which consist only of functions of the actions, are identified under identification of the coordinates $\mathcal{P}$, $\mathcal{Q}$ and $\mathcal{P}'$, $\mathcal{Q}'$.

## §7. The theorems of Darboux and Weinstein

One of the possible formulations of the classical theorem of Darboux asserts that locally all canonical manifolds of dimension $2n$ have the same structure. We shall prove this below, together with vastly more general facts discovered recently by A. Weinstein. We recommend the reader also to look at the classical proof of Darboux' theorem (see, for example, [2]).

7.1. We consider the following situation. Let $\Omega_0$ and $\Omega_1$ be two closed non-degenerate 2-forms on a manifold $M$, $\dim M = 2n$. By an *exact deformation* $\Delta = (\omega_t, \varphi_t)$ of $\Omega_0$ and $\Omega_1$ we mean two $t$-smooth families of forms $\omega_t \in \Lambda^2(M)$ $\varphi_t \in \Lambda^1(M)$, $0 \leqslant t \leqslant 1$, such that $\omega_0 = \Omega_0$, $\omega_1 = \Omega_1$, $d\omega_t = 0$, $d\varphi_t = \frac{d}{dt}\,\omega_t$ and $\omega_t$ is non-degenerate for all $t$. We call $x \in M$ a fixed point of such a deformation if $\varphi_t|_x = 0$, $0 \leqslant t \leqslant 1$. The set fix ($\Delta$) of fixed points of $\Delta$ is clearly closed.

THEOREM 7.1. *Suppose that two closed non-degenerate 2-forms $\Omega_0$ and $\Omega_1$ on $M$ are joined by an exact deformation $\Delta = (\omega_t, \varphi_t)$. Then there are domains $U_0$ and $U_1$ of $M$, $U_0 \cap U_1 \supset \text{fix}(\Delta)$, and a diffeomorphism $F$: $U_0 \to U_1$ such that $F^*(\Omega_1|_{U_1}) = \Omega_0|_{U_0}$ and $F|_{\text{fix}(\Delta)} = \text{id}$.*

We try to construct a family of diffeomorphisms $F_t$ such that the form $F_t^*(\omega_t)$ does not depend on $t$, that is $\frac{d}{dt} F_t^*(\omega_t) = 0$, $0 \leqslant t \leqslant 1$, and $F_0$ is the identity map. Except for the condition $F|_{\text{fix}(\Delta)} = \text{id}$, this would give us the result needed, because $\Omega_0 = \omega_0 = F_0^*(\omega_0) = F_1^*(\omega_1) = F_1^*(\Omega_1)$ and we can set $F = F_1$. But $0 = \frac{d}{dt} F_t^*(\omega_t) = F_t^*(Y^t(\omega_t) + \frac{d}{dt}\omega_t) = F_t^*(Y^t(\omega_t) + d\varphi_t)$, where $Y^t \in \mathscr{D}(M)$ is the family of vector fields corresponding to the family $F_t$ (see §0), from which it follows that $0 = Y^t(\omega_t) + d\varphi_t$. Since $d\omega_t = 0$, we have $Y^t(\omega_t) = d(Y^t \lrcorner\, \omega_t)$, and we arrive at the condition $d(Y^t \lrcorner\, \omega_t) + d\varphi_t = 0$, which is satisfied if we define $Y^t$ by $Y^t \lrcorner\, \omega_t = -\varphi_t$. Since $\omega_t$ is non-degenerate, this equation is uniquely soluble (see §2.2) and defines a smooth family of fields $Y^t \in \mathscr{D}(M)$. Since $\varphi_t|_{\text{fix}(\Delta)} = 0$, we see that $Y^t|_{\text{fix}(\Delta)} = 0$, therefore, if $U_0$ is the set of points where $F_1$ is defined, clearly $U_0$ is open, $F_1|_{\text{fix}(\Delta)} = \text{id}$, $U_0 \supset \text{fix}(\Delta)$ and $U_1 = F_1(U_0) \supset \text{fix}(\Delta)$.■

REMARK 7.1. The proof we have given shows that if $M$ is compact, then we can take for the domains $U_0$ and $U_1$ the whole of $M$.

COROLLARY 7.1. (Darboux' theorem). *Let $(K_i, \Omega_i)$ $(i = 0, 1)$ be $2n$-dimensional canonical manifolds, and $a_i \in K_i$. Then there are neighbourhoods $U_i$ of $a_i$ and a diffeomorphism $F$: $U_0 \to U_1$ such that $F^*(\Omega_1|_{U_1}) = \Omega_0|_{U_0}$, that is, all canonical manifolds of one and the same dimension locally have the same structure.*

Since, by §1.4, the 2-forms $\Omega_i|_{a_i}$ are isomorphic, we choose a linear map $\alpha$: $T_{a_0}(K_0) \to T_{a_1}(K_1)$ giving rise to some isomorphism of the forms $\Omega_i|_{a_i}$, and we extend it to a diffeomorphism $A$: $V_0 \to V_1$ of certain neighbourhoods $V_i$ of $a_i$ (that is, $dA|_{a_0} = \alpha$). Then $A^*(\Omega_1|_{a_1}) = \Omega_0|_{a_0}$, hence in some neighbourhood $\widetilde{V}_0 \subset V_0$ of $a_0$ the form $\omega_t = \Omega_0 + t(A^*(\Omega_1) - \Omega_0)$, $0 \leqslant t \leqslant 1$, is non-degenerate (because $\omega_t|_{a_0} \equiv \Omega_0|_{a_0}$) and closed. Shrinking $\widetilde{V}_0$, if necessary, we can find a $\varphi \in \Lambda^1(\widetilde{V}_0)$, $\varphi|_{a_0} = 0$, such that $(A^*(\Omega_1) - \Omega_0)|_{\widetilde{V}_0} = d\varphi$. Then, setting $\varphi_t \equiv \varphi$, we arrive at an exact deformation $\Delta = (\omega_t, \varphi_t)$ of $\Omega_0$ into $A^*(\Omega_1)$ on $\widetilde{V}_0$, with $a_0 \in \text{fix}(\Delta)$. When we now apply Theorem 7.1 to the situation in hand, we find a diffeomorphism $\widetilde{F}$: $U_0 \to U_1$, $a_0 \in (U_0 \cap \widetilde{U}_1) \subset \widetilde{V}_0$, such that $\widetilde{F}^*(A^*(\Omega_1)|_{\widetilde{U}_1}) = \Omega_0|_{U_0}$. We obtain the required diffeomorphism $F$ and the domain $U_1 \subset K_1$ by taking $F = A \circ \widetilde{F}$, $U_1 = A(\widetilde{U}_1)$.■

7.2. We now consider the problem of canonical coordinates (see §2.2). From the results of Section 7.1 we deduce the following proposition.

PROPOSITION 7.1. *Every point $x$ of an arbitrary canonical manifold $K$ has a neighbourhood furnished with a canonical system of coordinates.*

Let $\mathbf{R}^{2n}$ be the space of sequences $(q_1, \ldots, q_n, p_1, \ldots, p_n)$. We introduce a canonical structure on $\mathbf{R}^{2n}$ via the form $\Omega_1 = \sum_{i=1}^{n} dp_i \wedge dq_i$. Let $y \in \mathbf{R}^{2n}$. Then by Corollary 7.1 there is a diffeomorphism $F: U_0 \to U_1$, $x \in U_0 \subset K$, $y \in U_1 \subset \mathbf{R}^{2n}$, such that $F^*(\Omega_1|_{U_1}) = \Omega|_{U_0}$. Therefore, the functions $\mathcal{P}_i = F^*(p_i)$, $\mathcal{Q}_i = F^*(q_i)$ form a canonical system of coordinates in $U_0$.■

7.3. Our next aim is to investigate the structure of a canonical manifold $K$ near some Lagrangian manifold $L$. With the help of Theorem 7.2 we shall show that it is the same as that of $T^*(L)$ near $L = s_0(L)$, $0 \in \Lambda^1(L)$. In what follows we prepare the ground for the application of this theorem.

Let $L \subset K$ be a Lagrangian manifold, and $\mathcal{L}(x)$ the totality of all Lagrangian planes in $T_x(K)$, $x \in L$, transversal to the Lagrangian plane $T_x(L) \subset T_x(K)$. Then $\mathcal{L}(x)$ is a contractible manifold (Lemma 1.4). The set $\mathcal{L} = \bigcup_{x \in L} \mathcal{L}(x)$ can be naturally furnished with the structure of a smooth manifold. We have the obvious projection $\bar{p}: \mathcal{L} \to L$, $\bar{p}(\mathcal{L}(x)) = x$, which is, clearly, a smooth fibering with fibres $\mathcal{L}(x)$. The fact of elementary topology according to which there exist sections of a fibering with contractible fibres (see, for example, [7]), together with a procedure of smooth approximation proves the validity of the following assertion.

LEMMA 7.1. *There is a family of Lagrangian planes $l(x) \in \mathcal{L}(x)$, $x \in L$, depending smoothly on $x$.*

We remark that the map $\nu_x: l_x \to T_x^*(L)$ defined by the formula $(\nu_x(\alpha), \xi) = \Omega_x(\alpha, \xi)$, $\alpha \in l(x)$, $\xi \in T_x(L)$, is an isomorphism (see §1.5). The totality $N$ of all vectors $\alpha \in l(x)$, $x \in L$, is, clearly, a submanifold of $T(K)$, and the projection $\bar{\pi}|_N: N \to L$ ($\bar{\pi}: T(K) \to K$) turns it into a smooth vector bundle over $L$ with fibres $l(x)$. The map $\nu: N \to T^*(L)$ defined by the formula $\nu|_{l(x)} = \nu_x$ is an isomorphism of vector bundles.

The vector bundle $N \to L$ is, clearly, a fibering isomorphic to the normal fibering of $L$ in $K$. We identify $L$ with the zero section of $\bar{\pi}|_N: N \to L$ and $T_x(l(x))$ with $l(x)$, $x \in L$. This allows us to identify $T_x(N)$ with $T_x(K)$. The "tubular neighbourhood" theorem (see [11]) then guarantees that there is a diffeomorphism $a: U' \to K$, where $U'$ is some neighbourhood of $L$ in $N$, such that $a|_L: L \to L$ and $da_x: T_x(N) \to T_x(K)$, $x \in L$, are the identity maps. Let $U = \nu(U')$. Then $b = a \circ \nu^{-1}: U \to K$ is a diffeomorphism, and by construction $b^*(\Omega)_x = (\Omega_L)_x$ $\forall x \in L$, where $\Omega_L \in \Lambda^2(T^*(L))$ is a canonical 2-form

on $T^*(L)$. We set $\Omega_0 = \Omega|_{b(U)}$, $\Omega_1 = (b^*)^{-1}(\Omega_L)$. Then $\Omega_0$ and $\Omega_1$ are closed non-degenerate 2-forms on $b(U) \supset L$, $\Omega_0|_x = \Omega_1|_x$, $\forall x \in L$, and $\Omega_0|_L = \Omega_1|_L$, since $L$ is Lagrangian. The form $\omega_t = \Omega_0 + t(\Omega_1 - \Omega_0)$ is such that $\omega_t|_x = \Omega_0|_x$, $\forall x \in L$. Hence, there is a tubular neighbourhood $V_0$, $b(U) \supset V_0 \supset L$, such that $\omega_t|_{V_0}$ is non-degenerate for all $t$, $0 \leqslant t \leqslant 1$. Let $\beta'_t$: $T^*(L) \to T^*(L)$ be a "homothety" with coefficient $(1-t)$, that is, $\beta'_t(\lambda_x) = (1-t)\lambda_x$, $\lambda_x \in T^*_x(L) \subset T^*(L)$, $\beta_t =$
$= b \circ \beta'_t \circ (b^{-1})|_{V_0}$: $V_0 \to V_0$, and let $Y^t \in \mathscr{D}(V_0)$ be the family of vector fields on $V_0$ corresponding to the family $\beta_t$. If $\omega = \Omega_1 - \Omega_0$, then $d\omega = 0$ and $Y^t(\omega) = d(Y^t \lrcorner\, \omega)$. Therefore,

$$\frac{d}{dt}\,\beta_t^*(\omega) = \beta_t^*(Y^t(\omega)) = d\{\beta_t^*(Y^t \lrcorner\, \omega)\} \text{ and}$$

$$\beta_1^*(\omega) - \omega = d\left\{\int_0^1 \beta_t^*(Y^t \lrcorner\, \omega)\,dt\right\}.$$ Since $\beta_1 = i \circ (\bar{\pi}|_N)|_{V_0} \circ a^{-1}$, where $i$: $L \to K$ is the natural embedding, and $i^*(\omega) = \omega|_L = 0$, we see that $\beta_1^*(\omega) = 0$, hence, $\omega = d\varphi$, where $\varphi = -\int_0^1 \beta_t^*(Y^t \lrcorner\, \omega)\,dt$. Noting, finally, that $\frac{d\omega_t}{dt} = \omega = d\varphi$, we conclude that $\Delta = (\omega_t, \varphi_t \equiv \varphi)$ is an exact deformation of $\Omega_0$ into $\Omega_1$ on $V_0$, with $L \subset \text{fix}(\Delta)$, because $\omega_x = 0$, $\forall x \in L$. Applying Theorem 7.1 to this deformation, we find neighbourhoods $U_0$ and $U_1$, $V_0 \supset U_i \supset L$ $(i = 0, 1)$ and a diffeomorphism $F$: $U_0 \to U_1$ such that $F^*(\Omega_1|_{U_1}) = \Omega_0|_{U_0}$, $F|_L = \text{id}$. Then the composition $c = b^{-1} \circ F$ is such that $c^*(\Omega_L) = \Omega|_{U_0}$ and $c(L) = L$, so that we reach the following result.

THEOREM 7.2 (Weinstein [12]). *Let $L \subset K$ be some Lagrangian manifold. Then there are neighbourhoods $K \supset V \supset L$ and $T^*(L) \supset W \supset L$ and a diffeomorphism $c$: $V \to W$ such that $c^*(\Omega_L) = \Omega|_V$ and $c(L) = L$ (that is, the neighbourhood $V$ of $L$ in $K$ has the same canonical structure as the neighbourhood $W$ of $L = s_0(L)$ in $T^*(L)$).*

COROLLARY 7.2. *Every canonical diffeomorphism $f$: $K \to K$ sufficiently close to the identity can be included in a family of canonical diffeomorphisms $f_t$: $K \to K$, $f_0 = \text{id}$, $f_1 = f$.*

We do not make meaning of the words "sufficiently close" precise, and leave it to the reader to do this on the basis of the proof presented below. Let $\pi_i$: $K \times K \to K$ be the projection onto the $i$-th factor $(i = 1, 2)$ and $\omega = \pi_1^*(\Omega) - \pi_2^*(\Omega)$. Then, clearly, $d\omega = 0$ and $\omega$ is non-degenerate, that is, $\omega$ is a canonical structure on $K \times K$. Let $f$: $K \to K$ be a canonical diffeomorphism. Then for the "graph" $\bar{f}$: $K \to K \times K$ of $f$: $\bar{f}(x) = (x, f(x))$, we have $\bar{f}^*(\omega) = \bar{f}^*\pi_1^*(\Omega) - f^*\pi_2^*(\Omega) =$
$= (\pi_1 \circ \bar{f})^*(\Omega) - (\pi_2 \circ \bar{f})^*(\Omega) = 1^*(\Omega) - f^*(\Omega) = 0$, that is, $\bar{f}(K)$ is a Lagrangian manifold in $K \times K$. If $f = \text{id} = 1$, then $\bar{1}(K) = \Delta$ is the

diagonal in $K \times K$. By Theorem 7.2, there are a tubular neighbourhood $V$ of $\Delta \approx K$ in $K \times K$ and a diffeomorphism $F\colon V \to W \subset T^*(K)$ such that $F|_K = \mathrm{id}$ and $(F^{-1})^*(\omega) = \Omega_K|_W$, where $\Omega_K$ is a canonical form on $T^*(K)$. Now let $f\colon K \to K$ be a canonical diffeomorphism such that $\bar{f}(K) \subset V$. Then $F(f(K)) \subset W \subset T^*(K)$, and if $f$ together with its first derivatives is sufficiently close to the identity map, then since also $F(\bar{1}(K)) = s_0(K) \subset T^*(K)$, $0 \in \Lambda^1(K)$, it intersects the fibres of the projection $\pi_K\colon T^*(K) \to K$ transversally, that is, it is of the form $s_\lambda(K) = F(\bar{f}(K))$, $\lambda \in \Lambda^1(K)$. But $F(\bar{f}(K))$ is Lagrangian, therefore $d\lambda = 0$, and the canonical field $X = \Gamma^{-1}(\pi_K^*(\lambda))$ is such that $A_t \cdot s_\lambda = s_{(1-t)\lambda}$, where $\{A_t\} \Leftrightarrow X$ (Corollary 3.2). We consider the manifolds $F^{-1}(s_{t\lambda}(K))$. If $f$ together with its first derivatives is sufficiently close to 1, then $F^{-1}(s_{t\lambda}(K))$ has the form of the "graph" $\bar{f}_t(K)$, where $f_t\colon K \to K$ is some smooth canonical map, $0 \leqslant t \leqslant 1$, owing to the fact that, as before, the Lagrangian manifolds $F^{-1}(s_{t\lambda}(K))$ like $\Delta$, intersect the fibres of $\pi_1$ transversally. Finally, since $f$ is "sufficiently close" to 1, the $f_t$ are also diffeomorphisms. By construction, $f_0 = \mathrm{id}$, $f_1 = f$.■

7.4. We now show that all Hamiltonians have the same structure in neighbourhoods of their non-singular points.

PROPOSITION 7.2. *Let $H \in \mathcal{F}(K)$ and $dH_x \neq 0$. Then there are canonical coordinates $(p, q)$ near $x$ such that $H = q_1$.*

Let $(\mathcal{P}, \mathcal{Q})$ be some canonical coordinate system near $x$. By a linear (non-homogeneous) canonical transformation of the variables $(\mathcal{P}, \mathcal{Q})$ we can, clearly, achieve that $dH_x = d\mathcal{Q}_1|_x$, $H(x) = \mathcal{Q}_1(x)$. Assuming that is so, we consider the function $H_t = H + t(\mathcal{Q}_1 - H)$ and we try to find a family of canonical transformations $F_t$ such that $F_t^*(H_t) = \mathrm{const}$. $F_0 = \mathrm{id}$. If we can do this, then in view of the fact that $H = F_0^*(H_0) = F_1^*(\mathcal{Q}_1)$, $F_1$ is the transformation we are looking for. But

$$\frac{d}{dt} F_t^*(H_t) = F_t^* \,[Y^t(H_t) + \mathcal{Q}_1 - H],$$

where $\mathcal{D}(K) \ni Y^t \Leftrightarrow \{F_t\}$, therefore, the condition $\frac{d}{dt} F_t^*(H_t) = 0$ leads us to the condition

$$Y^t(H_t) + \mathcal{Q}_1 - H = 0, \qquad \text{or} \qquad X^t(h_t) = \mathcal{Q}_1 - H,$$

where $\Gamma(X^t) = dH_t$, and $h_t$ is a local Hamiltonian of $Y^t$. Let $N$ be a hypersurface in $K$ containing $x$ and transversal to all the trajectories of all the fields $X^t$, $0 \leqslant t \leqslant 1$. Since by construction $dH_t|_x = d\mathcal{Q}_1|_x = dH_x$, we have $X^t|_x = -\left.\frac{\partial}{\partial \mathcal{P}_1}\right|_x$, so that $N$ can be taken to be a sufficiently small neighbourhood of $x$ on a hypersurface $N'$ containing $x$ and not tangent to the vector $\left.\frac{\partial}{\partial \mathcal{P}_1}\right|_x$. Now when we give on $N$ an arbitrary function $\varphi_t$ as initial condition $h_t|_N = \varphi_t$, we can find $h_t$ itself by solving the equations $X^t(h_t) = \mathcal{Q}_i - H$. In view of the fact that the trajectories of $X^t$ do not touch $N$, this equation can be solved by the method of characteristics. Thus, the required fields $Y^t = \Gamma^{-1}(dh_t)$ exist and hence so does the required transformation $F_t$.■

COROLLARY 7.3. *Let $x_i \in K_i$ $(i = 1, 2)$, where $K_i$ are canonical manifolds of the same dimension. If $H_i \in \mathcal{F}(K_i)$, $H_1(x_1) = H_2(x_2)$, and $dH_{i}|_{x_i} \neq 0$ $(i = 1, 2)$, then there are neighbourhoods $U_i$ of $x_i$ and a canonical transformation $f\colon U_1 \to U_2$ such that $f^*(H_2) = H_1$.*

We introduce canonical coordinate systems $(p, q)$ and $(\mathcal{P}, \mathcal{Q})$ near $x_i$ such that $H_1 = q_1$, $H_2 = \mathcal{Q}_1$, and we identify them.■

## §8. Invariant Hamilton–Jacobi theory

8.1. In this section we discuss in an invariant way the theory of the Hamilton–Jacobi (H–J) partial differential equations and the connection between solutions of these and of the corresponding Hamiltonian system. Essentially, the whole H–J theory is covered by the following two almost obvious geometrical facts: 1) a solution of an H–J equation is a Lagrangian manifold lying on a level surface $\{H = c\}$; 2) if there is an $n$-parameter family of solutions of an H–J equation, that is, an $n$-parameter family of Lagrangian manifolds lying on level surfaces of $H$, then the parameters, regarded as functions on $\Phi$, form a family of $n$ pairwise commuting symmetries of $H$.

The usual method of developing the H–J theory achieves a certain mystery by the introduction of coordinates. We draw the reader's attention to the fact that all the lengthy passages in the following exposition are concerned only with harmonizing the invariant point of view with the usual coordinate one.

8.2. We recall that an H–J equation is a first order partial differential equation of the special form $H(q, \frac{\partial u}{\partial q}) = c$, where $q = (q_1, \ldots, q_n)$, $\frac{\partial u}{\partial q} = \left(\frac{\partial u}{\partial q_1}, \ldots, \frac{\partial u}{\partial q_n}\right)$ and $c$ is a constant.[1]

A special feature of this equation consists in the fact that $H$ does not depend on $u$. We interpret this invariantly, and to do this we remark that, as a consequence of the specific nature of the H–J equation, together with each function $u(q)$ that is a solution of it, all functions of the form $u(q)$ + const are also solutions. In view of the fact that the totality of such functions is uniquely characterized by the differential of each of them, we can say that the H–J equation is an equation in the differential $du$ of

[1] The H–J equations are usually written in the form $u_t = \varphi(t, x, u_x)$, $x = (x_1, \ldots, x_n)$, $u_x = (u_{x_1}, \ldots, u_{x_n})$. Setting $x_i = q_i$, $t = q_{n+1}$, $H(q, u_q) = u_t - \varphi(t, x, u_x)$, we arrive at the form indicated.

an unknown function $u(q)$. We consider next the equation $H(q, p) = c$ in $T^*(\mathbf{R}^n)$, $p \in T_q^*(\mathbf{R}^n)$. This is the equation of a level surface $\Gamma$ of $H$ on $T^*(\mathbf{R}^n)$. If $u(q)$ is a solution of the H–J equation, then the graph of its differential $L = \left\{(q, p) | \, p = \frac{\partial u}{\partial q}(q)\right\}$ lies on $\Gamma$. Clearly, the converse is also true: if $L$ is the graph of the differential of some function $u(q)$ and if $L$ lies on $\Gamma$, then $u(q)$ is a solution of the H–J equation. But, as we already know, the graphs of differentials of functions on $\mathbf{R}^n$ are Lagrangian manifolds in $T^*(\mathbf{R}^n)$ that project one-to-one onto $\mathbf{R}^n$, and vice versa. Therefore, the solutions of the H–J equation are Lagrangian manifolds lying on $\Gamma$ and projecting one-to-one onto $\mathbf{R}^n$. On the other hand, even the simplest H–J equation for the harmonic oscillator $\frac{1}{2}\left(\frac{\partial u}{\partial q}\right)^2 + \frac{1}{2} q^2 = c$ has no global solution, that is, a solution in single-valued functions defined on the whole of $\mathbf{R}^n$. Fig. 1 illustrates the graph of a many-valued function $u(q)$, whose differential has the circle $p^2 + q^2 = 2c$ as its graph.

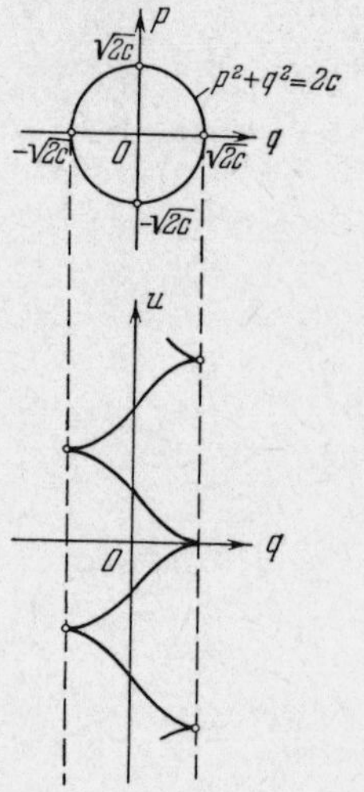


Fig. 1.

It is therefore sensible to give up asking for a one-to-one projection and to count as solutions of the H–J equation arbitrary Lagrangian manifolds lying on the surfaces $\Gamma$. These Lagrangian manifolds, which have no one-to-one projections, can be regarded as graphs of the differentials of many-valued functions (see §1.3).

Resuming what we were saying above, we find it natural to define an H–J equation as an arbitrary hypersurface $\Gamma$ (possibly with singularities) in $T^*(M)$ or, more generally, on an arbitrary canonical manifold $K$, and its solution as an arbitrary Lagrangian manifold $L$ lying on $\Gamma$. Next, by the H–J problem we mean the task of finding Lagrangian manifolds lying on $\Gamma$. Then the statement of the H–J problem is indeed invariant, since the terms in which it is expressed do not appeal to coordinates and are invariant under canonical transformations.

REMARK 8.1. Arguing in a similar way, one can develop a theory of many-valued solutions of arbitrary non-linear partial differential equations (see [13]).

**8.3.** We now turn to the H–J theory, which consists in the study of the connection between solutions of the H–J equations and those of the system of Hamilton's ordinary differential equations. This connection is given in an invariant way by the absorption principle.

Let $\Gamma$ be a hypersurface (that is, $\dim \Gamma = 2n - 1$) on a canonical manifold $K$. As we saw earlier (see §1.4), there is then a field of direction lines on $\Gamma$. Namely, at each point $x \in \Gamma$ there is defined a line (= one-dimensional subspace) $l_x(\Gamma) \subset T_x(\Gamma) \subset T_x(K)$, the direction line of the hypersurface $T_x(\Gamma)$ in $T_x(K)$ with respect to the skew-bilinear form $\Omega_x$.

**The absorption principle.** Every Lagrangian manifold $L$ lying wholly on $\Gamma$ is tangent to the direction field of $\Gamma$.

What we are saying means that $l_x(\Gamma) \subset T_x(L)$ for every point $x \in L$. But this follows from the local (linear) absorption principle (see §1.4). For the linear space $T_x(L)$ is a Lagrangian plane in $T_x(K)$ with respect to $\Omega_x$. Further, since $L \subset \Gamma$, we have $T_x(L) \subset T_x(\Gamma)$. Consequently, the Lagrangian plane $T_x(L)$ contains the direction line of $T_x(\Gamma)$, that is, $l_x(\Gamma)$. ■

Properly speaking, from now on we use the absorption principle in the following form: if a Lagrangian manifold $L$ lies on a non-singular part of a level surface of $H$, then the corresponding Hamiltonian vector field $X_H$ is tangent to $L$. This formulation of the absorption principle follows straight from the first. For if $\Gamma = \{H = c\}$ and $x \in L$, then $X_H|_x \in l_x(\Gamma) \subset T_x(L)$.

Since $X_H$ is tangent to $L$ under the preceding conditions, $L$ is woven out of the trajectories of $X_H$, that is, if $x \in L$, then the trajectory of $X_H$ passing through $x$ lies wholly in $L$, therefore, $L$ is invariant under the shift operators $\{A_t\} \Leftrightarrow X_H$. This can be regarded as a third formulation of the absorption principle.

**8.4.** A method of constructing solutions of an H–J problem follows immediately from the third formulation of the absorption principle. Namely, let $L = \bigcup_t A_t(L')$, where $\{A_t\} \Leftrightarrow X_H$, and let $L'$ be an arbitrary $(n-1)$-dimensional manifold lying on $\Gamma$, transversal to $X_H$ and such that $\Omega|_{L'} = 0$ (that is, $\Omega|_{L'}$ is pre-Lagrangian). For, if $x \in A_t(L')$, then $T_x(L) = T_x(A_t(L')) \oplus \{\lambda X_H|_x, \lambda \in \mathbf{R}\} = T_x(A_t(L')) \oplus l_x(\Gamma)$. Hence it follows that $T_x(L)$ is a Lagrangian plane in $T_x(K)$, that is, $L$ is a Lagrangian manifold.

In other words, $L$ is the union of the trajectories of $X_H$ issuing from the points of the submanifold $L'$. Hence it is clear that, although the field $X_H$ and the operators $\{A_t\}$ are connected with a specific choice of the function $H$ having $\Gamma$ as a level surface, the manifold $L$ above, which is a solution of the H–J problem, does not depend on this choice, that is, is determined only by $\Gamma$ and the pre-Lagrangian $(n-1)$-dimensional manifold $L'$ lying on $\Gamma$.

8.5. From what we have said above it follows that a solution of an H–J problem is everywhere locally of the form $L = \bigcup_t A_t(L')$, where $L'$ is an $(n-1)$-dimensional pre-Lagrangian manifold in $\Gamma$ (generally speaking, this is not true globally). Therefore, at least locally, a solution of an H–J problem reduces to 1) finding all possible $L'$ and 2) constructing $L$ from $L'$. Technically, 2) is connected with the solution of a system of ordinary Hamilton differential equations with initial conditions lying on $L'$, which are in principle always soluble. We now consider the question of constructing $(n-1)$-dimensional pre-Lagrangian manifolds $L' \subset \Gamma = \{H = c\}$.

We suppose first that $K = T^*(M)$ and $\pi|_{L'}$ is a diffeomorphism of $L'$ onto $\pi(L') = N$. Then $N$ is an $(n-1)$-dimensional submanifold of $M$. We note, next, that for any point $x \in N$ we have the natural embedding $\alpha_x = \alpha_x(N)\colon T_x(N) \to T_x(M)$ and the associated projection $\beta_x = \alpha_x^*\colon T_x^*(M) \to T_x^*(N)$. Let $T_N^*(M) = \bigcup_{x \in N} T_x^*(M) \subset T^*(M)$. Now $T_N^*(M)$ is clearly a $(2n-1)$-dimensional submanifold of $T^*(M)$ that maps naturally onto $T^*(N)$ via the map $\beta = \alpha^* = \alpha^*(N)\colon T_N^*(M) \to T^*(N)$, where $\beta|_{T_x^*(M)} = \beta_x$. Thus, $\beta$ is a linear fibre map, that is, the fibre $T_x^*(M)$ is mapped by $\beta$ onto the fibre $T_x^*(N)$. In particular, all the inverse images $\mathbf{R}_y = (\beta^{-1})(y) = (\beta_x^{-1})(y)$, $y \in T_x^*(N)$, are lines in the fibres $T_x^*(M)$. We call these lines *characteristic* with respect to $N$. Now let $\Gamma_N = \Gamma \cap T_N^*(M)$; then $\Gamma_N$ is a $(2n-2)$-dimensional manifold with singularities. We denote by $\Gamma_N^0$ the non-singular part of $\Gamma_N$, that is, $\Gamma_N$ with its singular points discarded. Then $\Gamma_N^0$ is a $(2n-2)$-dimensional manifold in $T_N^*(M) \subset T^*(M)$. We call a point $z \in \Gamma_N^0$ *characteristic* if the characteristic line $\mathbf{R}_y$, $y = \beta(z)$, touches $\Gamma_N^0$ at $z$, that is, $T_z(\mathbf{R}_y) \subset T_z(\Gamma_N^0)$. The point of the concept of a characteristic point is made clear by the following proposition.

PROPOSITION 8.1. *The map $\beta|_{\Gamma_N^0}$ is a local diffeomorphism everywhere except at the characteristic points.*

If a point $z \in \Gamma_N^0$ is not characteristic, then $\operatorname{Ker}(d\beta)_z = T_z(\mathbf{R}_y) \not\subset T_z(\Gamma_N^0)$, therefore, $(d\beta)|_z|_{T_z(\Gamma_N^0)} = d(\beta|_{\Gamma_N^0})|_z$ is an isomorphism. These same arguments show that $\operatorname{Ker}(d(\beta|_{\Gamma_N^0})|_z) = T_z(\mathbf{R}_y)$, if $z$ is characteristic.■

COROLLARY 8.1. *If $P \subset T^*(N)$ is some m-dimensional submanifold and $P \cap \beta(\Gamma_N^c) = \emptyset$, where $\Gamma_N^c$ is the totality of all characteristic points in $\Gamma_N^0$, then $\widetilde{P} = (\beta|_{\Gamma_N^0})^{-1}(P)$ is an m-dimensional submanifold of $\Gamma_N^0$.*

Clearly, $\beta$ has the following natural property.

PROPOSITION 8.2. $\beta^*(\rho_N) = \rho|_{T_N^*(M)}$, *where $\rho_N$ is the universal 1-form on $T^*(N)$ and $\rho$ that on $T^*(M)$.*

COROLLARY 8.2. *The line $T_z(\mathbf{R}_y) \subset T_z(T_N^*(M))$, $z \in T_N^*(M)$, $y = \beta(z)$, is a direction line of the hypersurface $T_N^*(M) \subset T^*(M)$.*

We must show that $\Omega_z(\xi, \eta) = 0$, $\xi \in T_z(\mathbf{R}_y)$, $\eta \in T_z(T_N^*(M))$. But $\Omega|_{T_N^*(M)} = d\rho|_{T_N^*(M)} = \beta^*(d\rho_N)$. Therefore, if $X$ is a vector field on $T_N^*(M)$ such that $X_a \in T_a(\mathbf{R}_{\beta(a)})$ for all $a \in T_N^*(M)$ and $\xi = X_z$, and if $Y \in \mathscr{D}(T_N^*(M))$, $Y_z = \eta$, then $d(\beta^*(\rho_N))(X, Y) = X(\beta^*(\rho_N))(Y) - Y[\beta^*(\rho_N)(X)]$. But $X$ is compatible with $\beta$, and $\beta(X) = 0$. Therefore, $\beta^*(\rho_N)(X) = \rho_N(\beta(X)) = 0$ and $X(\beta^*(\rho_N)) = \beta^*(\beta(X)(\rho_N)) = 0$, that is, $d(\beta^*(\rho_N))(X, Y) = 0$. In particular, $0 = d(\beta^*(\rho_N))|_z(X_z, Y_z) = \Omega_z(\xi, \eta)$.∎

COROLLARY 8.3. *The form $\Omega|_{\Gamma_N^0 \setminus \Gamma_N^c}$ is non-degenerate, and the map $\beta|_{\Gamma_N^0 \setminus \Gamma_N^c}$ is locally a canonical diffeomorphism of $(\Gamma_N^0 \setminus \Gamma_N^c, \Omega|_{\Gamma_N^0 \setminus \Gamma_N^c})$ into $(T^*(N), d\rho_N)$.*

Combining Corollaries 8.1 and 8.3, we came to the following result.

PROPOSITION 8.3. *Let $L'' \subset T^*(N)$ be some Lagrangian manifold. Then $L' = (\beta|_{\Gamma_N^0 \setminus \Gamma_N^c})^{-1}(L'') \subset (\Gamma_N^0 \setminus \Gamma_N^c)$ is a Lagrangian submanifold of the canonical manifold $(\Gamma_N^0 \setminus \Gamma_N^c, \Omega|_{\Gamma_N^0 \setminus \Gamma_N^c})$. In particular,* $\dim L' = n - 1$, *$\Omega|_{L'} = 0$, and $L' \subset \Gamma$.*

Proposition 8.3 provides us with the method of constructing $(n-1)$-dimensional pre-Lagrangian manifolds $L' \subset \Gamma$ that we were talking of at the beginning of this section. For, given some function $\varphi$ on $N$, we can take for $L''$ the graph of the differential of this function, that is, $L'' = s_{df}(N) \subset T^*(N)$, and then find the inverse image $(\beta|_{\Gamma_N^0 \setminus \Gamma_N^c})^{-1}(L) = L$, a task that is simplified by the fact that $\beta|_{\Gamma_N^0 \setminus \Gamma_N^c}$ is a local diffeomorphism. Since from the local point of view any Lagrangian manifold in $T^*(N)$ can be represented by means of a suitable canonical transformation in the form of the graph $s_{df}(N)$, the above procedure makes it possible to construct all the pre-Lagrangian manifolds $L'$ described in Proposition 8.3.

Finally, we remark that the method described above for constructing the manifolds $L'$ works not only when $K = T^*(M)$, but also when $K = U \subset T^*(M)$ is some domain of $T^*(M)$. To see this we need only restrict all the above arguments to $U$. Furthermore, every domain $V$ of an arbitrary canonical manifold $K$ in which canonical coordinates are introduced is canonically isomorphic to some domain $U \subset T^*(M)$, therefore, our method works inside such a domain. Finally, we recall that any point $x \in K$ has a canonical neighbourhood $V$. Thus, this method allows us to construct, at least locally, all possible pre-Lagrangian manifolds $L' \subset \Gamma$.

REMARK 8.2. Suppose that $N = \{q_1 = 0\}$ in suitable local coordinates $(q_1, \ldots, q_n)$. Then to specify a function $\varphi$ on $N$ is to specify $\varphi$ for $q_1 = 0$, that is, to give initial conditions by the usual procedure. Therefore, what we have set out above can be regarded as an invariant theory of giving initial conditions.

**8.6.** Thus, we have clarified how solutions of an H–J problem are constructed from solutions of Hamilton's equations. We now examine this procedure in the reverse direction.

To do this we consider the one-parameter family $\Gamma_c = \{H = c\}$, $c \in \mathbf{R}$, of level surfaces of $H$ and the corresponding family of H–J problems. For the sake of brevity, henceforth we call this family of problems the H–J problem.

We suppose that we have an $n$-parameter family of solutions of the H–J problem. More precisely, let $U$ be a domain in $K$, $V$ a domain in $\mathbf{R}^n$ (the parameter space) and $\varphi: U \to V$ a regular map onto, where $\varphi^{-1}(v) = L_v$ for all $v \in V$ is a Lagrangian manifold and a solution of the Hamilton–Jacobi problem, that is, lies on any one of the level surfaces of $H$. We denote by $Q_i: V \to \mathbf{R}$ ($i = 1, \ldots, n$) the coordinate functions on $V$. If $V \ni v = (c_1, \ldots, c_n)$, then $L_v = \bigcap_{i=1}^{n} \{\bar{Q}_i = c_i\}$, where $\bar{Q}_i = \varphi^*(Q_i)$ are the parameters of the Lagrangian manifolds $L_v$, regarded as functions on $K$. The following lemma shows that the $\bar{Q}_i$ form a family of $n$ commuting functions on $U$ whose differentials are linearly independent owing to the regularity of $\varphi$.

LEMMA 8.1. *Let $L_v$ be an $n$-parameter family of Lagrangian manifolds lying on level surfaces of both $f_1$ and $f_2$, where $f_1, f_2 \in \mathscr{F}(K)$. Then* $\{f_1, f_2\} = 0$.

In other words, $f_i = \varphi^*(h_i)$, where $h_i \in \mathscr{F}(V)$ $(i = 1, 2)$. Clearly, $\{\varphi^*(h_i) = c_i\} = \varphi^{-1}\{h_i = c_i\}$, so that all the manifolds $L_v$ lie wholly on the level surfaces of $\varphi^*(h_i)$. By the absorption principle the fields $X_{\varphi^*(h_i)}$ are tangent to the $L_v$. In particular, $X_{\varphi^*(h_1)}$ is tangent to all $L_v$, $v \in V$, hence, to the level surfaces of $\varphi^*(h_2)$. Therefore, $0 = X_{\varphi^*(h_1)}(\varphi^*(h_2)) = \{\varphi^*(h_1), \varphi^*(h_2)\} = \{f_1, f_2\}$. ∎

As the $\bar{Q}_i$ are constant along the $L_v$, $v \in V$, and the latter by the absorption principle, consist of trajectories of $X_H$, the $\bar{Q}_i$ are also constant on the trajectories of $X_H$, that is, they are symmetries of the Hamiltonian system with Hamiltonian $H$. So we have proved the following result.

PROPOSITION 8.4. *Suppose that the Hamilton–Jacobi problem has a non-degenerate $n$-parameter family of solutions in some domain $U$ of the canonical manifold $K$. Then the corresponding Hamiltonian system is completely integrable in $U$, and as a collection of commuting symmetries we can take the parameters, regarded as functions on $U$.*

## §9. Hamiltonians of mechanical type

In this section we define and study Hamiltonians of mechanical type. The reason for singling out these Hamiltonians from the rest is the following. Firstly, the kinetic energy of a mechanical system is a positive-definite quadratic form on its configuration space $M$, in other words, a Riemannian metric on $M$. Secondly, as is well known, the energy of a mechanical system is made up from its kinetic and its potential energy, which depends only on the position of the system. Therefore, the Hamiltonian (= energy) of a mechanical system has the form: Riemannian metric (regarded as a function on $T^*(M)$) plus $\pi^*(V)$, $V \in \mathcal{F}(M)$. Relativistic mechanics, which is concerned with the Minkowski metric, makes it necessary to consider also pseudo-Riemannian metrics. For this reason it is natural to call the sum of a pseudo-Riemannian metric and a function of position, that is, a function of the form $\pi^*(V)$, $V \in \mathcal{F}(M)$, a *Hamiltonian of mechanical type*.

Of course, not every Hamiltonian of mechanical type is the Hamiltonian of a real mechanical system. However, it is convenient to study the common properties of genuine mechanical Hamiltonians in a wider framework.

We remark that the possibility we pointed out above of regarding kinetic energy as a Riemannian metric leads to deep interconnections between geometry and mechanics, many aspects of which are still waiting to be studied. We give some examples of this kind in the next section.

**9.1.** A *metric* on $M$ is a non-degenerate symmetric $\mathcal{F}(M)$-bilinear form $G(X, Y)$ on the $\mathcal{F}(M)$-module $\mathcal{D}(M)$. Its non-degeneracy means that the map $g: \mathcal{D}(M) \to \Lambda^1(M)$, $g(X) = G(X, \ )$ is an isomorphism of $\mathcal{F}(M)$-modules. As well as by $G$, a metric can be specified by a non-degenerate symmetric $\mathcal{F}(M)$-bilinear form $\bar{G}$ on $\Lambda^1(M)$: $\bar{G}(\omega, \nu) = G(g^{-1}(\omega), g^{-1}(\nu))$, $\omega, \nu \in \Lambda^1(M)$. The corresponding $\mathcal{F}(M)$-quadratic form $\bar{G}(\omega, \omega)$ defines a function $\tilde{G} \in F(\Phi)$ by the equation $\bar{G}(\omega, \omega) = S_w^*(\tilde{G})$. Clearly, $\bar{G}$ is uniquely recoverable from $\tilde{G}$.

We call a function $H \in F(\Phi)$ of the form $\tilde{G} + \pi^*(V)$, $V \in F(M)$ a *Hamiltonian of mechanical type*, $\tilde{G}$ the *kinetic* and $\pi^*(V)$ the *potential energy* of the system. We show below that the representation of $H$ in the form $\tilde{G} + \pi^*(V)$ is unique.

**9.2.** To compare our approach to mechanics with the usual one based on variational principles (that is, with the Lagrangian approach), we need some new concepts. First of all, we have to understand the invariant meaning of a system of second-order ordinary differential equations on a manifold, which we need for an invariant treatment of the Euler-Langrange equations of motion. We recall that such a system in $\mathbf{R}^n$ has the form $\ddot{q}_i = f_i(q, \dot{q})$ $(i = 1, \ldots, n)$ or, equivalently,

$$\dot{q}_i = v_i, \ \dot{v}_i = f_i(q, v) \ (i = 1, \ldots, n).$$

The latter is a system of first-order differential equations, with which, according to §0, there is associated a vector field on a $2n$-dimensional manifold with local coordinates $(q, v)$. In view of the geometrical meaning of the equations $q_i = v_i$ it is natural to assume that this field is given on the tangent space to the configuration manifold $M$, and the equations $\dot{q}_i = v_i$ express the fact that it is *special*. More precisely, a field $Y \in \mathscr{D}(T(M))$ is called special if $d\bar{\pi}_{(q,v)}(Y_{(q,v)}) = v \quad \forall (q, v) \in T(M)$.

A more substantial global definition of a special field is based on the concept of a *universal derivative*. We call a function $\tau(f) \in \mathscr{F}(T(M))$ a universal derivative of $f \in \mathscr{F}(M)$ if for all $X \in \mathscr{D}(M)$

$$S_X^*(\tau(f)) = X(f). \tag{9.1}$$

Here $S_X: M \to T(M)$ is the graph of the vector field $X$, that is, $S_X(q) = X_q \in T_q(M) \subset T(M)$, $q \in M$. Clearly, $\tau$ defines an **R**-linear derivation of the ring $\mathscr{F}(M)$ into the $\mathscr{F}(M)$-module $F(T(M))$. Since (9.1) is satisfied for all $f \in \mathscr{F}(M)$, it can be written in operator form:

$$S_X^* \circ \tau = X. \tag{9.2}$$

Using $\tau$, we can define special fields on $T(M)$: a field $Y \in \mathscr{D}(T(M))$ is special if $Y \circ \bar{\pi}^* = \tau$. This definition is, clearly, equivalent to the first.

**9.3.** Hamiltonian vector fields on $T^*(M)$ are closely connected with special vector fields on $T(M)$. This connection is realized by means of the Legendre map, which we are now going to describe.

Let $X \in \mathscr{D}(T^*(M))$. The Legendre map associated with the field $X$ is the map $\alpha_X: T^*(M) \to T(M)$ that is defined by

$$\alpha_X(q, p) = (q, d\pi_{(q,p)}(X_{(q,p)})) \in T_q(M) \subset T(M). \tag{9.3}$$

We note that $\alpha_{X_1} = \alpha_{X_2}$ if and only if $(X_1 - X_2)$ is $\pi$-vertical, that is, $\pi(X_1 - X_2) = 0$. If the Hamiltonian field $X_H$ is vertical, then $X = \pi^*(h)$ for some $h \in \mathscr{F}(M)$; hence, we have the following result.

LEMMA 9.1. *$\alpha_{X_H} = \alpha_{X_{H'}}$ if and only if $H = H' + \pi^*(h)$, $h \in \mathscr{F}(M)$.*

Being clearly fibred, $\alpha_X$ takes sections of the cotangent bundle to sections of the tangent bundle, and we arrive at a map $\bar{\alpha}_X: \Lambda^1(M) \to \mathscr{D}(M)$. It follows from the definition that

$$\bar{\alpha}_X(\omega) = S_\omega^* \circ X \circ \pi^*. \tag{9.4}$$

When $X = X_H$, we write $\alpha_H$ and $\bar{\alpha}_H$ instead of $\alpha_X$ and $\bar{\alpha}_X$, respectively. It then follows from (9.4) that

$$\bar{\alpha}_H(\omega)(f) = S_\omega^*(\{H, \pi^*(f)\}). \tag{9.5}$$

THEOREM 9.1. *If $\alpha_H$ is a diffeomorphism, then $\alpha_H(X_H)$ is special.*

We verify that $\alpha_H(X_H) \circ \bar{\pi}^*(f) = \alpha_H^{*-1} \circ X_H \circ \alpha_H^* \circ \pi^*(f) = \tau(f)$, $\forall f \in \mathscr{F}(M)$.

Since $\bar{\pi} \circ \alpha_H = \pi$, this reduces to the equation $\alpha_H^{*-1} \circ X_H \circ \pi^*(f) = \alpha_H^{*-1}(\{H, \pi^*(f)\}) = \tau(f)$, which, by (9.1), is equivalent to the fact that $S_Y^* \circ \alpha_H^{*-1}(\{H, \pi^*(f)\}) = Y(f)$ for all $Y \in D(M)$. Setting $\omega = \bar{\alpha}_H^{-1}(Y)$ and noting that $Y = \bar{\alpha}_H(\omega)$, $\alpha_H^{-1} \circ S_Y = S_w$, we arrive at (9.5).■

**9.4.** Now we can give an invariant characterization of Hamiltonians of mechanical type.

THEOREM 9.2. *A Hamiltonian H is of mechanical type if and only if the map $\bar{\alpha}_H$ is an isomorphism of $\mathcal{F}(M)$-modules or, equivalently, $\alpha_H|_{T_q^*(M)}: T_q^*(M) \to T_q(M)$ is an isomorphism of linear spaces for all $q \in M$.*

The equivalence of the two formulations follows from the method of constructing $\bar{\alpha}_H$ from $\alpha_H$.

If $H$ is a Hamiltonian of mechanical type, then it follows from the definition that $\bar{\alpha}_H$ is an isomorphism. Let $\bar{\alpha}_H$ be an isomorphism. We define $\tilde{G} \in F(\Phi)$ by $2\pi^* \circ S_{df}^*(\tilde{G}) = \{\{H, \pi^*(f)\}, \pi^*(f)\}$, $f \in \mathcal{F}(M)$. It is not hard to see that $\alpha_H = \alpha_{\tilde{G}}$. By Lemma 9.1, $H - \tilde{G} = \pi^*(V)$, $V \in \mathcal{F}(M)$. ■

Let $V = 0$ (a free system) and $\tilde{G} = g^{ij}(q)p_i p_j$ (locally). A straightforward calculation shows that the equations corresponding to the field $\alpha_H(X_H)$ have the form

$$\ddot{q}_i + \Gamma^i_{kl}\dot{q}_k\dot{q}_l = 0 \qquad (i = 1, \ldots, n), \tag{9.6}$$

where $\Gamma^i_{kl}$ are the Christoffel symbols of the Riemannian connection associated with the metric $\| g_{ij} \| = \| g^{ij} \|^{-1}$. This is the equation of a geodesic of the metric $\| g_{ij} \|$ (see [1]). From this and the equation $\bar{\pi} \cdot \alpha_H = \pi$ we obtain the next result.

PROPOSITION 9.1. *The projection of a trajectory of $X_{\tilde{G}}$ on M is a geodesic of the metric G.*

REMARK 9.1. Proposition 9.1 admits the following interpretation: "a ball rolling by inertia" on a Riemannian manifold describes a geodesic.

**9.5.** The connection between the Lagrangian and the Hamiltonian approaches to mechanics consists in the following.

PROPOSITION 9.2. *If $\alpha_H$ is a diffeomorphism, then the system of second-order equations corresponding to the special field $\alpha_H(X_H)$ is equivalent to the system of Euler-Lagrange equations corresponding to the Lagrangian $\mathcal{L} = (\alpha_H^{-1})^*(\rho(X_H) - H)$.*

We cannot give here an invariant proof of this proposition because for this we would need an invariant account of the Lagrangian formalism (see [6], [8]). The reader can verify it by appealing to coordinates.

## §10. Mechanical systems. Examples

From the mathematical point of view, to study the motion of a given mechanical system whose configuration space is a manifold $M$ means to study the Hamiltonian system $X_H$ on $T^*(M)$, where $H$ is the total energy of the system. This approach to classical mechanics can be called Hamiltonian mechanics. If we call the theory expounded earlier the Hamiltonian formalism, then we can say that the language of Hamiltonian mechanics is the Hamiltonian formalism.

However, the Hamiltonian formalism, serves also as a language for a whole range of other theories. One example of this kind is geometrical optics. The equation describing the propagation of rays of light in an inhomogeneous and anisotropic medium is of the form

$$\sum_{i=1}^{3} \frac{1}{\alpha_i(x)} \left( \frac{\partial f}{\partial x_i} \right)^2 = 1, \tag{10.1}$$

where $\alpha_i(x)$ is the speed of propagation of light at the point $x$ along the $x_i$-axis. Now (10.1) is the Hamilton–Jacobi equation corresponding to the Hamiltonian $\frac{p_1^2}{\alpha_1} + \frac{p_2^2}{\alpha_2} + \frac{p_3^2}{\alpha_3}$. The corresponding mechanical system is of mechanical type, and as a result there arises an optical-mechanical analogy which has played an exceptionally important role in the development of a whole range of areas of physics and mathematics. In fact, it was just this analogy that led Hamilton to create his mechanics (see [14]). It also played a decisive role in the research of Schrödinger, leading to the discovery of the fundamental equation of non-relativistic quantum mechanics – the Schrödinger equation (see [15]).

Here is another example. The symbol of a linear differential operator $\mathscr{D}$ on a manifold $M$ can be regarded as a function on $T^*(M)$, that is, as a Hamiltonian (see [16]). Formally, to do this we must make the substitution $\frac{\partial}{\partial x_i} \Rightarrow p_i$ in the highest terms of $\mathscr{D}$. The properties of the Hamiltonian system arising in this way are closely connected, for example, with properties of local solubility of $\mathscr{D}$ (see [17]). Altogether, the Hamiltonian formalism is widely used in questions of various kinds in the theory of linear differential operators (see [18]–[20]). This is the most important domain of its non-mechanical application.

In this section we restrict ourselves to some specific mechanical questions connected with the use of the Hamiltonian method, and we discuss a number of examples.

**10.1.** Thus, to investigate the motion of a mechanical system by the methods of Hamiltonian mechanics, we have to find first the configuration space $M$ of this system, and then to specify on its phase space, that is, on $T^*(M)$, the appropriate Hamiltonian function, that is, the energy of the system. The principles that allow us to give an answer to these two questions form the physical basis of mechanics. As soon as these questions are answered, the relevant mechanical problem becomes purely mathematical and falls within the framework of the Hamiltonian formalism. Thus, Hamiltonian mechanics, as a physical theory, is made up of two parts: the physical, which consists of the principles referred to, and the mathematical, which is the Hamiltonian formalism.

To find the configuration space $M$ of a mechanical system $S$ is a purely geometrical question, because $M$ is, by definition, the totality of all possible positions that $S$ can take. In every mechanical system $S$ one can select a number of points, say $A_1, \ldots, A_m$, whose position in the physical space $\mathbf{R}^3$ determines the position of $S$ (we are not considering systems of the type of a string, which have infinitely many degrees of freedom; for these the Hamiltonian formalism is more complicated, see [59]). The points $A_1, \ldots, A_m$ cannot, generally, speaking, take arbitrary positions in $\mathbf{R}^3$ and in relation to one another, because, one says, of the presence of constraints. Analytically, these constraints are given by relations of the form $f_j(w_1, \ldots, w_n) = 0$ ($j = 1, \ldots, N$), where $w_k = (x_k, y_k, z_k)$ are the coordinates of $A_k$ in $\mathbf{R}^3$. Thus, the position of $S$ can be characterized by a point of the space $\mathbf{R}^{3m} = \{(w_1, \ldots, w_m)\}$ satisfying the system

$$f_j(w_1, \ldots, w_m) = 0 \qquad (j = 1, \ldots, N). \tag{10.2}$$

In other words, the configuration manifold of $S$ is the submanifold $M$ of $\mathbf{R}^{3m}$ given by the system (10.2). Here we assume that the constraints $f_1, \ldots, f_N$ are independent, which means analytically that the conditions of the implicit function theorem are satisfied for (10.2). In practice, systems of mechanical interest, apparently, all satisfy this assumption, from which it follows that the configuration space is a smooth manifold.

It is possible to have constraints that operate instantaneously, and these are included when the system arrives at an extreme position. There are all sorts of restrictions acting as constraints of this kind. For example, appropriate restrictions do not permit the pendulum of a clock to deviate from its equilibrium position by more than an angle $\varphi_0$. Mathematically, the restrictions are described by inequalities of the form $\varphi(w_1, \ldots, w_m) \geqslant 0$, which define a closed region on the manifold (10.2) defined by constraints constantly in force. In other words, the configuration space of a system with restrictions is a manifold with boundary. Constraints of instantaneous operation can also arise when a system arrives at a position such that some of its parts collide. The corresponding points of the configuration space are called exceptional. For example, if the system consists of two free material points, then $M = \mathbf{R}^3 \times \mathbf{R}^3 = \{(w_1, w_2)\}$, where $w_i$ is the radius vector of the $i$-th point ($i = 1, 2$). Then the diagonal $\Delta = \{(w, w)\} \subset \mathbf{R}^3 \times \mathbf{R}^3$ and it alone consists of exceptional points. In situations of a similar sort it is necessary, in describing the motion of a mechanical system, to add to the Hamiltonian formalism rules for the behaviour of the system in positions corresponding to boundary or exceptional points of the configuration space.

In the last example there must be such a rule for the collision of the particles.

A wheel able to roll without friction on a plane is a system with non-holonomic constraints, that is, constraints imposed on the speed (or momentum) of various parts of this system. In this example, the non-holonomic constraint has the form $V = \omega r$, where $V$ is the translational and $\omega$ the angular velocity of the wheel, and $r$ its radius.

To describe non-holonomic mechanical systems the Hamiltonian formalism must be enriched with the theory of reaction constraints.[1]

Thus, within the framework of the Hamiltonian formalism developed above we can study holonomic mechanical systems with configuration spaces without boundary and exceptional points (or their motion in interior domains without exceptional points). We call such systems admissible.

We remark that not every smooth manifold $M$ can play the role of the configuration space of some admissible system. The fact that $M$ is given by a system of equations of the form (10.2) shows that the normal bundle of $M$ is trivial (see [11]). This entails, in particular, the orientability of $M$. There are also other algebraic-topological properties that the configuration space must have.

10.2. Since all possible positions of a mechanical system are represented by points of its configuration space $M$, the motion of $S$ from the position $a_0 \in M$ from time $t_0$ up to time $t_1$ is represented by a curve $\gamma\colon [t_0, t_1] \to M$, $\gamma(t_0) = a_0$, where the point $\gamma(t) \in M$ represents the position of $S$ at time $t$, $t \in [t_0, t_1]$. The tangent vector to $\gamma$ at time $t$ is called the state of the system at this moment of time. Mathematically, the fact that the system in question is holonomic means that any vector tangent to $M$ can be realized as the state of $S$ in the process of one of its possible motions. Therefore, $T(M)$ is the state space of $S$.

The calculation of the kinetic energy of a mechanical system in a given state is based on the following two principles:

A) The kinetic energy of a material point of mass $m$ moving with velocity $V$ is equal to $mV^2/2$;

B) kinetic energy is additive, that is, if a system $S$ is mentally split into two parts $S_1$ and $S_2$, then its kinetic energy in some state is equal to the sum of the kinetic energies of $S_1$ and $S_2$ in those states that are defined by their motion as constituent parts of $S$.

If $S$ consists of finitely many material points, then the rules A) and B) allow us to find the kinetic energy by a simple summation process. In the case of a continuous distribution of mass (for example, in a rigid body) summation is obviously replaced by integration. The procedure indicated leads, by A), to some positive-definite quadratic function on $T_a(M)$ for all $a \in M$, that is, we obtain a Riemannian metric $G$ on $M$. Then for the kinetic part of the Hamiltonian of $S$ we must take the function

[1] See Appendix II.

$1/4\ \widetilde{G} \in \mathcal{F}(T^*(M))$ (see §9.1). The appearance of the factor 1/4 in front of $\widetilde{G}$ is explained by the fact that the kinetic energy is transferred from $T(M)$ to $T^*(M)$ by the Legendre map $\alpha_H^{-1} = \alpha_{\text{kin}}^{-1}$, where $H_{\text{kin}}$ is the kinetic part of the total mechanical energy $H \in \mathcal{F}(T^*(M))$ of $S$, and the linear Legendre map $\alpha_{H_{\text{kin}}}$ differs by a factor 2 from the usual metric identification of the fibres of $T_a^*(M)$ and $T_a(M)$ coming from the metric $(H_{\text{kin}})_a$ on $T_a^*(M)$.

The configuration space $M$ and the kinetic energy $G$ are determined solely by the internal structure of $S$, that is, by the geometry of the system. In contrast, the potential energy is determined from the external conditions (external field of force) in which the motion is proceeding. Since for a system with a given geometry the external conditions can be of a very diverse physical nature, there can be no general principles for finding the potential energy. Nevertheless, in view of the fact that the potential energy depends only on the position of the system, finding the potential energy never presents difficulties in practice if the physics of the interaction of the system with the external force field is known.

The remarks made in the preceding paragraphs are, as a rule, sufficient to construct without difficulty the manifold $M$ and the appropriate Hamiltonian $H$ on $T^*(M)$ for a given specific system $S$. If $\gamma: I \to T^*(M)$ is a trajectory of $X_H$, then $\pi \circ \gamma: I \to M$ describes one of the possible motions of $S$ and, moreover, every curve in $M$ describing the motion of $S$ has the form shown.

It seems to us that this approach to solving specific problems of classical mechanics is most simple and elegant.

**10.3.** It is natural to call the motion of a mechanical system free when external influences are absent, that is, when the motion is determined just by its geometry; the system itself is also called free. The Hamiltonian of a free mechanical system is of the form $\widetilde{G}/4$, where $G$ is the Riemannian metric on $M$ determined by its kinetic energy. As we have seen, the geodesics of $G$ are the trajectories of the motion of the system in question. We show now that very often the study of non-free motion of a mechanical system can be reduced to that of free motion.

Let $H = \widetilde{G} + \pi^*(V)$ be some Hamiltonian of mechanical type and $P_c = \{H = c\}$ one of its level surfaces. Let $U_c \subset M$ be the domain consisting of those $x \in M$ where $V(x) \neq c$. Then the function $\pi^*(V) - c$ vanishes nowhere in $\widetilde{U}_c = \pi^{-1}(U_c)$, and so the hypersurface $P_c \cap \widetilde{U}_c$ in $\widetilde{U}_c$ can be given by the equation $\frac{1}{c-\pi^*(V)}\ \widetilde{G} = 1$. Since $G_c = \frac{1}{c-\pi^*(V)}\ G$ is, clearly, a metric on $U_c$ and $\widetilde{G}_c = \frac{1}{c-\pi^*(V)}\ \widetilde{G}_{|\widetilde{U}_c}$, we see that $P_c \cap \widetilde{U}_c = \{\widetilde{G}_c = 1\}$. This leads to the next lemma.

LEMMA 10.1. *The hypersurface $P_c \cap \widetilde{U}_c$ in $\widetilde{U}_c = T^*(U_c)$ coincides with the unit level surface of the metric $G_c$.*

Owing to this lemma, the fields $X = X_H$ and $Y = Y_{\widetilde{G}_c}$ are tangent to $P_c \cap \widetilde{U}_c$, and, by the uniqueness of the field of direction lines of a hypersurface, we have $Y_a = f(a)X_a$, where $a \in P_c \cap \widetilde{U}_c$ and $f \in \mathcal{F}(P_c \cap \widetilde{U}_c)$. We can find the function $f$ by using the linearity of $\Gamma$: $(d\widetilde{G}_c)_a = f(a)dH_a$. But $d\widetilde{G}_c = \frac{[c-\pi^*(V)]\,d\widetilde{G} + \widetilde{G}\,d\pi^*(V)}{[c-\pi^*(V)]^2}$, so that at the points $a \in P_c \cap \widetilde{U}_c$ where $\widetilde{G}_a = c - \pi^*(V)(a)$, we have

$$(d\widetilde{G}_c)_a = \left[\frac{d\widetilde{G}+d\pi^*(V)}{c-\pi^*(V)}\right](a) = \frac{dH_a}{c-\pi^*(V)(a)},$$ that is, $f = [c-\pi^*(V)]^{-1}|_{P_c \cap \widetilde{U}_c}$.

Let $\gamma\colon I \to T^*(M)$ be a trajectory of $Y$ lying in $P_c \cap \widetilde{U}_c$, and $\gamma'\colon I \to I$ a trajectory of the field $\gamma^*(f^{-1})\frac{d}{dt}$ on $I$. Then $\bar{\gamma} = \gamma \circ \gamma'$ is a trajectory of $X$, that is, $\bar{\gamma}^* \circ X = \frac{d}{dt'} \circ \bar{\gamma}^*$, where $t'$ is a coordinate on $I'$. For

$$\bar{\gamma}^* \circ X = \gamma'^* \circ \gamma^* \circ X = \gamma'^* \circ \gamma^* \circ (f^{-1}Y) = \gamma'^*\left[\gamma^*(f^{-1})\frac{d}{dt} \circ \gamma^*\right] = \frac{d}{dt'} \circ \gamma'^* \circ \gamma^* =$$
$$= \frac{d}{dt'} \circ \bar{\gamma}^*.$$

It follows from this fact that im $\gamma$ = im$\bar{\gamma}$, that is, the trajectories of the Hamiltonian systems with Hamiltonians $H$ and $\widetilde{G}_c$ on $P_c \cap \widetilde{U}_c$ coincide if we regard them as submanifolds of $T^*(M)$. As $f \neq 0$ on $P_c \cap \widetilde{U}_c$, $\gamma'$ is a diffeomorphism of $I'$ into $I$, and we can transfer the parameter $t'$ on $I'$ to $I$ by means of $(\gamma'^*)^{-1}$, that is, reparametrize $I$. Then $\gamma$ is a trajectory of $X$ relative to the parameter $t'$ on $I$. So we can say that the trajectories of the system $X_H$ lying on $P_c \cap \widetilde{U}_c$ are obtained from the trajectories of the system $X_{\widetilde{G}_c}$ lying on $P_c \cap \widetilde{U}_c$ by a reparametrization. To find the parameter $t'$ on $I$ we must solve the equations corresponding to the field $\gamma^*(f^{-1})\frac{d}{dt}$. Therefore, $t'$, as a function of $t$, is given by the equation $t' = \int^t \gamma^*(f)dt$. Bearing in mind what we have said and the results of §9. we obtain the following theorem.

THEOREM 10.1. *Every motion of an admissible mechanical system $S$ with energy $H = c$ is described in the domain $U_c \subset M$ by a geodesic $\nu\colon I \to M$ of the metric $4G_c$, where $G$ is the kinetic energy of $S$, reparametrized by $t' = \int^t \frac{dt}{c-\nu^*(V)}$, where $t$ is the arc length of $\nu(I)$ in the metric $4G_c$.*

The points of the hypersurface $\{V = c\} = M \setminus U_c$, where the metric $G_c$ is not defined, clearly correspond to those positions where the system moving with energy $H = c$ stops instantaneously on arrival (the momentum (or velocity) in this position is zero). If the function $V$ is bounded, $V \leqslant c_0$, then $U_c = M$ for energy values $c > c_0$, hence, the motion of $S$ with energy $c$ is everywhere described by the geodesics of the metric $4G_c$. If $M$ is compact, this is always so.

Theorem 10.1 allows us to use various facts of Riemannian geometry to obtain mechanical results. For example, the theorem of Lyusternik and Fet that on every closed Riemannian manifold (that is, compact without boundary) there is a closed geodesic (see [21]) leads to the following assertion.

THEOREM 10.2. *Every admissible mechanical system with a compact configuration space $M$ for energy values $H >$ max $V$ has at least one periodic motion.*

**10.4.** In this subsection we present some facts concerning symmetries of systems of mechanical type.

When do such systems have non-standard infinitesimal symmetries? To answer this question we find the points $y \in T^*(M)$ where $dH_y = d(\widetilde{G} + \pi^*(V))_y = 0$. Clearly, we have the following result.

LEMMA 10.2. $dH_y = 0$ *if and only if* $y \in M$ *and* $dV_y = 0$.

Using now a remark in §4.2, we obtain the next proposition.

PROPOSITION 10.1. *A system of mechanical type can have a non-standard infinitesimal symmetry only when the differential of $V$ does not vanish anywhere.*

We turn now to non-hidden symmetries, that is, symmetries of the form $\hat{X}$, where $X \in \mathscr{D}(M)$. Let $X \Leftrightarrow \{A_t\}$ and $\hat{X} \Leftrightarrow \{\bar{A}_t\}$. Then directly from the definition of $\bar{A}_t$ and $\hat{X}$ we have $\bar{A}_t^*(H) = \bar{A}_t^*(\widetilde{G} + \pi^*(V)) = \widetilde{A_t^*(G)} + \pi^*(A_t^*(V))$. Thus, $\bar{A}_t^*(H)$ is once more a Hamiltonian of mechanical type, and so its decomposition into the components $\widetilde{A_t^*(G)}$ and $\pi^*(A_t^*(V))$ is unique. If $\hat{X}$ is a symmetry of $X_H$, then $\bar{A}_t^*(H) = H +$ const, hence, $\widetilde{A_t^*(G)} = \widetilde{G}$ and $\pi^*(A_t^*(V)) = \pi^*(V) +$ const, from which it is obvious that $A_t^*(G) = G$, $A_t^*(V) = V +$ const or, equivalently, $X(G) = 0$, $X(V) =$ const.

Thus, $A_t$ is an isometry of the Riemannian manifold $(M, G)$ and $X$ is an infinitesimal symmetry of it, also known as the Killing field of $G$. This leads to the next result.

THEOREM 10.3. *Every non-hidden infinitesimal symmetry of the Hamiltonian $H = \widetilde{G} + \pi^*(V)$ is the Killing field of the metric $G$ for which* $X(V) =$ const.

To conclude this subsection, we clarify the connection between standard infinitesimal symmetries of the Hamiltonians $H = \widetilde{G} + \pi^*(V)$ and $\widetilde{G}_c = \dfrac{1}{c-\pi^*(V)} G$. Let $h \in \mathscr{F}(T^*(M))$. Then $\{h, \widetilde{G}_c\} = \dfrac{1}{c-\pi^*(V)}\{h, \widetilde{G}\} + \dfrac{\widetilde{G}}{[c-\pi^*(V)]^2}\{h, \pi^*(V)\}$. In view of the fact that

$\widetilde{G}|_{P_c} = (c - \pi^*(V))|_{P_c}$, we have

$$\{h, \widetilde{G}_c\}|_{P_c \cap \widetilde{U}_c} = \left[\frac{1}{c-\pi^*(V)}(\{h, \widetilde{G}\} + \{h, \pi^*(V)\})\right]_{P_c \cap \widetilde{U}_c} = \frac{\{h, H\}}{c-\pi^*(V)}\Big|_{P_c \cap \widetilde{U}_c}.$$

It is clear from this that a function $h \in \mathscr{F}(T^*(M))$ that is constant on the trajectories of $X_H$ lying in $P_c \cap \widetilde{U}_c$ is constant on the trajectories of $X_{\widetilde{G}_c}$ lying on $\{\widetilde{G}_c = 1\} = P_c \cap \widetilde{U}_c$, and vice versa. Loosely speaking, the standard infinitesimal symmetries of $X_H$ coincide with the symmetries of $X_{\widetilde{G}_c}$ on $\{\widetilde{G}_c = 1\}$. This follows from the fact that the geometrical shapes of the trajectories of $X_H$ and $X_{\widetilde{G}_c}$ in $P_c \cap \widetilde{U}_c$ are the same (see § 10.3).

In conclusion, we consider three classical examples, which have played a prominent part in the history of mechanics.

**10.5. Kepler's problem.** This is the problem of the motion of a particle $A$ in the gravitational field of a fixed point mass. The position of this mechanical system is uniquely determined by the position of $A$, that is, its configuration space is $\mathbf{R}^3$. Suppose that the fixed attracting mass is situated at $\mathbf{O} \in \mathbf{R}^3$. This point is clearly exceptional. Let $q_1, q_2, q_3$ be Cartesian coordinates in $\mathbf{R}^3$, $q = \sqrt{q_1^2 + q_2^2 + q_3^2}$, $\mathbf{q} = (q_1, q_2, q_3)$; $p_1, p_2, p_3$ be the respective momenta, $p^2 = \sum_{i=1}^{3} p_i^2$ and $\mathbf{p} = (p_1, p_2, p_3)$. For an appropriate choice of units, the Hamiltonian function of the system in question has the form $H = \frac{p^2}{2} - \frac{\alpha}{q}$, where $\alpha > 0$ is some constant. Here $\widetilde{G} = p^2/2$ and $\mathbf{G}$ is the standard Euclidean metric on $\mathbf{R}^3$. Rotations about $\mathbf{O}$ are isometries of $\mathbf{R}^3$ that preserve the potential energy $-\alpha/q$. An infinitesimal rotation about the $q_i$ axis is of the form $X_i = -q_{i+1}\frac{\partial}{\partial q_{i+2}} + q_{i+2}\frac{\partial}{\partial q_{i+1}}$, where the indices are taken mod 3. Then the Hamiltonian of the field $\hat{X}_i$ is equal to $M_i = -q_{i+1}p_{i+2} + q_{i+2}p_{i+1}$. By Lemma 4.2, $\{H, M_i\} = 0$. We note that $\{M_i, M_{i+1}\} = M_{i+2}$. This shows that the linear span of the functions $M_i$ $(i = 1, 2, 3)$ forms a Lie algebra under Poisson brackets, which is isomorphic to the Lie algebra of the group $SO(3)$ (see [2]). Let $M^2 = \sum_{i=1}^{3} M_i^2$. Then, clearly, $\{H, M^2\} = \{M_i, M^2\} = 0$. Therefore, $f_1 = H$, $f_2 = M^2$, $f_3 = M_1$ is a complete collection of pairwise commuting symmetries of $H$, $s(H) = 3$, and Kepler's problem is completely integrable.

REMARK 10.1. We have exploited here the following useful method of finding pairwise commuting symmetries. Suppose that the Hamiltonian system admits a finite-dimensional algebra $L$ of symmetries, generated by, say, the functions $h_1, \ldots, h_k$. Then every invariant of $L$ (a polynomial over $L^*$ that is invariant under the adjoint action of $L$ on $L^*$), regarded as a polynomial in the $h_i$, commutes with all the $h_i$, and, clearly, also with $H$.

We recommend the reader starting from this to find an exact solution of Kepler's problem, using the methods developed above. We remark that the preceding arguments are valid for any central potential, that is, a potential of the form $\varphi(q)$. The Newtonian potential $-\alpha/q$ has, over and above this, the following remarkable symmetry property, Let $\mathbf{M} = (M_1, M_2, M_3)$ and $\mathbf{A} = \mathbf{M} \times \mathbf{p} - \frac{\alpha}{q}\mathbf{q}$ (we use the standard vector notation). Then, as can be verified immediately, $\{H, A_i\} = 0$, where $\mathbf{A} = (A_1, A_2, A_3)$. The functions $N_i = 2|H|^{1/2} A_i$ are also, clearly, symmetries of $H$. Furthermore, a straightforward calculation shows that

$$\text{(10.3)} \qquad \left\{ \begin{array}{l} \{M_i, N_{i+1}\} = \{N_i, M_{i+1}\} = N_{i+2}, \\ \{M_i, N_i\} = 0, \quad \{N_i, N_{i+1}\} = \pm M_{i+2}, \end{array} \right.$$

where the plus sign corresponds to the case $H < 0$ and the minus sign to the case $H > 0$. The formulae (10.3) show that the linear span of the functions $M_i$, $N_i$ ($i = 1, 2, 3$) forms a Lie algebra of dimension 6, isomorphic to the Lie algebra of the group $O(4)$ for $H > 0$ and of $O(3, 1)$ for $H < 0$. The triple of pairwise commuting functions $H$, $N_1$ and $\sum_{i=1}^{3} N_i^2$ forms a collection of commuting symmetries of $H$ that cannot be expressed functionally in terms of $H$, $M$, and $M^2$. This gives us yet one more example in connection with the problem discussed in §5.3.

**10.6. Geodesics on an ellipsoid.** This example demonstrates the fruitfulness of the "mechanical" approach to a geometrical problem. We want to find the geodesics on the triaxial ellipsoid $M = \left\{\frac{x^2}{a} + \frac{y^2}{b} + \frac{z^2}{c} = 1\right\}$, $a > b > c > 0$. Let $G$ be the metric on $M$ induced by the embedding $M \subset \mathbf{R}^3$. From the mechanical point of view it is natural to examine the Hamiltonian $H = \widetilde{G}/2$ of the corresponding free mechanical system and investigate it for symmetry. If we could find a symmetry of this Hamiltonian, independent of it, this would mean that it is completely integrable. We note that, as the ellipsoid has no infinitesimal symmetries, the symmetries of $H$, if there are any at all, must be hidden.

The geometrically natural coordinate system on some surface $P \subset \mathbf{R}^3$ is the system of its lines of curvature. A practically useful way of finding the lines of curvature is based on a theorem of Dupin, which consists in the following. Let $F_i(x, y, z)$ ($i = 1, 2, 3$) be functions on some domain $U \subset \mathbf{R}^3$, where the level surfaces of the $F_i$ intersect at right angles at all points of $U$ (a tri-orthogonal system). If $P = \{F_1 = c\}$ and $u = F_2|_P$, $v = F_3|_P$, then $(u, v)$ is a coordinate system on $P$ for which the lines $u$ = const and $v$ = const are lines of curvature of $P$ (see [22] p.334).

The functions $\lambda_i(x, y, z)$, where $\lambda_i (i = 1, 2, 3)$ are the roots of the equation $\frac{x^2}{a-\lambda} + \frac{y^2}{b-\lambda} + \frac{z^2}{c-\lambda} = 1$, form, as is not hard to check, a tri-orthogonal system and $M = \{\lambda_1 = 0\}$. The resulting coordinates $(u, v)$ on $M$ are connected with the Cartesian ones by the equations

$$x^2 = \frac{a(u-a)(v-a)}{(b-a)(c-a)}, \quad y^2 = \frac{b(u-b)(v-b)}{(a-b)(c-b)}, \quad z^2 = \frac{c(u-c)(v-c)}{(a-c)(b-c)}.$$

In the coordinates $(u, v)$ the metric $G$ can be written in the form

$$ds^2 = \frac{u-v}{4}[\varphi(u)du^2 - \varphi(v)dv^2], \quad \text{where} \quad \varphi(w) = \frac{w}{(a-w)(b-w)(c-w)}.$$

Therefore,

$$H=\frac{1}{u-v}\left[\frac{p_1^2}{\varphi(u)}-\frac{p_2^2}{\varphi(v)}\right], \tag{10.4}$$

where $(u, v, p_1, p_2)$ is a special canonical coordinate system on $T^*(M)$. We rewrite (10.4) in the form

$$\frac{p_1^2}{\varphi(u)}-Hu=\frac{p_2^2}{\varphi(v)}-Hv. \tag{10.5}$$

Then this equation gives $H$ implicitly. We now make use of the following general lemma, which lies behind the method of separation of variables.

LEMMA 10.3. *Let $\alpha(\varphi_1, \ldots, \varphi_k, H) = \beta(\psi_1, \ldots, \psi_l, H)$ be an equation giving $H$ implicitly, where $\varphi_i, \psi_j \in \mathcal{F}(K)$, $K$ is a canonical manifold, and $\{\varphi_i, \psi_j\} = 0$, $1 \leqslant i \leqslant k$, $1 \leqslant j \leqslant l$. Then* $\{\alpha(\varphi_1, \ldots, \varphi_k, H), H\} = 0$.

From the equations $\alpha(\varphi, H) = \beta(\psi, H)$ we express $H$ as a function of the $\varphi_i$ and $\psi_j$. Then $H_{\varphi_i} = \dfrac{\alpha_{\varphi_i}}{\beta_H-\alpha_H}$. Further

$$\{\alpha(\varphi_1, \ldots, \varphi_k, H), H\}=\sum_{i=1}^{k}\alpha_{\varphi_i}\{\varphi_i, H\}+\alpha_H\{H, H\}=\sum_{i=1}^{k}\alpha_{\varphi_i}\{\varphi_i, H\}=$$

$$=\sum_{i=1}^{k}\alpha_{\varphi_i}\sum_{j=1}^{k}H_{\varphi_j}\{\varphi_i, \varphi_j\}=\frac{1}{\beta_H-\alpha_H}\sum_{i, j=1}^{k}\alpha_{\varphi_i}\alpha_{\varphi_j}\{\varphi_i, \varphi_j\}.$$

But the expression $\sum_{i,j=1}^{k}\alpha_{\varphi_i}\alpha_{\varphi_j}\{\varphi_i, \varphi_j\}$, being both symmetric and skew-symmetric in $i$ and $j$, is zero.■

Applying this lemma to (10.5), we find that the function $h = \dfrac{p_1^2}{\varphi(u)} - Hu = \dfrac{v}{v-u}\dfrac{p_1^2}{\varphi(u)} + \dfrac{u}{u-v}\dfrac{p_2^2}{\varphi(v)}$ is a symmetry of the Hamiltonian $H$ and clearly independent of it. The two-parameter family of Lagrangian manifolds $M_{c_1,c_2} = \{H=c_1\} \cap \{h=c_2\}$ can be put in the following form by expressing $p_1$ and $p_2$ in terms of $H = c_1$ and $h = c_2$, $p_1 = \pm\sqrt{\varphi(u)(c_1u+c_2)}$, $p_2 = \pm\sqrt{\varphi(v)(c_1v+c_2)}$. From this we find an explicit expression for the two-parameter family of solutions of the corresponding Hamilton–Jacobi problem:

$$S(u, v, c_1, c_2)= \tag{10.6}$$

$$=\pm\int^{u}\sqrt{\varphi(w)(c_1w+c_2)}\,dw\pm\int^{v}\sqrt{\varphi(w)(c_1w+c_2)}\,dw.$$

We take $H = \mathcal{Q}_1$ and $h = \mathcal{Q}_2$ as one half of new canonical variables and complement them by the corresponding variables $\mathcal{P}_1$ and $\mathcal{P}_2$. Then in these coordinates Hamilton's equations take the form $\dot{\mathcal{Q}}_1=\dot{\mathcal{Q}}_2=0$, $\dot{\mathcal{P}}_1=-1$, $\dot{\mathcal{P}}_2=0$, so that $\mathcal{Q}_i=c_i$ $(i=1, 2)$, $\mathcal{P}_1=-t+c_3$, $\mathcal{P}_2=c_4$, $c_i \in \mathbf{R}$ $(i = 1, \ldots, 4)$.

But $\mathcal{P}_i = -\frac{\partial S}{\partial Q_i} \left(= -\frac{\partial S}{\partial c_i}\right)$, therefore,

$$\pm\frac{1}{2}\int^u w\sqrt{\frac{\varphi(w)}{c_1w+c_2}}\,dw \pm\frac{1}{2}\int^v w\sqrt{\frac{\varphi(w)}{c_1w+c_2}}\,dw = -t+c_3, \tag{10.7}$$

$$\pm\frac{1}{2}\int^u \sqrt{\frac{\varphi(w)}{c_1w+c_2}}\,dw \pm\frac{1}{2}\int^v \sqrt{\frac{\varphi(w)}{c_1w+c_2}}\,dw = c_4. \tag{10.8}$$

For fixed values of the constants $c_1$, $c_2$ and $c_4$, (10.8) is a relation between $u$ and $v$, that is, it gives an implicit equation of the geodesics. A parametrization of the geodesics can be derived from (10.7).

**10.7. A rigid body. The Hamiltonian formalism on Lie groups.** By a rigid body we mean in mechanics a system of material points (finite or infinite), the distance between any pair of which is invariable. An orthonormal frame of reference rigidly connected to a rigid body uniquely defines the position of the points of this body in $\mathbf{R}^3$. We consider a free rigid body. The configuration space $M$ of this mechanical system clearly coincides with the totality of all orthonormal frames of reference $\mathscr{E} = \{a; \varepsilon_1, \varepsilon_2, \varepsilon_3\}$ in $\mathbf{R}^3$, where $a \in \mathbf{R}^3$ and the ordered triple of vectors $\varepsilon_i$ is orthonormal. Therefore, $M = \mathbf{R}^3 \times SO(3)$. We fix one such frame $\mathscr{E}_0$. Then every frame $\mathscr{E} \in M$ uniquely determines a motion $g(\mathscr{E})\colon \mathbf{R}^3 \to \mathbf{R}^3$ carrying $\mathscr{E}_0$ to $\mathscr{E}$. This lets us identify $M$ with the Lie group $\mathscr{G}(3)$ of all motions of $\mathbf{R}^3$. Let $\mathscr{E} \in M$. Then $T_{\mathscr{E}}(M) = T_a(\mathbf{R}^3) \oplus T_\varepsilon(SO(3))$, where $\mathscr{E} = (\varepsilon_1, \varepsilon_2, \varepsilon_3)$, and we can represent every vector $\xi \in T_{\mathscr{E}}(M)$ uniquely in the form $\xi = (v, \omega)$, $v \in T_a(\mathbf{R}^3)$, $\omega \in T_\varepsilon(SO(3))$,[1] which means that the state of a rigid body is described by the velocity vector $v$ of its translational motion and the velocity vector $\omega$ of its angular motion. It is convenient to situate $\mathscr{E}_0$ at the so-called centre of gravity of the body, which is defined by the condition that in the metric $G$ given by the kinetic energy of the body the tangent spaces $T_a(\mathbf{R}^3)$ and $T_\varepsilon(SO(3))$ are orthogonal for all $\mathscr{E} \in \mathscr{G}(3)$. The existence and uniqueness of the centre of gravity are affirmed by an elementary theorem of König, which we do not prove (see [24]). Taking account of what was said above, the kinetic energy $G$ of a rigid body is the sum of the kinetic energies of its translational and its rotational motion. The first is equal to $m \sum_{i=1}^{3} v_i^2/2$, where $m$ is the mass of the body and $v = (v_1, v_2, v_3)$, and the second is a quadratic form in the $\omega_i$ $(i = 1, 2, 3)$, where $\omega = \sum_{i=1}^{3} \omega_i \varepsilon_i$. By obvious symmetry arguments, it does not depend

[1] We recall that $\omega \in \mathbf{R}^3$ denotes the rotation about the axis of the vector $\omega$ in the positive direction with angular velocity $|\omega|$.

on $\mathscr{E}$. Therefore, $\mathscr{E}_0$ can be affixed to the body (at the centre of gravity) in such a way that this quadratic form is diagonal: $\sum_{i=1}^{3} I_i\omega_i^2/2$. The constants $I_i$ are called the principal moments of inertia of the body in question. Thus, $G_{\mathscr{E}} = m\sum_{i=1}^{3} v_i^2/2 + \sum_{i=1}^{3} I_i\omega_i^2/2$. We now turn our attention to the fact that the formula for $G$ is not the usual way of writing the metric in local coordinates. Essentially, we have introduced a coordinate system $(v_i, \omega_i)$ on $T_{\mathscr{E}_0}(M)$ and then carried it over via $g(\mathscr{E})$ to $T_{\mathscr{E}}(M)$. This method of introducing coordinates is not linked with a choice of local coordinates on $M$, and just because of this the expression given for $G$ has constant coefficients. It can be shown that, with the usual way of writing the metric in terms of a choice of local coordinates, it is impossible to obtain constant coefficients for $G$.

The point of carrying out this procedure is explained by the theory of Lie groups, an elementary knowledge of which we take for granted (see [2]). Let $\mathscr{G}$ be a Lie group, $e \in \mathscr{G}$ its unit element, and $l_g\colon \mathscr{G} \to \mathscr{G}$ left multiplication by $g \in \mathscr{G}$. If $(e_1, \ldots, e_n)$ is a basis of $T_e(\mathscr{G})$, then it can be carried over to any $T_g(\mathscr{G})$ via the isomorphism $(dl_g)_e\colon T_e(\mathscr{G}) \to T_g(\mathscr{G})$, which leads to the basis $e_i^g = (dl_g)_e(e_i)$. The vector fields $X_i$ such that $X_i|_g = e_i^g$ are clearly left-invariant, that is, $l_g(X_i) = X_i \;\; \forall g \in G$, and form a basis over $\mathbf{R}$ of the space $L = L(\mathscr{G})$ of all left-invariant fields on $\mathscr{G}$. If $X, Y \in L$, then clearly $[X, Y] \in L$; $L$ is called the Lie algebra of $G$. Let $[X_i, X_j] = \sum_{k=1}^{n} c_{ij}^k X_k$; the $c_{ij}^k$ are called the structure constants of $\mathscr{G}$ (or of $L$). Similarly, let $(\bar{e}_1, \ldots, \bar{e}_n)$ be the basis of $T_e^*(\mathscr{G})$ dual to $(e_1, \ldots, e_n)$ and $\bar{e}_i^g = [(dl_g)_e^*]^{-1}(\bar{e}_i)$. Then the forms $\omega_i \in \Lambda^1(\mathscr{G})$, $\omega_i|_g = \bar{e}_i^g$ are left-invariant and form a basis of $L^*$, $\mathbf{R}$-dual to the basis $(X_1, \ldots, X_n)$ of $L$. Furthermore, since the $X_i$ (respectively, $\omega_i$) form an $\mathscr{F}(\mathscr{G})$-basis of $\mathscr{D}(\mathscr{G})$ (respectively, of $\Lambda^1(\mathscr{G})$), every vector field (respectively, 1-form) on $\mathscr{G}$ can be written in the form $X = \sum_{i=1}^{n} \alpha_i X_i$ (respectively, $\omega = \sum_{i=1}^{n} \alpha_i\omega_i$), where $\alpha_i \in \mathscr{F}(\mathscr{G})$. In particular, every metric $G$ on $\mathscr{G}$ can be written in the form

$G(X, Y) = \sum_{i,j=1}^{n} \mu_{ij}\alpha_i\beta_j$, where $X = \sum_{i=1}^{n} \alpha_i X_i$, $Y = \sum_{i=1}^{n} \beta_i X_i$, $\mu_{ij} = \mu_{ji} \in \mathscr{F}(\mathscr{G})$.

If $G$ is left-invariant, then the $\mu_{ij}$ are constants. The formula for the kinetic energy of a rigid body is, as is now clear, an expression of a similar kind when $\mathscr{G} = \mathscr{G}(3) = M$.

Since $G$ is left-invariant, the Hamiltonian $H = \widetilde{G}/4$ of a rigid body is left-invariant. If $(p_i, \nu_i)$ are coordinates in $T^*(M)$ dual to $(v_i, \omega_i)$, then clearly $H = \sum_{i=1}^{3} \left(\frac{p_i^2}{2m} + \frac{v_i^2}{2I_i}\right)$. Thus, the theory of the motion of a rigid body leads us to the necessity of considering left-invariant Hamiltonian systems on Lie groups.

The theory of left-invariant Hamiltonian systems on a Lie group $\mathcal{G}$ is simplified considerably if we discuss only the left-invariant part of the Hamiltonian formalism on $T^*(\mathcal{G})$. This is justified, because $l_g^*(\rho) = \rho \ \forall\, g \in G$ (Corollary 2.2), that is, the fundamental forms $\rho$ and $d\rho$ on $T^*(G)$, as well as the map $\Gamma$, are left-invariant. This allows us to make full use of the property of being left-invariant, which clearly entails the fact that $\mathcal{G}$ is a group of (non-hidden) symmetries of the left-invariant Hamiltonian $H$ on $T^*(\mathcal{G})$.

Thus, we consider the left-invariant Hamiltonian formalism on $T^*(\mathcal{G})$, which means that we deal only with functions, forms, fields, etc. on $T^*(\mathcal{G})$ that are invariant under the diffeomorphisms $l_g$, $g \in \mathcal{G}$. Every left-invariant function $\varphi$ on $T^*(\mathcal{G})$ can be put into one-to-one correspondence with the function $l(\varphi) = \varphi|_{T_e^*(\mathcal{G})}$ on the linear space $T_e^*(\mathcal{G})$, which is naturally identified with $L^*$; $l$ is an isomorphism of the ring of smooth left-invariant functions on $T^*(\mathcal{G})$ and $\mathcal{F}(L^*)$. If $\varphi \in \mathcal{F}(T^*(\mathcal{G}))$ and $X \in \mathcal{D}(T^*(\mathcal{G}))$ are left-invariant, then so is the function $X(\varphi)$. Therefore, the field $X^l \in \mathcal{D}(L^*)$ is well defined by $X^l(\psi) = l(X(l^{-1}(\psi)))$, $\psi \in \mathcal{F}(L^*)$.

If $H \in \mathcal{F}(T^*(\mathcal{G}))$ is left-invariant, then so is $X_H$. Therefore, the field $X_H^l$ and the $\mathbf{R}$-linear map $\Gamma_l$: $\mathcal{F}(L^*) \to \mathcal{D}(L^*)$, $\Gamma_l(\psi) = X^l_{l^{-1}(\psi)}$ are defined. We compute $\Gamma_l$. The functions $\pi_i = l(\Pi_i)$, where $\Pi_i = \rho(\hat{X}_i)$, form a coordinate system on $L^*$. The structure formula $[X_i, X_j] = \Sigma\, c_{ij}^k X_k$ leads to $[\hat{X}_i, \hat{X}_j] = \sum_{k=1}^{n} c_{ij}^k \hat{X}_k$. Applying the form $\rho$ to both sides of the equation

$\hat{X}_i(\Pi_j) = \Pi_i, \Pi_j = d\rho(\hat{X}_i, \hat{X}_j) = \hat{X}_i(\rho(\hat{X}_j)) - \hat{X}_j(\rho(\hat{X}_i)) - \rho([\hat{X}_i, \hat{X}_j])$,

we obtain $\hat{X}_i(\Pi_j) = \sum_{k=1}^{n} c_{ij}^k \Pi_k$, hence $X_i^l = \sum_{j=1}^{n} \left(\sum_{k=1}^{n} c_{ij}^k \pi_k\right) \frac{\partial}{\partial \pi_j}$. If $H = f(\pi_1, \ldots, \pi_n)$ is some left-invariant Hamiltonian, then

$l(H) = f(\pi_1, \ldots, \pi_n)$ and $X_H = \sum_{i=1}^{n} \frac{\partial f}{\partial \pi_i} X_{\Pi_i} = \sum_{i=1}^{n} \frac{\partial f}{\partial \pi_i} \hat{X}_i$. Hence

$$(10.9) \qquad \Gamma_l(f) = X_H^l = \sum_{i=1}^{n} \frac{\partial f}{\partial \pi_i} \left(\sum_{j=1}^{n} \left(\sum_{k=1}^{n} c_{ij}^k \pi_k\right) \frac{\partial}{\partial \pi_j}\right) = \sum_{j=1}^{n} \left(\sum_{i,k=1}^{n} \frac{\partial f}{\partial \pi_i} c_{ij}^k \pi_k\right) \frac{\partial}{\partial \pi_j}.$$

To finish this brief excursion into the left-invariant theory, we remark that any function $h$ invariant under the adjoint representation of $\mathcal{G}$ on $L^*$ is a symmetry of any left-invariant function if we regard it as a function on $T^*(\mathcal{G})$ (it is obvious that $h$ is an invariant if and only if $\Gamma_l(h) = 0$). For more details about this, see [23].

Let us return, however, to the rigid body. Above, we took as a basis of $T_e(\mathcal{G}(3)) = T_{\mathcal{E}_0}(M) = T_{a_0}(\mathbf{R}^3) \oplus T_e(SO(3))$ the vectors $e_1, e_2, e_3$ of unit translations along the inertial axes of the body in question (a basis of $T_{a_0}(\mathbf{R}^3)$) and the vectors $e_4, e_5, e_6$ of unit rotations about the axes $e_1, e_2, e_3$ (a basis of $T_e(SO(3))$). Then

$$
(10.10)\quad \begin{cases} [e_i, e_j] = [e_i, e_{i+3}] = 0, \quad 1 \leqslant i,\ j \leqslant 3, \quad [e_4, e_5] = e_6, \\ [e_5, e_6] = e_4, \quad [e_6, e_4] = e_5, \quad [e_4, e_2] = e_3, \quad [e_4, e_3] = -e_2, \\ [e_5, e_1] = -e_3, \quad [e_5, e_3] = e_1, \quad [e_6, e_1] = e_2, \quad [e_6, e_2] = -e_1. \end{cases}
$$

When we now apply (10.9) to the Hamiltonian $H = \sum_{i=1}^{3} \left(\frac{p_i^2}{2m} + \frac{\nu_i^2}{2I_i}\right)$, and calculate the values of the structure constants, which are given by (10.10), we obtain

$$
(10.11)\quad X_H^l = \left(\frac{\nu_3 p_2}{I_3} - \frac{\nu_2 p_3}{I_2}\right)\frac{\partial}{\partial p_1} + \left(\frac{\nu_1 p_3}{I_1} - \frac{\nu_3 p_1}{I_3}\right)\frac{\partial}{\partial p_2} + \left(\frac{\nu_2 p_1}{I_2} - \frac{\nu_1 p_2}{I_1}\right)\frac{\partial}{\partial p_3} + \nu_3\nu_2\left(\frac{1}{I_3} - \frac{1}{I_2}\right)\frac{\partial}{\partial \nu_1} + \nu_1\nu_3\left(\frac{1}{I_1} - \frac{1}{I_3}\right)\frac{\partial}{\partial \nu_2} + \nu_2\nu_1\left(\frac{1}{I_2} - \frac{1}{I_1}\right)\frac{\partial}{\partial \nu_3}.
$$

Thus, the left-invariant Hamiltonian system for a rigid body is:

$$
(10.12)\quad \begin{cases} \dot p_1 = \frac{\nu_3 p_2}{I_3} - \frac{\nu_2 p_3}{I_2}, \quad \dot p_2 = \frac{\nu_1 p_3}{I_1} - \frac{\nu_3 p_1}{I_3}, \quad \dot p_3 = \frac{\nu_2 p_1}{I_2} - \frac{\nu_1 p_2}{I_1}, \\ \dot\nu_1 = \left(\frac{1}{I_3} - \frac{1}{I_2}\right)\nu_2\nu_3, \quad \dot\nu_2 = \left(\frac{1}{I_1} - \frac{1}{I_3}\right)\nu_1\nu_3, \\ \dot\nu_3 = \left(\frac{1}{I_2} - \frac{1}{I_1}\right)\nu_2\nu_1. \end{cases}
$$

The Legendre map $\alpha_H$ is clearly left-invariant and gives rise to a map $\alpha_H^l\colon L^* \to L$. In the case in question, $\alpha_H^l$ is given by the formulae $v_i = p_i/m$, $\omega_i = \nu_i/I_i$. Therefore, the system of equations on $L$ associated with the field $\alpha_H^l(X_H^l)$ has the form

$$
(10.13)\quad \begin{cases} \dot v_1 = \omega_3 v_2 - \omega_2 v_3, \quad \dot v_2 = \omega_1 v_3 - \omega_3 v_1, \quad \dot v_3 = \omega_2 v_1 - \omega_1 v_2, \\ \dot\omega_1 = \frac{I_2 - I_3}{I_1}\omega_2\omega_3, \quad \dot\omega_2 = \frac{I_3 - I_1}{I_2}\omega_3\omega_1, \quad \dot\omega_3 = \frac{I_1 - I_2}{I_3}\omega_1\omega_2. \end{cases}
$$

The first three equations of this system are, as is not hard to see, the laws of conservation of momentum along the coordinate axes, written in the moving frame $\mathcal{E}$, and are therefore of no interest. But the second three equations are the famous Euler equations of motion of a rigid body. For this reason, the field $\alpha_H^l(X_H^l)$ on $L$ in the case of an arbitrary left-invariant Hamiltonian $H$ on $T^*(\mathcal{G})$ can be regarded as a generalization of Euler's equations for a rigid body (provided, of course, that $\alpha_H$ is a diffeomorphism). The functions $p^2 = \sum_{i=1}^{3} p_i^2$ and $\nu^2 = \sum_{i=1}^{3} \nu_i^2$ are invariants of $G(3)$, so that $X^l_{p^2} = X^l_{\nu^2} = 0$; hence they are symmetries of $H$. The projection of the manifold $\{H = c_1\} \cap \{p^2 = c_2\} \cap \{\nu^2 = c_3\}$ on the $\nu$-space $(= T_e^*(SO(3)) \subset T_{\mathscr{E}_0}(\mathcal{G}(3)))$ is given by the equations

$$\sum_{i=1}^{3} \frac{\nu_i^2}{2I_i} = c_1 - \frac{c_2}{2m}, \quad \nu^2 = c_3. \tag{10.14}$$

Thus, the motion of a rigid body is described in non-parametric form by the curves (10.11) or, after applying $\alpha_H$, by the curves

$$\sum_{i=1}^{3} I_i\omega_i^2 = \text{const}, \qquad \sum_{i=1}^{3} I_i^2\omega_i^2 = \text{const}, \tag{10.15}$$

which are, therefore, the solutions of Euler's equations.

The following connection between the trajectories of $X_H$ and of the "Euler field" $\alpha_H^l(X_H^l)$ is an immediate consequence of the definitions: if $\omega(t)$ is a trajectory of the Euler field, then the projection $\gamma(t)$ of the trajectory $\bar{\gamma}(t)$ of $X_H$ on $\mathcal{G}$ is of the form $\gamma(t) = \exp \omega(t)$, and vice versa. Here $\exp\colon L \to \mathcal{G}$ denotes the exponential map (see [2]). The case of a rigid body fixed at one or two points can be treated similarly.

The theory expounded above is, as we have seen, the restriction of the Hamiltonian formalism to the subring of $\mathscr{F}(T^*(\mathcal{G}))$ consisting of left-invariant functions. A similar theory can be constructed, if instead of this subring we take any other subring of $\mathscr{F}(T^*(\mathcal{G}))$ that is invariant under the operators $\bar{l}_g$, $g \in \mathcal{G}$. The smaller this subring, the simpler the resulting theory. The study of the motion of a rigid body in, say, a gravitational field can most conveniently be brought within the framework of such a theory.

# APPENDIX I

## Bifurcations

Here we want to describe how the manifolds $P_c$ vary with a change of $c \in \mathbf{R}^k$. It is not hard to give to the loose arguments presented below a rigorous meaning in the framework of the theory of singularities of smooth maps (see, for example, the survey [6]). In particular, this refers to the concept of "general position", which lies at the basis of our discussion and which nonetheless we leave undefined.

We begin with the following general situation: $K$ is some $m$-dimensional manifold, $f_1, \ldots, f_k \in \mathscr{F}(K)$, $k < m$, is a collection of functions in "general position", $P_c = \{x \in K \mid f_i(x) = c_i,\ i = 1, \ldots, k\}$, $c = (c_1, \ldots, c_k) \in \mathbf{R}^k$. Then we make modifications, taking into account the fact that $\{f_i, f_j\} = 0$. Let

$$\text{Bif}\,(f_1, \ldots, f_k) = \{x \in K \mid df_1|_x, \ldots, df_k|_x \text{ are linearly independent}\}.$$

Generally speaking, Bif $(f_1, \ldots, f_k)$ is a manifold with singularities, whose points in "general position" are the $x$ for which rank $(df_1|_x, \ldots, df_k|_x) = k - 1$, $df_i|_x \neq 0$ $(i = 1, \ldots, k)$. We restrict ourselves to the analysis of these points. Let $P'_{c'} = \{x \in K \mid f_i(x) = c_i,\ i = 1, \ldots, k-1\}$, $c' = (c_1, \ldots, c_{k-1}) \in \mathbf{R}^{k-1}$. For almost all $c' \in \mathbf{R}^{k-1}$ the manifold $P'_{c'}$ can be assumed to be non-singular, that is, rank $(df_1|_x, \ldots, df_{k-1}|_x) = k - 1$. A point $x$ lies in $P'_{c'} \cap \text{Bif}\,(f_1, \ldots, f_n)$ only when $d(f_k|_{P'_{c'}})|_x = 0$, that is, when $x$ is an extremum point of $f_k|_{P'_{c'}}$ on $P'_{c'}$. The extremum points of a function in "general position" form a discrete set (see [6]), so that the set $P'_{c'} \cap \text{Bif}\,(f_1, \ldots, f_k)$ can be regarded as such. When the $(k-1)$-dimensional parameter $c'$ varies, each of the $x \in P'_{c'} \cap \text{Bif}\,(f_1, \ldots, f_k)$ also varies and describes a $(k-1)$-dimensional manifold. Thus, $\text{Bif}\,(f_1, \ldots, f_k) = \bigcup_{c' \in \mathbf{R}^{k-1}} (P'_{c'} \cap \text{Bif}\,(f_1, \ldots, f_k))$ is a $(k-1)$-dimensional manifold with singularities.

We now consider the map $F\colon K \to \mathbf{R}^k$, $F(x) = (f_1(x), \ldots, f_k(x))$. Then the set bif $(f_1, \ldots, f_k) = F(\text{Bif}\,(f_1, \ldots, f_k))$, is, in general, also a $(k-1)$-dimensional manifold in $\mathbf{R}^k$ with singularities and self-intersections. It is called the *bifurcation diagram* of the system $f_1, \ldots, f_k$ and it splits $F(K) \subset \mathbf{R}^k$ into a number of connected domains, say $B_i$. These domains have the property that under some natural conditions of completeness all the manifolds $P_c$, $c \in B_i$, are diffeomorphic, while $P_c$ and $P_{\bar{c}}$ are not diffeomorphic if $c$ and $\bar{c}$ belong to different domains $B_i$. Thus, the manifold $P_c$ changes on passing through the bifurcation diagram from one domain $B_i$ to another. Let us prove this.

Let $\widetilde{B}_i = F^{-1}(B_i)$, $x \in \widetilde{B}_i$, $c(x) = (f_1(x), \ldots, f_k(x))$. Then $\text{Ker}\, dF_x = T_x(P_{c(x)})$. We introduce a Riemannian metric on $\widetilde{B}_i$ (this is always possible (see [1])) and consider an orthogonal complement $O_x$ to Ker $dF_x$ in $T_x(K)$ in the sense of this metric.

Now $dF_x|_{O_x}\colon O_x \to T_{F(x)}(\mathbf{R}^k)$ is an isomorphism, since $dF_x$ is an epimorphism because $x \in \widetilde{B}_i$. Every field $Z \in \mathscr{D}(B_i)$ can be lifted to a field $\widetilde{Z} \in \mathscr{D}(\widetilde{B}_i)$, by setting $\widetilde{Z}_x = (dF_x|_{O_x})^{-1}(Z_{F(x)})$. Here, clearly, $\widetilde{Z} \circ F^* = F^*Z$. It follows from this equality that $F \circ \widetilde{C}_t = C_t \circ F$, where $\{C_t\} \Leftrightarrow Z$, $\{\widetilde{C}_t\} \Leftrightarrow \widetilde{Z}$ at places where $\widetilde{C}_t$ is defined. In this context we call $F$ *full* if $\widetilde{C}_t$ is defined on $F^{-1}(V)$ whenever $C_t$ is defined on $V \subset B_i$.

LEMMA. *Let $F$ be full. Then all the fibres $P_c$, $c \in B_i$, are diffeomorphic.*

PROOF. Let $c$, $\bar{c} \in B_i$ and let $Z \in \mathscr{D}(B_i)$ be a field such that one of its trajectories passes through the points $c$ and $\bar{c}$, and $C_t(c) = \bar{c}$ for some $t \in \mathbf{R}$. If $V$ is a neighbourhood of $c$ where $C_t$ is defined, then $\widetilde{C}_t$ is defined on $F^{-1}(V) \supset P_c$. Since $F \circ \widetilde{C}_t = C_t \circ F$, we have $\widetilde{C}_t(P_c) \subset P_{\bar{c}}$, and $\widetilde{C}_{-t}(P_{\bar{c}}) \subset P_c$, from which it is clear that $\widetilde{C}_t|_{P_c}: P_c \to P_{\bar{c}}$ is a diffeomorphism.

We give the following almost obvious proposition without proof.

PROPOSITION. 1) *A map $F$ is full if $\forall Z \in \mathscr{D}(B_i)$ and $\forall x \in \widetilde{B}_i$, the trajectory of $Z$ starting at $x$ is defined for those $t$ for which the trajectory of $Z$ starting at $F(x)$ is defined.*

2) *A map $F$ is full if the preceding condition is satisfied for some system of fields $\{Z_\alpha\}$ that generates the $\mathscr{F}(B_i)$-module $\mathscr{D}(B_i)$.*

3) *The property of a map $F$ of being full does not depend on the choice of the Riemannian metric in $B_i$.*

4) *If $F$ is full, then $F|_{\widetilde{B}_i}: \widetilde{B}_i \to B_i$ is a smooth fibering with the fibres $P_c = F^{-1}(c)$, $c \in B_i$* (see [7]).

We turn now to the case of interest to us, that of a canonical manifold $K$ and a commuting system of essentially independent functions $f_1, \ldots, f_k$. Then a point $x$ lies in $P'_{c'} \cap \mathrm{Bif}(f_1, \ldots, f_k)$, as before, if $d(f_k|_{P'_{c'}})|_x = 0$. But $x$ cannot be isolated, because of the conditions $X_i(f_k) = \{f_i, f_k\} = 0$, $X_i = X_{f_i}$. For it follows from $X_i(f_k) = 0$ that $d(f_k|_{P'_{c'}})|_{A^i_{t(x)}} = 0$, where $\{A^i_t\} \Leftrightarrow X_i$. More precisely, there is some $(k-1)$-dimensional manifold $\bar{N}_{x'} \subset P'_{c'}$ passing through $x$ and consisting of points where $d(f_k|_{P'_{c'}}) = 0$. The manifold $\bar{N}'_x$ is constructed from $f_1, \ldots, f_{k-1}$ just as $\bar{N}_x$ from $f_1, \ldots, f_k$. Thus, $\mathrm{bif}(f_1, \ldots, f_k)$ is a $(k-1)$-parameter family of $(k-1)$-dimensional manifolds, that is, a manifold of dimension $2k-2$. Nevertheless, $\mathrm{Ker}\, dF_x \cap T_x(\mathrm{Bif}(f_1, \ldots, f_k)) = T_x(N_x)$ is of dimension $k-1$, so that $\mathrm{Bif}(f_1, \ldots, f_k)$ is, as before, a $(k-1)$-dimensional manifold with singularities. For the rest, everything proceeds as before.

We consider one simple example that illustrates the theory of bifurcation: a "particle in a central field". In this case $M = \mathbf{R}^2 \setminus \{0\}$, $T^*(M) \approx M \times \mathbf{R}^2$ and $H = \dfrac{p_1^2 + p_2^2}{2} + U(x)$, where $(q_1, q_2)$ are Cartesian coordinates in $\mathbf{R}^2 \supset M$, $(q_i, p_i)$ $(i = 1, 2)$ is the corresponding special coordinate system in $T^*(M)$, $x = \sqrt{(q_1^2 + q_2^2)}$. The Hamiltonian $H$ is, clearly, symmetric with respect to rotations about the origin, and the field $X \in \mathscr{D}(M)$ corresponding to the one-parameter group of turns about the origin is $q_1 \dfrac{\partial}{\partial q_2} - q_2 \dfrac{\partial}{\partial q_1}$. Therefore, the Hamiltonian $f_2$ of $\hat{X}$ is $q_1 p_2 - p_1 q_2$, and $\{f_2, H\} = 0$. We consider the bifurcations of the pair $f_1 = H$, $f_2$. Since

$$df_1 = p\,dp_1 + p_2\,dp_2 + \frac{U'(x)}{x}(q_1\,dq_1 + q_2\,dq_2),$$

$$df_2 = -q_2\,dp_1 + q_1\,dp_2 + p_2\,dq_1 - p_1\,dq_2,$$

the set $\mathrm{Bif}(f_1, f_2)$ consists of the points where the rank of the matrix

$$\begin{pmatrix} p_1 & p_2 & V(x)\,q_1 & V(x)\,q_2 \\ -q_2 & q_1 & p_2 & -p_1 \end{pmatrix},$$

where $V(x) = U'(x)/x$, does not exceed 1. An elementary calculation shows that $\mathrm{Bif}(f_1, f_2)$ consists of "pieces of two types:

$\mathrm{Bif}_1 = \{p_1 = \pm\sqrt{(V(x))}q_2, p_2 = \mp\sqrt{(V(x))}q_1, V(x) > 0\}$ – points in "general position";
$\mathrm{Bif}_2 = \{p_1 = p_2 = 0, V(x) = 0\}$ – points "not in general position".
The projections of these pieces on the plane $\mathbf{R}^2 = \{f_1, f_2\}$ are

$$\mathrm{bif}_1 = F(\mathrm{Bif}_1) = \left\{ f_1 = U(y) + \frac{1}{2}U'(y)\,y,\ f_2 = \pm\sqrt{(U'(y))}y^{3/2} \right\},$$

$U'(y) > 0, y > 0$ is a parameter;

$$\mathrm{bif}_2 = F(\mathrm{Bif}_2) = \{f_1 = a_i,\ f_2 = 0\},$$

where $a_i = U(x_i)$ and the $x_i$ are roots of the equation $U'(x) = 0$. Thus, $\mathrm{bif}(f_1, f_2) = F(\mathrm{Bif}_1) \cup F(\mathrm{Bif}_2)$.

For example, let $U(x) = 1/x$ ("repelling potential"). Then $\mathrm{Im}\,F = \{f_1 > 0\}$, $\mathrm{bif}(f_1, f_2) = \phi$. But if $U(x) = -1/x$ ("attracting potential"), then $\mathrm{Im}\,F = \mathbf{R}^2$, $\mathrm{bif}(f_1, f_2) = \{2f_1f_2^2 = -1\}$.

For other examples, see [26].

## APPENDIX II

## The Hamiltonian form of mechanics with friction, non-holonomic mechanics, invariant mechanics, the theory of refraction and impact

A. V. Bocharov and A. M. Vinogradov

In this appendix we present the Hamiltonian formulation of the branches of classical mechanics indicated in the heading, about which nothing is said in the main text. Here, in contrast to the traditional approach, we are in no way guided by the Lagrangian point of view. The invariant Hamiltonian formulation of the branches of mechanics discussed in subsections 2, 3, and 5 is apparently new.

**1. Mechanics with friction.** The action of frictional forces is manifested by a decrease of the momentum of a mechanical system in "the direction opposite to its direction of motion". Therefore, the resistive action of frictional forces can be described by adding to the non-resistive Hamiltonian field of the system a vertical vector field on $T^*(M)$, in which all the vectors that are tangent to the fibres $T^*_x(M)$ "face" the point $(x, 0) \in T^*(M)$. In particular, $X_\rho$ is such a field. Hence, any field of frictional forces is of the form $\lambda X_\rho$, where $\lambda$ is a non-negative function on $T^*(M)$. Thus, if $\mathscr{H}$ is the usual Hamiltonian of the system in question, then its motion, when friction is taken into account, is described by a vector field on $T^*(M)$ of the form $X = \Gamma^{-1}(d\mathscr{H} + \lambda\rho)$. The function $\lambda$ is the coefficient of friction, and its form is determined by the concrete physical conditions.

**2. Mechanics of non-holonomic systems.**[1] The kinematic constraints imposed on a system by their very meaning must be understood as a

[1] For a Lagrangian treatment of non-holonomic systems see [28]. See also [29].

certain submanifold $\Pi \subset T^*(M)$ on which the phase trajectories of the system have to lie. Thus, it is necessary to indicate the principles on which a vector field $Z \in \mathscr{D}(\Pi)$ is constructed for a given Hamiltonian of mechanical type. It is convenient to formulate it, by introducing the concept of a partial symplectic structure.

Let $Q$ be a linear subspace of a linear space $V$, and $P = p_0 + Q$ an affine subspace. We say that a bilinear form $\Omega: V \times V \to \mathbf{R}$ is non-degenerate on $P$ if it is non-degenerate on its associated linear subspace $Q$.

If $\Omega$ is such a form, then $P$ can be uniquely written in the form $p_1 + Q$, where $p_1$ is a vector skew-orthogonal to $Q$ (with respect to $\Omega$), the so-called *canonical displacement* (with respect to $\Omega$) of the affine space $P$.

If a skew-symmetric form $\Omega$ is non-degenerate on $P$, then the map $\gamma_\Omega: P \to Q^*$, $\gamma_\Omega(x) = x \lrcorner \Omega$ is an "isomorphism". We consider the isomorphism $\Gamma_\Omega: Q \to Q^*$, $\Gamma_\Omega(y) = y \lrcorner \Omega$. The maps $\gamma_\Omega$ and $\Gamma_\Omega$ are connected by the relations

$$(1) \qquad \begin{cases} \Gamma_\Omega(x) = \gamma_\Omega(p_1 + x), \\ \gamma_\Omega^{-1}(\omega) = p_1 + \Gamma_\Omega^{-1}(\omega), \\ \forall x \in Q, \quad \omega \in Q^*, \end{cases}$$

in which $p_1$ is the canonical displacement of $P$.

Let $L$ be a smooth manifold. We say that a *partial symplectic structure* is given on $L$ if there are specified a distribution of affine subspaces $T_x^0(L) \subset T_x(L)$ $\forall x \in L$, and a differential 2-form $\Omega \in \Lambda^2(L)$, with 1) $\Omega$ non-degenerate on $T_x^0$ for any $x \in L$, 2) $d\Omega = 0$.

Together with $\{T_x^0\}$ we consider the distribution of linear spaces $T_x^1 \subset T_x(L)$ and the field of canonical displacements $\mathscr{P}$ $(T_x^0 = \mathscr{P}_x + T_x^1)$.

The family of maps $\gamma_{(-\Omega_x)}^{-1}: (T_x^1)^* \to T_x^0$, $\Gamma_{(-\Omega_x)}^{-1}: (T_x^1)^* \to T_x^1$ obviously defines linear operators

$$Z\colon\ \Lambda^1(L) \to \mathscr{D}(L), \qquad Y\colon\ \Lambda^1(L) \to \mathscr{D}(L).$$

By (1), $Z(\omega) = \mathscr{P} + Y(\omega)$, where $\mathscr{P}$ is the field of canonical displacements.

DEFINITION. *We call the* 1*-form* $\omega$ *a generalized Hamiltonian of the vector field* $Z(\omega)$. *When* $\omega = df$, *we call the function* $f$ *a Hamiltonian of the field* $Z(\omega)$.

The existence of the map $C^\infty(L) \xrightarrow{d} \Lambda^1(L) \xrightarrow{Y} \mathscr{D}(L)$ allows us to define on a manifold with a partial symplectic structure a pairing $\langle f, g\rangle = Y_{(df)}(g)$ satisfying all the axioms of the Poisson bracket (see [16]), except the Jacobi identity.

Let $M$ be the configuration manifold of a mechanical system and $G$ its kinetic energy (a Riemannian metric on $M$). Let $P_x \subset T_x(M)$ be a distribution of $k$-dimensional affine subspaces, realizing the kinematic constraints imposed on the system. We introduce the following objects:

1°. $\alpha_x: T_x(M) \to T_x^*(M)$ – the Legendre map associated with the metric $G$;

2°. a $k$-dimensional distribution $\Pi_x = \alpha_x(P_x) \subset T_x^*(M)$;

3°. a $2k$-dimensional distribution $T_b^0(\Pi) \subset T_b(\Pi)$ on the manifold

$\Pi = \bigcup_{x \in M} \Pi_x$:

$$T_b^0(\Pi) = \{\xi \in T_b(\Pi) \mid d\pi|_b(\xi) \in P_{\pi(b)}\}$$

is the inverse image of $P_x$ under the natural projection $\pi: \Pi \to M$.

FUNDAMENTAL ASSERTION. *The distribution $T^0$ and the 2-form $(d\rho)|_\pi$ give a partial symplectic structure on $\Pi$.*

If we now understand by *non-linear* kinematic constraints an arbitrary submanifold $\Pi \subset T^*(M)$, fibred over $M$ into smooth surfaces $\Pi_x \subset T_x(M)$, then the partial symmetric structure on $\Pi$ must be given "backwards". Namely, let $b \in \Pi$, $\pi(b) = x$; we replace the surface $\Pi_x \subset T_x^*(M)$ by its linearization $\bar{\Pi}(b)$ at $b$. We construct the element $P_x(b) = \alpha_x^{-1}(\bar{\Pi}(b)) \subset T_x(M)$. We set $T_b^0(\Pi) = \{\xi \in T_b(\Pi) | d\pi|_b(\xi) \in P_x(b)\}$ etc.

Let $F \in \Lambda^1(M)$ be the "field" of force in which the system is situated, regarded as a differential 1-form, and let $\widetilde{G}$ be the quadratic function on $T^*(M)$ corresponding to the metric $G$. By a generalized Hamiltonian of the system we mean the 1-form $E = d\widetilde{G} - \pi^*(F)$, where $\pi: T^*(M) \to M$ is the natural projection.

DEFINITION. Suppose that a manifold $\Pi$ realizes the kinematic constraints imposed on a mechanical system, $\omega \in E|_\Pi$, $\Omega = (d\rho)|_\Pi$. Then the trajectories of the vector field $Z(\omega)$ are called the *phase trajectories* of this mechanical system. We assume that the forces admit a force function $F = dU$, $\mathscr{H} = \widetilde{G} - \pi^*(U)$. Since a free mechanical system would move along the trajectories of the Hamiltonian vector field $X_{\mathscr{H}}$, the vector $R_{\mathscr{H}} = Z_{d\mathscr{H}} - X_{\mathscr{H}}$ must be taken to be the force of reaction of the non-holonomic constraints.

Let us describe the objects introduced above in special coordinates on $T(M)$ and $T^*(M)$, assuming that the kinematic constraints are linear, and let us verify that the well-known Chaplygin–Kanjel equations (see [27]) and the formulae for the force of reaction follow from the principle we have formulated. Let $\dim M = n$, $n = k + m$.

1°. The constraints $P_x = \left\{ \sum_{i=1}^{n} a_{si}(q)\dot{q}_i + a_{s0}(q) = 0,\ s = 1, \ldots m \right\}$, where $q_1, \ldots, q_n, \dot{q}_1, \ldots, \dot{q}_n$ are special coordinates on $T(M)$.

2°. The Hamiltonian $E = \sum_{i=1}^{n} \frac{\partial \widetilde{G}}{\partial p_i} dp_i + \sum_{i=1}^{n} g_i(q, p) dq_i$.

3°. The inverse Legendre map $q_i = \frac{\partial \widetilde{G}}{\partial p_i}$ $(i = 1, \ldots, n)$, where $q_1, \ldots, q_n, p_1, \ldots, p_n$ are special coordinates on $T^*(M)$.

4°. $\Pi = \left\{ \sum_{i=1}^{n} a_{si} \frac{\partial \widetilde{G}}{\partial p_i} + a_{s0} = 0,\ s = 1, \ldots, m \right\}$.

PROPOSITION. *For $\omega = E|_\Pi$ the vector field $Z(\omega)$ is of the form*

$$(2) \qquad -\sum_{i=1}^{n}\Big(g_i-\sum_{s=1}^{m}\lambda_s a_{si}\Big)\frac{\partial}{\partial p_i}+\sum_{i=1}^{n}\frac{\partial \widetilde{G}}{\partial p_i}\frac{\partial}{\partial q_i},$$

*where the* $\lambda_s$ $(s = 1, \ldots, m)$ *are functions on* $\Pi$.

In particular, when $F = dU$, $\mathscr{H}=\widetilde{G} - \pi^*(U)$, we have the classical Chalygin–Kanjel equations

$$\begin{cases} \dfrac{dq_i}{dt}=\dfrac{\partial \mathscr{H}}{\partial p_i}, \quad \dfrac{dp_i}{dt}=-\Big(\dfrac{\partial \mathscr{H}}{\partial q_i}-\displaystyle\sum_{s=1}^{m}\lambda_s a_{si}\Big), \\ \displaystyle\sum_{i=1}^{n} a_{si}\dfrac{\partial \mathscr{H}}{\partial p_i}+a_{s0}=0 \quad (s=1, \ldots, m), \end{cases}$$

$$R_{\mathscr{H}}=\sum_{i=1}^{n}\Big(\sum_{s=1}^{m}\lambda_s a_{si}\Big)\frac{\partial}{\partial p_i}.$$

If the configuration manifold of the system is a Lie group $\mathscr{G}$, and the distribution $P_g \subset T_g(\mathscr{G})$ and the kinetic energy of the system are left-invariant, then so is the surface $\Pi \subset T^*(\mathscr{G})$. In this case the phase trajectories of the system are projected naturally onto the linear space $\Pi_e \subset L^* = T^*_e(\mathscr{G})$, where $e$ is the unit element of the group. Here the equations of the projected phase flow can be uniquely recovered from the element of the metric $G_e$: $T_e \times T_e \to \mathbf{R}$ and the structure constants of the Lie algebra $L = T_e(\mathscr{G})$, similar to the way this happens in the holonomic case (see §10). In particular, the classical equations "in quasicoordinates" describing the rocking of a body on a rough plane are equations of such a flow on the appropriate Lie algebra.

Bringing mechanics with friction and non-holonomic mechanics together, we come to the following general principle.

If a force field $F$ and a coefficient of friction are given on the configuration manifold $M$ of a non-holonomic mechanical system, then the non-holonomic constraints on the system must be interpreted as a submanifold $\Pi \subset T^*(M)$, and the phase trajectories of this system are the trajectories of the vector field $Z(\omega)$, where $\omega = (d\widetilde{G} - \pi^*(F) + \lambda p)|_{\Pi}$.

**3. The Hamiltonian formalism with values in a representation of a Lie group.** The formalism developed below is suitable for the description of mechanical systems whose configuration manifold is a Lie group $\mathscr{G}$, whose kinetic energy is left-invariant, and whose potential energy is "only slightly non-invariant". We show below that the study of such a system reduces to that of a system with left-invariant Hamiltonian on a Lie group $\mathscr{G}_1$, an extension of $\mathscr{G}$.

We consider the right representation of $\mathscr{G}$ on $C^\infty(\mathscr{G})$: $g$: $f \to R_g^*(f)$, where $R_g$ is right multiplication by $g$. Let $V \subset C^\infty(\mathscr{G})$ be an invariant subspace of this representation, $A(V) \subset C^\infty(\mathscr{G})$ the subalgebra generated by $V$, and $\pi^*(A) = \{\pi^*(f) \mid f \in A(V)\}$, ($\pi : T^*(\mathscr{G}) \to \mathscr{G}$ is the projection). We denote by $C^\infty_L(T^*(\mathscr{G}))$ the algebra of left-invariant Hamiltonians (see §10).

FUNDAMENTAL ASSERTION. $C_L^\infty(T^*)\cdot\pi^*(A)$ is a subalgebra of $C^\infty(T^*(\mathcal{G}))$ and is closed under the Poisson bracket.

We indicate at once the main source of invariant subspaces of the right representation in $C^\infty(\mathcal{G})$. Let $\Phi$: $\mathcal{G} \to \mathrm{Aut}(W)$ be a linear representation. We fix a vector $e \in W$. For each $\xi \in W^*$ we define a function $f_\xi \in C^\infty(\mathcal{G})$: $f_\xi(g) = \xi(\Phi_{(g)}[e])$, $g \in \mathcal{G}$. The space $V$ of these functions is right-invariant, and if $W$ is finite-dimensional, so is $V$.

EXAMPLE. The potential energy of a rigid body, fastened at one fixed point and situated in a uniform gravitational field, is a function on $SO(3)$, but is not right-invariant. The orbit of this function under the right representation has a three-dimensional linear span.

From now on we assume that $\dim V < \infty$. Let $L = T_e(\mathcal{G})$ be the Lie algebra of $\mathcal{G}$. Regarding $V$ as a commutative Lie algebra, we examine the action of $L$ on $V$ induced by the right representation, and the corresponding semidirect product $L_1$ of the algebras $L$ and $V$. Now the ground ring $K = C_L^\infty(T^*)\cdot\pi^*(A)$ can be regarded as a ring of functions on $L_1^* = L^* \oplus V^*$ or, equivalently (§10), as a ring of left-invariant Hamiltonians on the corresponding Lie group $\mathcal{G}_1$. In particular, the Hamiltonian of the original mechanical system $\tilde{G} + \pi^*(V)$ is of this kind if $v \in V$.

It is not hard to check that the canonical structure arising on $K$ from the Lie algebra $L_1$ (see [16]) is the same as the structure induced on it by the Poisson bracket, by the fundamental assertion (see above).

EXAMPLE. For a heavy rigid body fastened at a fixed point, $\mathcal{G}_1$ is the group of motions of three-dimensional Euclidean space. The classical equations of motion of a rigid body are equations on a trajectory of the Hamiltonian vector field $X_{\mathcal{H}}$ if the Hamiltonian

$$\mathcal{H} = \sum_{i=1}^{3} A_i \nu_i^2 + \sum_{i=1}^{3} x_i \gamma_i$$

of the heavy rigid body is regarded as an element of the ring of functions on $T_e^*(\mathcal{G}_1)$.

**4. Invariant mechanics on a uniform space.** Generalizing the left-invariant theory, we can construct an invariant Hamiltonian formalism on a uniform space. The analogue of the ring $C_L^\infty(T^*(\mathcal{G}))$ in this case is the ring of functions on $T_{x_0}^*(M)$ ($x_0$ is some fixed point) that are invariant under the action of the stabilizer of $x_0$.

From results of Gel'fand (see [30]) it follows, for example, that the Poisson bracket of any two $O(n)$-invariant functions on $T^*(G_n^k)$, where $G_n^k$ is a Grassman manifold, vanishes.

**5. The theory of refraction and impact.** We suppose that the phase space of a mechanical system is divided by a hypersurface $\Gamma$ into two domains and that the behaviour of the system in these domains is described by *distinct* Hamiltonians. We discuss below the rules of "refraction and reflection" of a phase trajectory of such a system on $\Gamma$.

Let $M$ be a smooth manifold and $\Gamma \subset M$ a hypersurface splitting $M$ into two connected components:

$$M \setminus \Gamma = M_- \cup M_+.$$

By a function *divided by M* we mean a pair of functions $f_\mp : M_\mp \cup \Gamma \to \mathbf{R}$, each of which is infinitely differentiable at interior points and is extended to a smooth function $\tilde{f}_\mp : M \to \mathbf{R}$. (In particular, the restriction of a divided function to $\Gamma$ is two-valued.) We denote the ring of divided functions by $\mathscr{F}_\Gamma(M)$. If $M$ is a symplectic manifold, $\mathscr{H} \in \mathscr{F}_\Gamma(M)$, then a Hamiltonian vector field $X_{\mathscr{H}}$ and the notion of a trajectory of this field are defined on each of the components.

A point $x \in \Gamma$ is called *minus-opposing* for $\mathscr{H}$ if the vector $(X_{\mathscr{H}_-})_x$ points into $M_+$, and called *plus-opposing* for $\mathscr{H}$ if $(X_{\mathscr{H}_+})_x$ points into $M_-$.

There is a natural requirement, which we now impose on a trajectory: the trajectory must pass from opposing points into non-opposing points along the characteristics of the hypersurface (see §8). Let $\varepsilon, E \in \mathbf{R}$, $\varepsilon > 0$ and let $\gamma_- : (-\varepsilon, 0] \to \Gamma \cup M_-$ be a trajectory of the vector field $X_{\mathscr{H}_-}$ lying on the level surface $\{\mathscr{H}_- = E\}$, where $x = \gamma_-(0) \in \Gamma \cap \{\mathscr{H}_- = E\}$. We consider the characteristic $l$ passing through $x$. A point $y \in l$ is called *decisive* for $x$ if $y$ lies in $l \cap \{\mathscr{H}_- = E\}$ and is not minus-opposing or $y$ lies in $l \cap \{\mathscr{H}_+ = E\}$ and is not plus-opposing.

PRINCIPLE OF TRANSITION. A trajectory $\gamma_-$ can be extended from any decisive point as the trajectory of the corresponding Hamiltonian vector field (that is, the field $X_{\mathscr{H}_+}$ in the case of non-plus-opposition and $X_{\mathscr{H}_-}$ in the case of non-minus opposition).

Having formulated the principle of transition in this form, we deliberately refrain from quoting the theorem about the uniqueness of the trajectory, which has an obvious physical justification.

We indicate (in the spirit of [16]) an algebraic interpretation of the principle of transition. Instead of $\mathscr{F}_\Gamma(M)$ we consider a fairly arbitrary $R$-algebra $K$ with a canonical structure $\{\ ,\ \}$. Then the analogue of $\Gamma$ in $M$ is a self-stabilizing ideal $\mathfrak{J} \subset K$, and the analogue of the field of characteristic directions is a family of derivations $\chi_g : K \to K/\mathfrak{J}$, $g \in \mathfrak{J}$, $\chi_g(f) = \pi(\{g, f\})$, where $\pi: K \to K/\mathfrak{J}$ is the natural homomorphism. Let $I \subset \mathbf{R}^1$ be an interval, $0 \in I$, and let $F_0(I)$ be the ring of functions divided by the point 0.

We denote by $\overline{\mathscr{F}_0(I)} \subset \mathscr{F}_0(I)$ the subring of functions continuous at zero. Let $\mathscr{H} \in K$.

DEFINITION. We call a homomorphism of $R$-algebras $\gamma: K \to \mathscr{F}_0(I)$ a *trajectory* of the Hamiltonian $\mathscr{H}$, *divided by the ideal* $\mathfrak{J}$, if

1°. the diagram

$$\begin{array}{ccc} K & \xrightarrow{\{\mathscr{H},\,\cdot\}} & K \\ \gamma\downarrow & \frac{d}{dt} & \downarrow\gamma \\ \mathscr{F}_0(I) & \longrightarrow & \mathscr{F}_0(I) \end{array}$$

is commutative;

2°. $\gamma(\mathscr{H}) \in \overline{\mathscr{F}_0(I)}$;

3°. it follows from $\chi_g(f) = 0 \;\; \forall g \in \mathscr{I}$, that $\gamma(f) \in \overline{\mathscr{F}_0(I)}$.

We examine the case $M = T^*(L)$; $N \subset L$ is a submanifold of codimension 1, $\Gamma = \bigcup_{x \in N} T_x^*(L)$. The axioms 2° and 3° in the definition above have a simple mechanical meaning; 2° is the law of conservation of energy on passing through $N$, and 3° implies the continuity of the projection of the trajectory on $L$ and the law of conservation of "the component of momentum tangential to $N$".

Let $U$ be a domain in $L$, and $q_1, \ldots, q_n, p_1, \ldots, p_n$ be special coordinates on $T^*(U)$ such that $N = \{q_n = 0\}$, $M_+ \cap T^*(U) = \{q_n > 0\}$, $M_- \cap T^*(U) = \{q_n < 0\}$.

Let $\mathscr{H} = \mathscr{H}_\mp(q_1, \ldots, q_n, p_1, \ldots, p_n)$. Then the set of plus-opposing points of $H$ is described by the inequality

$$\frac{\partial \mathscr{H}_+}{\partial p_n} < 0, \tag{3}$$

and the domain of minus-opposing points by the inequality

$$\frac{\partial \mathscr{H}_-}{\partial p_n} > 0. \tag{4}$$

EXAMPLE. We suppose that the branches $\mathscr{H}_\mp$ of $\mathscr{H}$ are positive-definite quadratic functions on each fibre $T_x^*(L)$.

We consider in $T_x^*(L)$, $x \in N$ the pair of surfaces $S_+ = \{\mathscr{H}_+ = E\}$ and $S_- = \{\mathscr{H}_- = E\}$. Let $l$ be a characteristic passing through a minus-opposing point on $S_-$. It is easy to verify that a) $l$ intersects $S_-$ again at a minus-opposing point, therefore, there is always a "reflected" trajectory: b) if $l$ intersects $S_+$ at two points, then exactly one of them is plus-opposing and there is a "refracted" trajectory: c) if $l$ touches $S_+$, then the point of contact is non-opposing.

The cases b) and c) correspond to *refraction* of a trajectory of the system on $N$, and the case when $l$ does not intersect $S_+$ at all to a *completely internal reflection.*

EXAMPLE. (Capture of a trajectory). We consider on $T^*(\mathbf{R}^2)$ and $\Gamma = \{q_2 = 0\}$, the Hamiltonian $\mathscr{H}_\mp = \mp p_1 p_2$. The domains of plus- and minus-opposition of this Hamiltonian coincide: $\{p_1 < 0,\ q_2 = 0\}$. Therefore, there are no non-opposing points of $\mathscr{H}$ on the characteristic $p_1 = \text{const} < 0$, and trajectories falling on such a characteristic cannot be extended. It would be interesting to clarify whether there is any physical effect corresponding to this mathematical situation.

In conclusion we point out that impulsive motion of systems can also be studied by means of the formalism we have constructed. For example, elastic impact of a rigid body at a fixed barrier without friction is described by reflection of the trajectory of the rigid body from a hypersurface $\Gamma \subset T^*(SO(3) \times \mathbf{R}^3)$. Here we must postulate that the Hamiltonian $\mathscr{H}_+$ is "identically equal to $\infty$", that is, that $\{\mathscr{H}_+ = E\} \cap \Gamma = \emptyset$ for any $E$.

## References

[1] R. L. Bishop and R. J. Crittenden, Geometry of manifolds, Pure and Appl. Math. **14**, Academic Press, New York–London 1964. MR **29** #6401.
Translation: *Geometria mnogoobrazii*, Mir, Moscow 1967. MR **35** #4833.

[2] S. Sternberg, Lectures on differential geometry, Prentice–Hall, Englewood Cliffs, N.J., 1964. MR **33** #1797.
Translation: *Lektsii po differential'noi geometrii*, Mir, Moscow 1970. MR **52** #15255.

[3] R. Narasimhan, Analysis on real and complex manifolds, North-Holland, Amsterdam 1968. MR **40** #4972.
Translation: *Analiz na deistvitel'nykh i kompleksnykh mnogoobraziyakh*, Mir, Moscow 1971.

[4] M. F. Atiyah and I. G. Macdonald, Introduction to commutative algebra, Addison–Wesley, Reading, Mass. –London 1969. MR **39** #4129.
Translation: *Vvedenie v kommutativnyu algebru,* Mir, Moscow 1972. MR **50** #2138.

[5] Sze-tsen Hu, Homotopy theory, Pure and Appl. Math. **18**, Academic Press, New York–London 1959. MR **21** #5186.
Translation: *Teoria gomotopii*, Mir, Moscow 1964.

[6] H. Goldschmidt and S. Sternberg, The Hamilton–Cartan formalism in the calculus of variations, Ann. Inst. Fourier (Grenoble) **23** (1973), 203–267. MR **49** #6279.

[7] N. E. Steenrod, The topology of fibre bundles, Princeton Mathematical Series **14**, Princeton University Press, Princeton, N.J., 1951. MR **12**–522.
Translation: *Topologia kosykh proizvedenii,* Izdat, Inost. Lit., Moscow 1953.

[8] B. A. Kupershmidt, The Lagrangian formalism in the calculus of variations, Funktsional. Anal. i Prilozhen. **10**:2 (1976), 77–78.
= Functional Anal. Appl. **10** (1976).

[9] J. Milnor, Morse theory, Ann. of Math. Studies **51**, Princeton University Press, Princeton, N.J., 1963. MR **29** #634.
Translation: *Teoriya Morsa,* Mir, Moscow 1965. MR **31** #6249.

[10] V. V. Lychagin, A local classification of first order non-linear partial differential equations, Dokl. Akad. Nauk. SSSR **210** (1973), 525–528. MR **50** #1295.
= Soviet Math. Dokl. **14** (1973) 761–764. Local classification of non-linear first order partial differential equations, Uspekhi Mat. Nauk **30**:1 (1975), 101–171.
= Russian Math. Surveys **30**:1 (1975), 105–175.

[11] R. Thom, Quelques propriétés globales des variétés différentiables, Comment. Math. Helv. **28** (1954), 17–86. MR **15**–890.
Translation: in the coll. "Fibre spaces and their applications", Izdat. Inost. Lit, Moscow 1958.

[12] A. Weinstein, Symplectic manifolds and their Lagrangian submanifolds, Advances in Math. **6** (1971) 329–346. MR **44** #3351.

[13] A. M. Vinogradov, Many-valued solutions, and a principle for the classification of non-linear differential equations, Dokl. Akad. Nauk. SSSR **210** (1973), 11–14. MR **50** #1294.
= Soviet Math. Dokl. **14** (1973), 661–665.

[14] F. Klein, Vorlesungen über die Entwicklung der Mathematik im 19ten Jahrhundert, Springer–Verlag, Berlin 1926.
Translation: *Lektsii o razvitii matematiki v XIX stoletii*, ONTI, Moscow–Leningrad 1937.

[15] E. Fermi, Quantum mechanics, Univ. of Chicago Press, Chicago 1961. MR **24** #1534.
Translation: *Kvantovaya mekhanika*, Mir, Moscow 1968. MR **40** #3803.

[16] A. M. Vinogradov and I. S. Krasil'shchik, What is the Hamiltonian formalism? Uspekhi Math. Nauk. **30**:1 (1975), 173–198.
= Russian Math. Surveys **30**:1 (1975), 177–202.

[17] Yu. V. Ergorov, On the solubility of differential equations with simple characteristics, Uspekhi Mat. Nauk. **26**:2 (1972), 183–199. MR **45** #5541.
= Russian Math. Surveys **26**:2 (1971), 113–130.

[18] A. M. Vinogradov, The algebra of logic of the theory of linear differential operators, Dokl. Akad. Nauk. SSSR **205** (1972), 1025–1028. MR **46** #3498.
= Soviet Math. Dokl. **13** (1972), 1057–1062.

[19] L. Hörmander, Linear partial differential operators, Springer–Verlag, Berlin–Heidelberg–New York 1963. MR **28** #4221.
Translation: *Lineinye differentsial'nye operatory s chastnymi proizvodnymi*, Mir, Moscow 1965. MR **37** #6595.

[20] V. P. Maslov, *Teoriya vozmushchenii i asimptoticheskie metody* (Perturbation theory and asymptotic methods), Izdat. Moskovsk. Gos. Univ., Moscow 1965.

[21] L. A. Lyusternik and A. I. Fet, Variational problems on closed manifolds, Dokl. Akad. Nauk. SSSR **81** (1951), 17–18. MR **13**–474.

[22] J. Favard, Cours de géométrie différentiale locale, Gauthier–Villars, Paris 1957. MR **18**–668.
Translation: *Kurs lokal'noi differentsial'noi geometrii*, Mir, Moscow 1966.

[23] A. A. Kirillov, *Elementy teorii predstavlenii,* Nauka, Moscow 1972.
Translation: Éléments de la théorie des représentations, Mir, Moscow 1974. MR **52** #14134.

[24] A. Appel', *Teoreticheskaya mekhanika* (Theoretical mechanics), Izdat, Fizmatgiz, Moscow 1960.

[25] L. S. Pontryagin, Nepreryvnye gruppy (third ed.), Nauka, Moscow 1973. MR **50** #10141.
Translation of second ed.: Continuous groups, Gordon and Breach, New York 1966. MR **34** #1439.

[26] S. Smale, Topology and mechanics, I, Invent. Math. **10** (1970), 305–331; II, ibid, **11** (1970), 45–64. MR **46** #8263, **47** #9671.
= Uspekhi Mat. Nauk **27**:2 (1972), 77–133.

[27] Yu. I. Neimark and N. A. Fufaev, *Dinamiki negolonomnykh sistem* (The dynamics of non-holonomic systems), Nauka, Moscow 1967.

[28] A. V. Vershik and L. D. Faddeev, *Lagranzheva mekhanika v invariantnom izlozhenii* (Lagrangian mechanics in an invariant presentation), in the coll. "*Teoreticheskaya i matematicheskaya fizika*" ("Theoretical and mathematical physics"), vol.II, Izdat. Leningrad. Gos. Univ., Leningrad 1974.

[29] F. A. Berezin, The Hamiltonian formalism in the general Lagrangian problem, Uspekhi Mat. Nauk. **29**:3 (1974), 183–184.

[30] I. M. Gel'fand, Spherical functions on symmetric Riemannian spaces, Dokl. Akad. Nauk. SSSR **70** (1950), 5–8. MR **11**–498.

[31] R. Abraham, Foundations of mechanics, Benjamin, New York–Amsterdam 1967. MR **36** #3527.

[32] V. I. Arnol'd, *Matematicheskie metody klassicheskoi mekhaniki* (Mathematical methods of classical mechanics), Nauka, Moscow 1974.

[33] C. Godbillon, Géométrie différentielle et mécanique analytique, Hermann, Paris 1969. MR **39** #3416.
Translation: *Differentsial'naya geometriya i analiticheskaya mekhanika*, Mir, Moscow 1973.

[34] Dynamical Systems (Proc. Sympos. Univ. of Bahia, San Salvador 1971), Academic Press, New York–London 1973.

[35] Dynamical Systems, Theory and Applications, Lecture Notes in Physics **38**. MR **52** #14236.

[36] J. K. Moser, Lectures on Hamiltonian systems, Mem. Amer. Math. Soc. **81**, Amer. Math. Soc., Providence, R.I., 1968. MR **37** # 6060.
Translation: *Lektsii o gamil'tonovykh sistemakh*, Mir, Moscow 1973.

[37] J. M. Souriau, Structure des systèmes dynamiques, Dunod, Paris 1970. MR **41** #4866.

[38] Symposia Mathematica vol. XIV (Convegno di Geometria Simplettica), INDAN, Roma, January 1973.

[39] A. Avez, A. Lichnerowicz, and A. Diaz-Miranda, Sur l'algèbre des automorphismes infinitésimaux d'une variété symplectique. J. Differential Geometry **9** (1974), 1–40. MR **50** #8602.

[40] W. B. Gordon, A theorem on the existence of periodic solutions to Hamiltonian systems with convex potentials, J. Differential Equations **10** (1971), 324–335. MR **45** #1200.

[41] S. M. Graff, On the conservation of hyperbolic invariantttori for Hamiltonian systems, J. Differential Equations **15** (1974), 1–69. MR **51** #1878.

[42] P. B. Guest, A non-existence theorem for realizations of semisimple Lie algebras, Nuovo Cimento A (10) **61** (1969), 593–604. MR **40** #2254.

[43] R. Hermann, Spectrum generating algebras and symmetries in mechanics, I, II, J. Math. Phys. **13** (1972), 833–842. MR **47** #2930.

[44] A. A. Kirillov, Local Lie algebras, Uspekhi Mat. Nauk. **31**:4 (1976), 57–76. = Russian Math. Surveys **31**:4 (1976), 55–76.

[45] A. Lichnerowicz, Cohomologie 1-différentiable des algèbres de Lie attachées à une variété symplectique ou de contact, J. Math. Pures Appl. **53** (1974), 459–483. MR **51** #4315.

[46] E. Lacomba Zamora, Mechanical systems with symmetry on homogeneous spaces, Trans. Amer. Math. Soc. **185** (1973), 477–491. MR **46** #9759.

[47] L. Losco, Intégrabilité en mécanique céleste, J. Mech. **13** (1974), 197–223. MR **51** #14154.

[48] S. V. Manakov, A remark on the integration of the Euler equations of the dynamics of a rigid body, Funktsional. Anal. i Prilozhen. **10**:4 (1976), 93–94. = Functional Anal. Appl. **10** (1976).

[49] A. S. MIshchenko and A. T. Fomenko, On the integration of the Euler equations on semisimple Lie algebras, Dokl. Akad. Nauk. SSSR **231** (1976), 536–538. = Soviet Math. Dokl. **17** (1976), 1591–1593.

[50] M. A. Ol'shanetskii and A. M. Perelomov, Geodesic flows on symmetric spaces of zero curvature and explicit solutions of the generalized Calogero model for the classical case, Funktsional. Anal. i Prilozhen. **10**:3 (1976), 86–87. = Functional Anal. Appl. **10** (1976).

[51] R. M. Santilli, Dissipative and Lie admissible algebras, Meccanica **4** (1963), 3–11 (Italian).

[52] A. Weinstein, Lagrangian submanifolds and Hamiltonian systems, Ann. of Math. (2) **98** (1973), 377–410. MR **48** #9761.

[53] M. Verge, La structure de Poisson sur l'algèbre symmétrique d'une algèbre de Lie nilpotente, Bull. Soc. Math. France **100** (1972), 301–335. MR **52** #657.

[54] P. A. M. Dirac, Lectures on quantum mechanics Yeshiva Univ., New York 1964. Translation: *Lektsii po kvantovoi mekhaniki*, Mir, Moscow 1973.

[55] L. M. C. Coelho de Souza and P. R. Rodrigues, Field theory with higher derivatives-Hamiltonian structure, J. Phys. A, Ser.2, **2** (1962), 304–310.

[56] M. Flato, A. Lichnerowicz, and D. Sternheimer, Deformations of Poisson brackets, Dirac brackets and applications, J. Math. Phys. **17** (1976), 1754–1762.

[57] P. L. Garcia, Symplectic geometry in the classical theory of fields, Collect. Math. **19** (1968), 73–134 (Spanish). MR **39** #5089.

[58] B. Kostant, Quantization and unitary representations, Lecture Notes in Math. **170**. MR **45** #3638.

[59] B. A. Kupershmidt, *Geometriya mnogoobraziya dzhetov i struktura lagranzheva i gamil'tonova formalizmov* (The geometry of a jet manifold and the structure of the Lagrangian and Hamiltonian formalisms), Trudy Moscov. Mat. Obshch., Moscow 1977.

[60] L. D. Faddeev, The Feynman integral for singular Lagrangians, Teoret. Mat. Fiz. **1** (1969), 3–18.

[61] E. Shmutser, *Osnovnye printsipy klassicheskoi mekhaniki i klassicheskoi teorii polya* (The basic principles of classical mechanics and classical field theory), Mir, Moscow 1976.

[62] O. I. Bogoyavlenskii and S. P. Novikov, On a connection between the Hamiltonian formalisms of stationary and non-stationary problems, Funktsional. Anal. i. Prilozhen. **10** (1976), 9–13. = Functional Anal. Appl. **10** (1976).

[63] L. J. F. Broer and J. A. Kobussen, Canonical transformations and generating functionals, Physica **61** (1972), 275–288.

[64] A. M. Vinogradov, I. S. Krasil'shchik, and V. V. Lychagin, *Primenenie nelineinykh differentsial'nykh uravnenii*. I (Applications of non-linear differential equations. I), MIIGA, Moscow 1977.

[65] A. M. Vinogradov, I. S. Krasil'shchik, and V. V. Lychagin, *Geometriya prostranstv dzhetov i nelineinye differentsial'nye uravneniya* (The geometry of jet spaces and non-linear differential equations), MIEM, Moscow 1977.

[66] C. S. Gardner, Korteweg-de Vries equation and generalizations. IV. The Korteweg-de Vries equation as a Hamiltonian system. J. Mathematical Phys. **12** (1971), 1548–1551. MR **44** #3615.

[67] I. M. Gel'fand and L. A. Dikii, Asymptotic behaviour of the resolvent of Sturm–Liouville equations and the algebra of the Korteweg-de Vries equations, Uspekhi Math. Nauk. **30**:5 (1975), 67–100.
= Russian Math. Surveys **30**:5 (1975), 77–113.

[68] I. M. Gel'fand and L. A. Dikii, Fractional powers of operators and Hamiltonian systems, Funktsional. Anal. i Prilozhen. **10**:4 (1976), 13–29.
= Functional Anal. Appl. **10** (1976).

[69] I. M. Gel'fand, Yu. I. Manin, and M. A. Shubin, Poisson brackets and the kernel of the variational derivative in the formal calculus of variations, Funktsional. Anal. i Prilozhen **10**:4 (1976), 30–34.
= Functional Anal. Appl. **10** (1976).

[70] B. A. Dubrovin, V. B. Matveev, and S. P. Novikov, Non-linear equations of Korteweg-de Vries type, finite-zone linear operators, and Abelian varieties, Uspekhi Mat. Nauk. **31**:1 (1976), 55–136.
= Russian Math. Surveys **31**:1 (1976), 59–146.

[71] V. E. Zakharov and L. D. Faddeev, The Korteweg-de Vries equation is a fully integrable Hamiltonian system, Funktsional. Anal. i Prilozhen. **5**:4 (1971), 18–27. MR **46** # 2270.
= Functional Anal. Appl. **5** (1971), 280–287.

[72] B. A. Kupershmidt and Yu. I. Manin, Equations of long waves with a free surface, conservation laws and solutions, Funktsional. Anal. i Prilozhen. **11**:3 (1977),
= Functional Anal. Appl. **11** (1977).

[73] N. V. Nikolenko, On the complete integrability of the non-linear Schrödinger equation, Funktsional. Anal. i Prilozhen. **10**:3 (1976), 55–69.
= Functional. Anal. Appl. **10** (1976).

[74] Y. Watanabe, Note on gauge invariance and conservation laws for a class of non-linear partial differential equations, J. Mathematical Phys. **15** (1974), 453–457. MR **49** # 5588.

Received by the Editors 20 October 1975

Translated by A. J. McIsaac

# WHAT IS THE HAMILTONIAN FORMALISM?

A. M. Vinogradov and I. S. Krasil'shchik

In this paper the basic concepts of the classical Hamiltonian formalism are translated into algebraic language. We treat the Hamiltonian formalism as a constituent part of the general theory of linear differential operators on commutative rings with identity. We take particular care in motivating the concepts we introduce. As an illustration of the theory presented here, we examine the Hamiltonian formalism in Lie algebras. We conclude by presenting a version of the "orbit method" in the theory of representations of Lie groups, which is a natural corollary of our view of the Hamiltonian formalism.

*"The present essay does not pretend to treat fully of this extensive subject, – a task which may require the labours of many years and many minds; but only to suggest the thought and propose the path to others."*
(From the introduction to "On a general method in dynamics" by W. R. Hamilton.)

A curious fact that has come to light in the course of the last ten or fifteen years is the appearance of the Hamiltonian formalism of classical mechanics as an essential component of a number of mathematical theories, far removed from mechanics as such. Examples of this kind are the orbit method in the theory of representations of Lie groups (see [10] and [11]), the theory of local solubility of linear differential operators (see the survey [8]), the theory of a canonical operator (see [13] and [14]), and others. It is of interest that the Hamiltonian formalism plays different functional roles in these theories; this suggests that the Hamiltonian formalism is part of a broader mathematical system. We believe that this is the "general theory of linear differential operators" in the sense of [2], and that this circumstance, although demonstrated in the corresponding specific investigations completely explains the many-sidedness of the Hamiltonian formalism and allows us to hope for a deeper understanding of the possibilities of Hamiltonian method, as well as for new domains of its applicability.

Our paper is concerned with the necessary first step in this direction: translating the basic concepts of the Hamiltonian formalism into its natural algebraic language (here we arrive at some new "mechanical" invariants of commutative rings).

The new facts, if they are new, are simple and illustrative. We present our exposition in two versions, "general" and "Lie", mainly to show the degree of unification our approach yields. Throughout we try to motivate the concepts we introduce.

For an understanding of many of them, apart from the "standard" facts, an acquaintance with the elementary parts of the theory of affine schemes (see [12], for example) is needed.

The authors are very grateful to V. I. Arnol'd, who read this work in manuscript, and to the referee for many useful remarks.

1. **Linear differential operators over a commutative ring K with identity.** We introduce here the necessary information from the theory of linear differential operators over a ring K. An outline of this theory, which will help to understand the role of the "Hamiltonian formalism", can be found in [2] and [3]. For surveys of the theory of differential operators from the point of view of the theory of schemes, see [4] and [7].

Let $A$ and $B$ be $K$-modules and let $\Delta: A \to B$ be an additive mapping. For each element $k \in K$ we define the mappings $k\Delta$, $k^+\Delta$ and $\delta_k(\Delta)$ by putting $(k\Delta)(a) = k\Delta(a)$, $(k^+\Delta)(a) = \Delta(ka)$, $a \in A$, and $\delta_k(\Delta) = k{+}\Delta - k\Delta$. We also set $\delta_{k_0, k_1, \ldots, k_s} = \delta_{k_0} \circ \delta$

FUNDAMENTAL DEFINITION. An additive mapping $\Delta: A \to B$ is called a *linear differential operator of order* $\leqslant s$ *over* $K$ if for each choice of $k_0, k_1, \ldots, k_s \in K$ we have

$$\delta_{k_0, k_1, \ldots, k_s}(\Delta) = 0.$$

If the ring $K$ has additional structure, we can consider differential operators that respect this structure. If, for example, $K$ is a $C$-algebra, we can speak of $C$-linear differential operators over $K$.

When $K = C^\infty(M)$ is the ring of all smooth (real or complex) functions on a smooth manifold $M$, and $A$ and $B$ are $K$-modules of smooth sections of vector bundles over $M$, then our concept of an **R**- or **C**-linear differential operator coincides with the classical one (see [15]).

The set of differential operators of order $\leqslant s$ from $A$ to $B$ is an Abelian group under the natural addition.

We can make this Abelian group into a $K$-module in two different ways. The first is by the multiplication $(k, \Delta) \to k\Delta$, and the second by $(k, \Delta) \to k^+\Delta$. We denote the corresponding modules by $\mathrm{Diff}_s(A, B)$ and $\mathrm{Diff}_s^+(A, B)$ and we write $\mathrm{Diff}_s^{(+)}(A, B)$ for the bimodule obtained by considering both structures simultaneously. It is clear that $\mathrm{Diff}_m^{(+)}(A, B) \subset \mathrm{Diff}_s^{(+)}(A, B)$ when $m \leqslant s$, so that we arrive at the bimodule $\mathrm{Diff}_*^{(+)}(A, B) = \bigcup_{s \geqslant 0} \mathrm{Diff}_s^{(+)}(A, B)$, filtered by the bimodules $\mathrm{Diff}_s^{(+)}(A, B)$, $s \geqslant 0$.

Let $A$, $B$, and $C$ be $K$-modules, $\Delta_1 \in \mathrm{Diff}_s^{(+)}(B, C)$ and $\Delta_2 \in \mathrm{Diff}_t^{(+)}(A, B)$.

The obvious relation $\delta_k(\Delta_1 \circ \Delta_2) = \delta_k(\Delta_1) \circ \Delta_2 + \Delta_1 \circ \delta_k(\Delta_2)$, $k \in K$, shows that composition of differential operators defines a multiplication

$$\mathrm{Diff}_s^{(+)}(A, B) \otimes_{\mathbf{Z}} \mathrm{Diff}_t^{(+)}(B, C) \to \mathrm{Diff}_{s+t}^{(+)}(A, C).$$

In particular, $\mathrm{Diff}_*^{(+)}(A, A)$ is a filtered ring. We consider the ring $\mathrm{Diff}_*^{(+)} K = \mathrm{Diff}_*^{(+)}(K, K)$. The graded ring $\mathrm{Smbl}_* K = \sum_{s \geqslant 0} \mathrm{Smbl}_s K$ corresponding to it, where $\mathrm{Smbl}_s K = \mathrm{Diff}_s^{(+)} K / \mathrm{Diff}_{s-1}^{(+)} K$, is called the *ring of symbols*; it is obviously commutative.

For each $K$-module $A$ we introduce a module $\mathrm{Diff}_s^+ A$ and a differential operator Д: $\mathrm{Diff}_s^+ A \to A$ of order $\leqslant s$ such that for any $K$-module $B$ the correspondence $f \to Д \circ f$, $f \in \mathrm{Hom}_K(B, \mathrm{Diff}_s^+ A)$ is an isomorphism of the modules $\mathrm{Hom}_K(B, \mathrm{Diff}_s^+ A)$ and $\mathrm{Diff}_s^+(B, A)$.

It is easy to see that by setting $\mathrm{Diff}_s^+ A \simeq \mathrm{Diff}_s^+(K, A)$, and $Д(\Delta) = \Delta(1)$, $\Delta \in \mathrm{Diff}_s^+ A$, $1 \in K$, we obtain a realization of the functor $A \to \mathrm{Diff}_s^+ A$.

2. **The functors $\mathbf{D}_i$ and $\mathbf{\Lambda}_i$.** The reader will see below that, perhaps somewhat unexpectedly, the Hamiltonian formalism cannot, in general (that is, for an arbitrary ring $K$), be stated in the language of differential forms. In this section we provide enough information on the calculus of differential forms over a ring $K$ to explain this circumstance. When $K = C^\infty(M, \mathbf{R}$ or $\mathbf{C})$, the concepts introduced here coincide with the classical ones. We recall that when $K = C^\infty(M, \mathbf{R})$, each smooth vector field $X$ on $M$ induces a derivation $d_X : K \to K$, that is, an additive mapping such that $d_X(f_1 f_2) = f_1 d_X(f_2) + f_2 d_X(f_1)$, $f_1, f_2 \in K$, and vice versa. Since the concept of a derivation is defined for any ring $K$, we naturally use the phrase "vector field on $K$" to mean a derivation of $K$. We write $D(K)$ for the $K$-module of derivations of $K$.

Now we recall that a derivation of $K$ with values in a $K$-module $A$ is an additive mapping $\Delta: K \to A$ such that $\Delta(k_1 k_2) = k_1 \Delta(k_2) + k_2 \Delta(k_1)$. We write $D(A)$ for the $K$-module of all such derivations. It is obvious that each of these derivations is a differential operator of order $\leqslant 1$ and that there is an embedding $D(A) \subset \mathrm{Diff}_1^+ A)$, which is a first order differential operator.

Let $R$ be a subset of a module $A$. We write $D(R \subset A)$ for the set of all derivations $\Delta: K \to A$ such that $\Delta(K) \subset R$; we define $\mathrm{Diff}_s^{(+)}(R \subset A)$ similarly. In particular $D(R \subset A) = D(R)$ if $R$ is a submodule of $A$.

We now define a sequence of functors $D_i$ and embeddings $D_i(A) \subset (\mathrm{Diff}_1^+)^i(A)$ inductively. We put $D_0 = id$, $D_1 = D$, $D_i(A) = D\{D_{i-1}(A) \subset (\mathrm{Diff}_1^+)^{i-1}(A)\}$. The composition

$D_i(A) = D\{D_{i-1}(A) \subset (\mathrm{Diff}_1^+)^{i-1}(A)\} \subset D(\mathrm{Diff}_1^+)^{i-1}(A) \subset (\mathrm{Diff}_1^+)^i(A)$ of the natural embeddings gives an embedding $D_i(A) \subset (\mathrm{Diff}_1^+)^i(A)$, which is a differential operator of order $\leqslant 1$. The embeddings $\alpha_i(A)\colon D_i(A) \subset D_{i-1}(\mathrm{Diff}_1^+ A)$ are constructed inductively from the natural embedding $\alpha_1(A)\colon D(A) \subset \mathrm{Diff}_1^+(A)$ and can be shown to be differential operators of order $\leqslant 1$. Since

$$D_{i-1}(\mathrm{Diff}_1^+ A) = D\{D_{i-2}(\mathrm{Diff}_1^+ A) \subset (\mathrm{Diff}_1^+)^{i-2}(\mathrm{Diff}_1^+ A)\} = \\ = D\{D_{i-2}(\mathrm{Diff}_1^+ A) \subset (\mathrm{Diff}_1^+)^{i-1}(A)\},$$

we see that

$$D_i(A) = D\{D_{i-1}(A) \subset (\mathrm{Diff}_1^+)^{i-1}(A)\}$$

can evidently be identified with

$$D\{\mathrm{Im}\, \alpha_{i-1}(A) \subset (\mathrm{Diff}_1^+)^{i-1}(A)\} \subset D\{D_{i-2}(\mathrm{Diff}_1^+ A) \subset (\mathrm{Diff}_1^+)^{i-1}(A)\},$$

which defines $\alpha_i(A)$.

We remark that the functor embedding $D_i(D_j) \subset D_{i+j}$ can be established directly by induction.

The module $\Lambda_i = \Lambda_i(K)$ $(i = 0, 1, 2, \ldots)$ of $i$-dimensional differential forms over $K$ is defined as a representing object for the functor $D_i$; that is, so that for every $K$-module $A$ there is an isomorphism $D_i(A) = \mathrm{Hom}_K(\Lambda_i, A)$. As we know, $\Lambda = \Lambda_1$ is the factor module of the free $K$-module generated by elements $dk$ $(k \in k)$ subject to the relations $d(k_1 k_2) = k_1 d(k_2) + k_2\, dk_1$, $d(k_1 + k_2) = dk_1 + dk_2$, $k_1, k_2 \in K$. It can be shown that the module $\Lambda_i$ exists and is the $i^{\text{th}}$ exterior power of $\Lambda$. In this context we mention that the homomorphisms $\Lambda_i \otimes \Lambda_j \to \Lambda_{i+j}$ of representing objects generated by the functor embedding $D_i(D_j) \subset D_{i+j}$ is nothing but exterior multiplication of differential forms.

We note that every differential operator $\Delta\colon \Lambda_{i-1} \to \Lambda_i$ of order $\leqslant 1$ determines for every $K$-module $A$ a mapping $\varphi_\Delta(A)\colon \mathrm{Hom}_K(\Lambda_i, A) \to \mathrm{Hom}_K(\Lambda_{i-1}, \mathrm{Diff}_1^+ A)$ by the formula $\varphi_\Delta(A)(\lambda) = \psi_{\Delta+\lambda}$, where by $\psi_{\Delta'}\colon \Lambda_{i-1} \to \mathrm{Diff}_1^+ A$, $\Delta' \in \mathrm{Diff}_1^+(\Lambda_{i-1}, A)$, we denote the homomorphism corresponding to $\Delta'$ according to the definition of the functor $\mathrm{Diff}_1^+$. We define the operator $d = d_i\colon \Lambda_{i-1} \to \Lambda_i$ of exterior differentiation so that the diagram

$$\begin{array}{ccc} \mathrm{Hom}(\Lambda_i, A) & \xrightarrow{\varphi_d(A)} & \mathrm{Hom}_K(\Lambda_{i-1}, \mathrm{Diff}_1^+ A) \\ \| & & \| \\ D_i(A) & \xrightarrow{\alpha_i(A)} & D_{i-1}(\mathrm{Diff}_1^+ A) \end{array}$$

commutes. In particular, for $i = 1$ the operator $d_1$ is the derivation of $K$ with values in $\Lambda$, and $d(k) = dk$, $k \in K$, and the isomorphism $D(A) = \mathrm{Hom}_k(\Lambda, A)$ associates with each derivation $\Delta \in D(A)$ a homomorphism $h_\Delta : \Lambda \to A$ so that the diagram

$$\begin{array}{ccc} K & \xrightarrow{\Delta} & A \\ & {\scriptstyle d}\searrow \quad \nearrow{\scriptstyle h_\Delta} & \\ & \Lambda & \end{array}$$

commutes.

The operator $d$ defined here can be shown to have many of the properties of the classical exterior differentiation of forms. For example, $d^2 = 0$ and $d(\omega_1 \wedge \omega_2) = d\omega_1 \wedge \omega_2 + (-1)^i \omega_1 \wedge d\omega_2$, $\omega_1 \in \Lambda_i$, $\omega_2 \in \Lambda_j$.

The representation of the module $\Lambda_i$ as the $i$th exterior power of $\Lambda$ allows us to regard the elements $\omega \in \Lambda_i$ as $i$-linear skew-symmetric functions on $D(K)$, which corresponds to the classical treatment of a differential form on a smooth manifold. The classical formula for the exterior differential:

$$(d\omega)(X_1, \ldots, X_{i+1}) = \sum_{p=1}^{i+1} (-1)^{p-1} X_p \omega(X_1, \ldots, X_{p-1}, X_{p+1}, \ldots, X_{i+1}) +$$
$$+ \sum_{p<q} (-1)^{p+q} \omega([X_p, X_q], X_1, \ldots, X_{p-1}, X_{p+1}, \ldots, X_{q-1}, X_{q+1}, \ldots, X_{i+1}),$$

where $X_1, \ldots, X_{i+1} \in D(K)$, $\omega \in \Lambda_i$, remains valid.

It is important to realize that a form $\omega \in \Lambda_i(K)$ need not, in general, be uniquely determined by its values on vector fields; that is, it can happen that $\omega(X_1, \ldots, X_i) = 0$ for all choices of $X_1, \ldots, X_i \in D(K)$, and yet $\omega \neq 0$; there is an example of this on p. 000. This cannot happen when the module $\Lambda = \Lambda(K)$ is projective and locally finite, for example, when $K = C^\infty(M, \mathbf{R}$ or $\mathbf{C})$. In 4. we shall use the fact that if a 3-form is determined by its values on elements of $D(K)$, then to prove that $\omega$ is closed we need only verify that it vanishes on the expression above.

**3. The cotangent (phase) space of a commutative ring K.** We recall that the arena of action of the Hamiltonian formalism of classical mechanics, the phase space, is the cotangent space $T^*(M)$ the manifold $M$ of configurations. The algebraic theory of differential operators, which forms the foundation of our constructions, is based on the transition from smooth manifolds to commutative rings. Here $M$ is replaced by $C^\infty(M, \mathbf{R})$ or $C^\infty(M, \mathbf{C})$.

REMARK. The theory of affine schemes (see [12], for example) permits us to recover the manifold $M$ from $C^\infty(M, \mathbf{R})$ or $C^\infty(M, \mathbf{C})$. But by now it is clear that the ring view-point is preferable, because two distinct rings correspond to one and the same manifold.

We must, therefore, find out, purely in terms of the algebraic structure of $K$, what this $T^*(M)$ should be, more accurately, what ring "corresponds" to $T^*(M)$.

Answer: the ring "corresponding" to $T^*(M)$ is $\mathrm{Smbl}_*(K)$, the ring of symbols of $K$.

What is the basis for this statement? Let us consider the classical case $K = C^\infty(M, \mathbf{C})$. Then, as is well known, the ring $\mathrm{Smbl}_*(K)$ can be identified naturally with the ring of those smooth functions on $T^*(M)$ that are polynomial along the fibres of the projection $\pi\colon T^*(M) \to M$. The "manifold" Specm $\{\mathrm{Smbl}_* K\}\setminus$ "bad" points (we do not define "bad" precisely) is then the complexification of the fibering $\pi$, so that $T^*(M) \subset \mathrm{Spec}\{\mathrm{Smbl}_* K\}$. Further, by what we have said above, each element of $\mathrm{Smbl}_*$ K, regarded as a function on $\mathrm{Spec}\{\mathrm{Smbl}_* K\}$, in accordance with the theory of affine schemes, can be uniquely recovered from its restriction to the subspace $T^*(M) \subset \mathrm{Spec}\{\mathrm{Smbl}_* K\}$. Thus, the points of the complement $\mathrm{Spec}\{\mathrm{Smbl}_* K\}\setminus T^*(M)$ are inessential from the point of view of the "geometrical realization" of the ring $\mathrm{Smbl}_* K$, that is, its representation as a ring of "functions on something". This shows why $T^*(M)$ can be regarded as the geometrical equivalent of $\mathrm{Smbl}_* K$.

NOTE 1. In quite a number of problems (the complex germs of Maslov [13], for example) it is useful, and even necessary, to consider the "complex" points of $\mathrm{Spec}\{\mathrm{Smbl}_* K\}$.

NOTE 2. What we have said above can be understood as our justification for taking $\mathrm{Spec}\{\mathrm{Smbl}_* K\}$ as the cotangent space of the "manifold" Spec $K$. Later we shall write briefly $T^*(K)$ instead of $\mathrm{Spec}\{\mathrm{Smbl}_* K\}$ (on this point see also [6]).

**4. The Hamiltonian mapping.** Let $\Delta \in \mathrm{Diff}_m\, K$ and let us write $\mathrm{smbl}_m(\Delta)$ for the image of $\Delta$ under the quotient homomorphism $\mathrm{Diff}_m\, K \to \mathrm{Smbl}_m K$. If $\Delta_i \in \mathrm{Diff}_{m_i}(K)$ $(i = 1, 2)$, then $[\Delta_1, \Delta_2] = \Delta_1\Delta_2 - \Delta_2\Delta_1 \in \mathrm{Diff}_{m_1+m_2-1}\, K$. This is an immediate consequence of the equality $\delta_k(\Delta_1\Delta_2) = \delta_k(\Delta_1)\cdot\Delta_2 + \Delta_1\cdot\delta_k(\Delta_2)$, which was established in 1. If $s_i = \mathrm{smbl}_{m_i}(\Delta_i)$ $(i = 1, 2)$, then it follows from what we have said that the element $\mathrm{smbl}_{m_1+m_2-1}[\Delta_1, \Delta_2]$, depends only on $s_i (i = 1, 2)$. We call this element the *Poisson bracket* of $s_1$ and $s_2$ and denote it by $\{s_1, s_2\} \in \mathrm{Smbl}_{m_1+m_2-1} K$. The following properties (analogous to those of the classical Poisson bracket) can be derived immediately from the definition:

1) $\{s_1 + s_2, s\} = \{s_1, s\} + \{s_2, s\}$,

2) $\{s_1, s_2\} = -\{s_2, s_1\}$,

3) $\{s, s_1s_2\} = s_1\cdot\{s, s_2\} + s_2\cdot\{s, s_1\}$,

4) $\{s_1, \{s_2, s_3\}\} + \{s_2, \{s_3, s_1\}\} + \{s_3, \{s_1, s_2\}\} = 0$.

As 3) shows, the mapping $Hs_0\colon s \mapsto Hs_0(s) = \{s_0, s\}$ for fixed $s_0 \in S = \mathrm{Smbl}_*\, K$ is a derivation of $S$. Since for us the concepts "derivation of $S$" and "vector field on $S$" are synonyms, we can regard $Hs_0$ as a "vector field" on $T^*(K)$; more precisely, $Hs_0$ is the "Hamiltonian vector field" on $T^*(K)$ corresponding to the "Hamiltonian" $s_0$. As 3) shows, the mapping $H.\ S \to D(S)$, $s \mapsto Hs$ is a derivation of $S$ with values in the $S$-module $D(S)$. There is, therefore, an $S$-homomorphism $h$: $\Lambda(S) \to D(S)$, such that the diagram

$$\begin{array}{ccc} S & \xrightarrow{H} & D(S) \\ & {\scriptstyle d}\searrow & \nearrow{\scriptstyle h} \\ & & \Lambda(S) \end{array}$$

commutes. We note that specifying the derivation $H$ is equivalent to specifying the homomorphism $h$; the Poisson bracket can, therefore, be expressed in terms of $h$. Indeed, $\{s_1, s_2\} = \langle h\, ds_1, ds_2\rangle$, where the brackets $\langle,\rangle$ denote the natural pairing of $\Lambda(S)$ and $D(S)$ induced by the isomorphism $D(S) = \mathrm{Hom}_S(\Lambda(S), S)$. Instead of $h$ we can consider equivalently the bilinear form $\Omega_h(\lambda_1, \lambda_2) = \langle h(\lambda_1), \lambda_2\rangle$ on $\Lambda(S)$. The form $\Omega_h$ is skew-symmetric because so is the Poisson bracket. When $h$ is an isomorphism, we can consider the skew form $\hat{\Omega}_h(D_1, D_2) = \Omega_h(h^{-1}(D_1), h^{-1}(D_2))$ on $D(S)$. If $\hat{\Omega}_h \in \Lambda_2(S)$, it is natural to interpret it as a differential 2-form on Spec $S = T^*(K)$.

PROPOSITION 1. *If the form $\hat{\Omega}_h$ exists and is defined by its values on $D(S)$, then it is closed, that is, $d\hat{\Omega}_h = 0$.*

PROOF. We need only prove that for any three derivations $D_i \in D(S)$ $(i = 1, 2, 3)$

$$\begin{aligned}(1)\qquad d\hat{\Omega}_h(D_1, D_2, D_3) = {} & D_1\hat{\Omega}_h(D_2, D_3) - D_2\hat{\Omega}_h(D_1, D_3) + \\ & + D_3\hat{\Omega}_h(D_1, D_2) - \hat{\Omega}_h([D_1, D_2], D_3) + \\ & + \hat{\Omega}_h([D_1, D_3], D_2) - \hat{\Omega}_h([D_2, D_3], D_1) = 0\end{aligned}$$

(see 2). Since $\Lambda(S)$ as a module over $S$ is generated by the elements of the form $ds$, and since $h$ in our case is an isomorphism, the module $D(S)$ is generated by derivations of the form $Hs$ with the same relations. Therefore, since (1) is linear in all three arguments, we need verify (1) only for $D_i = Hs_i$ $(i = 1, 2, 3)$. But then

$$\begin{aligned}d\hat{\Omega}_h(Hs_1, Hs_2, Hs_3) = {} & Hs_1\hat{\Omega}_h(Hs_2, Hs_3) - Hs_2\hat{\Omega}_h(Hs_1, Hs_3) + \\ & + Hs_3\hat{\Omega}_h(Hs_1, Hs_2) - \hat{\Omega}_h([Hs_1, Hs_2], Hs_3) + \\ & + \hat{\Omega}_h([Hs_1, Hs_3], Hs_2) - \hat{\Omega}_h([Hs_2, Hs_3], Hs_1) = \\ & = 2(\{s_1, \{s_2, s_3\}\} + \{s_3, \{s_1, s_2\}\} + \{s_2\ \{s_3, s_1\}\}) = 0\end{aligned}$$

by the Jacobi identity for the Poisson brackets.

In the classical case $K = C^\infty(M, \mathbf{R})$ we have $\hat{\Omega}_h = \sum_i dp_i \wedge dq_i$ in the corresponding coordinates on $T^*(M)$. So we see that when $h$ is an isomorphism, $\hat{\Omega}_h$ is a "geometrical realization" of $h$ (or of $H$, or of the Poisson bracket). Thus, the mapping $H$ (or $h$) is a more fundamental object than $\hat{\Omega}_h$.

Thus, although according to the conventional point of view the Hamiltonian (= canonical) structure is interpreted as a 2-form on $T^*(M)$

(which in this case is equivalent to specifying $H$ or $h$), in the general situation we must understand by the Hamiltonian (or canonical) structure one of the three objects: the operator $H$, the homomorphism $h$, or the form $\Omega_h$. It stands to reason that each of these three objects must satisfy the relevant conditions. For example, the operator $H$ must be a derivation from $S$ into $D(S)$ for which $Hs_1(s_2) + Hs_2(s_1) = 0$.

From now on we mean by a *canonical structure* on a ring $S$ (which is not, in general, the ring of symbols $\mathrm{Smbl}_* K$ for some ring $K$) a derivation $H: S \to D(S)$ satisfying the condition just stated. The pair $(S, H)$ is then called a *canonical ring.* We note that when $M$ is a canonical (= symplectic) manifold, then there is a natural canonical structure on the ring of functions $C^\infty(M, \mathbf{R})$ or $C^\infty(M, \mathbf{C})$. Thus, the concept of a canonical ring is a generalization of that of a symplectic manifold.

We remark again that the equation $Hs_1(s_2) = \{s_1, s_2\}$ defines a "Poisson bracket" on the canonical ring $S$ and that this in turn determines the canonical structure $H: S \to D(S)$ uniquely.

The concept of a canonical ring (canonical manifold), being structural, taken by itself is void of a "functional" meaning. However, the study of canonical rings is both justified and necessary, because they arise naturally as factor rings and subrings of "functional-valued" rings $S$ of the form $\mathrm{Smbl}_* K$. For example, the 2-dimensional sphere $S_2$, equipped with its standard canonical structure, is a factor object of a Hamiltonian formalism in the Lie algebra of the rotations of 3-dimensional space (see 11) and this, in turn, is a "left-invariant subobject" of a Hamiltonian formalism in $T^*(G)$, where $G$ is the group of rotations of 3-dimensional space.

NOTE. In classical mechanics the fundamental form $\Omega = \sum_i dp_i \wedge dq_i$ is exact and arises purely geometrically as the differential of the canonical 1-form $\rho = \sum_i p_i\, dq_i$ on the phase space $T^*(M)$; and this 1-form is determined naturally by the following universal property: $s_\omega^*(\rho) = \omega$, where $\omega$ is an arbitrary 1-form on $M$ and $s_\omega$ is the section corresponding to $\omega$ of the cotangent fibration $\pi: T^*(M) \to M$. The following arguments show that in the general situation this construction is not possible.

Suppose that $K$ is an algebra over the field of rational numbers and let $\omega = dk \in \Lambda_1(K)$, $k \in K$.

Further, let $\sigma = \mathrm{smbl}_m \Delta, \Delta \in \mathrm{Diff}_m K$. We define

$$s_\omega^*(\sigma) = \frac{1}{m!}\, \delta_{\underbrace{k, \ldots, k}_{m \text{ times}}}(\Delta) \in K = \mathrm{Diff}_0\, K.$$

The resulting mapping $s_\omega^*: \mathrm{Smbl}_* K \to K$ is a ring homomorphism and in the case $K = C^\infty(M, \mathbf{R} \text{ or } \mathbf{C})$ corresponds to the homomorphism

$s_\infty^*: C^\infty(T^*(M)) \to C^\infty(M)$, generated by the section $s_\omega$. Therefore, the homomorphism $s_\omega^*$ is an algebraic equivalent to the concept of a section of $T^*(M)$. However, for an arbitrary ring $K$ the existence of a mapping of the type $s_\omega^*$ does not, in general, enable us to construct a form $\rho$. The following example illustrates this.

Let $K$ be the $\mathbf{R}$-algebra of all continuous functions on the line $\mathbf{R}^1$ that are once differentiable at 0. We denote by $\mu_a \subset K, a \in \mathbf{R}$, the maximal ideal consisting of the functions that vanish at $a$. Clearly, $\Lambda_1(K) \neq (0)$, therefore there are derivations of $K$ with values in $K/\mu_0$. On the other hand, the modules $D(K/\mu_a)$ for all $a \neq 0$ are trivial (because $\mu_a^2 = \mu_a$, $a \neq 0$). Hence, the module $D(K)$ is trivial. For if there is a non-zero derivation $\Delta: K \to K$, then its composition with the projection $K \to K/\mu_a$ for some $a \neq 0$ is also a non-zero derivation owing to the continuity of the functions in $K$, which is a contradiction to what was said above. These arguments show that in our example the mapping $s_\omega^*$ does not depend on $\omega \in \Lambda_1(K)$; that is, there cannot be a universal form $\rho$. Thus, the classical mechanism, which guarantees that $\hat{\Omega}_h$ is exact, does not work, in general, on $T^*(M)$.

5. **Canonical transformations.** Since in the category of smooth manifolds smooth mappings are in one-to-one correspondence with homomorphisms of the corresponding rings of smooth functions, it is natural to understand by a diffeomorphism (smooth mapping) of the "manifold" Spec $S$ an automorphism (homomorphism into itself) of the ring $S$. This leads us to the following intuitive definition: a canonical transformation is an automorphism $\varphi$ of $S$ that respects the "Hamiltonian structure". This has to be made more precise.

DEFINITION 1. An automorphism $\varphi$ of $S$ is called *canonical* if the diagram

$$\begin{array}{ccc} \Lambda_1(S) & \xrightarrow{h} & D(S) \\ {\scriptstyle \Lambda\varphi}\downarrow & & \uparrow{\scriptstyle D\varphi} \\ \Lambda_1(S) & \xrightarrow{h} & D(S) \end{array} \tag{2}$$

commutes, that is, if $h = D\varphi \circ h \circ \Lambda\varphi$.

Here we understand by $\Lambda\varphi$ and $D\varphi$ the $S$-module morphisms:

$$\Lambda\varphi(s\,ds') = \varphi(s)\,d\varphi(s'),$$
$$D\varphi(\partial) = \varphi^{-1} \circ \partial \circ \varphi.$$

The morphisms $\Lambda\varphi$ and $D\varphi$ are obviously related with respect to the natural pairing of $\Lambda_1(S)$ with $D(S)$, in the following sense:

$$\langle \partial, \Lambda\varphi(\lambda) \rangle = \varphi\,\langle D\varphi(\partial), \lambda \rangle. \tag{3}$$

PROPOSITION 2. *The diagram* (2) *commutes if and only if* $\Omega_h$ *is invariant under* $\varphi$.

PROOF.

$$\Omega_h(\Lambda\varphi(\lambda_1), \Lambda\varphi(\lambda_2)) = \langle h\Lambda\varphi(\lambda_1), \Lambda\varphi(\lambda_2)\rangle =$$
$$= \varphi\,\langle D\varphi\circ h\circ\Lambda\varphi(\lambda_1), \lambda_2\rangle = \varphi\,\langle h(\lambda_1), \lambda_2\rangle = \varphi\Omega_{h},(\lambda_1\ \lambda_2)$$

(the second equality is just (3), and the third is a consequence of the commutativity of (2) in this chain). The chain of equalities proves our proposition in both directions.

PROPOSITION 3. *The automorphism* $\varphi$ *of the ring of symbols* $S$ *is canonical if and only if the diagram*

$$\begin{array}{ccc} S & \xrightarrow{H} & D(S) \\ \varphi\downarrow & & \uparrow D\varphi \\ S & \xrightarrow{H} & D(S) \end{array}$$

*commutes, that is, if and only if* $H(\varphi s) = \varphi\circ Hs\circ\varphi^{-1}$ *for all* $s\in S$.

The proof is obvious.

Of course, the concept of a canonical transformation introduced above corresponds in the classical case $K = C^\infty(M, \mathbf{C})$ or $K = C^\infty(M, \mathbf{C})$ to the "usual" canonical transformation in $T^*(M)$.

**6. Infinitesimal canonical transformations.** The analogue of an infinitesimal canonical transformation, that is, a vector field that respects the Hamiltonian structure, must, in general, be taken to be a derivation of the ring $S$ that respects the Hamiltonian structure. We give a precise definition.

DEFINITION 2. A derivation $\psi\in D(S)$ of the ring of symbols $S$ is called *canonical* if it satisfies the relation

$$\widetilde{D\psi}\circ h + h\circ\widetilde{\Lambda\psi} = 0,$$

where, by definition, $\widetilde{D\psi}(\partial) = [\partial, \psi]$, and $\widetilde{\Lambda\psi}(s\,ds') = (\psi s)ds' + sd(\psi s')$, $s$, $s'\in S$.

PROPOSITION 4. *A derivation* $\psi$ *is canonical if and only if it satisfies the following two conditions*:

1) $\Omega_h(\widetilde{\Lambda\psi}(\lambda_1), \lambda_2) + \Omega_h(\lambda_1, \widetilde{\Lambda\psi}(\lambda_2)) = \psi\Omega_h(\lambda_1, \lambda_2)$;
2) $H(\psi s) = [\psi, Hs]$.

The proof is obvious.

We introduce some more notation. As above, let $K$ be a commutative ring with identity and $S = \mathrm{Smbl}_* K$ its ring of symbols. We write Ham $S$ for the set of Hamiltonian vector fields on $T^*(K)$ (that is, derivations of $S$ of the form $Hs$, $s\in S$), and Can $S$ for the set of canonical vector fields on $T^*(K)$ (that is, canonical derivations of $S$).

Condition 2) of Proposition 4, applied to an arbitrary element $s'\in S$ indicates that $\psi\{s, s'\} = \{\psi s, s'\} + \{s, \psi s'\}$, and is, therefore, satisfied for all derivations $\psi = Hs''$, $s''\in S$ (in this case it is simply the Jacobi identity for Poisson brackets). Thus, we have the embedding Ham $S \subset$ Can $S$ (as sets). Further, the modules Ham $S$ and Can $S$ are Lie rings (and when $K$ is provided with the structure of an $A$-algebra they are Lie algebras over $A$). Condition 2) of Proposition 4 shows that Ham $S$ is an ideal in Can $S$. In the classical case $K = C^\infty(M, \mathbf{R})$ the "fundamental theorem of mechanics" asserts that Ham $S$ = Can $S$ locally (and globally, if $M$ is simply-connected). This is not so, in general, as is shown by examples below.[1] Thus, we must understand by the "fundamental theorem of mechanics" for a ring $K$ a theorem describing the structure of the Lie ring ($A$-algebra) Can $S$/Ham $S$, that is, describing by how much Can $S$ is "bigger" than Ham $S$).

As in **5.**, the concepts introduced here correspond in the classical case completely to the usual ones.

**7. Localization.** Definition 1 of **5.** admits of some generalization. We say that a ring homomorphism $\varphi\colon S_1 \to S_2$ of canonical rings is a *canonical homomorphism* if it preserves the Poisson brackets, that is, if $\varphi\{s_1, s_2\}_1 = \{\varphi s_1, \varphi s_2\}_2$, where $\{,\}_i$ are the Poisson brackets of $S_i$. If $s_1 \in \operatorname{Ker} \varphi$ then $0 = \{\varphi s_1, \varphi s_2\}_2 = \varphi\{s_1, s_2\}_1$, that is, $\{s_1, s_2\} \in \operatorname{Ker} \varphi$. Thus, $H s_1(\operatorname{Ker} \varphi) \subset \operatorname{Ker} \varphi$, $s_1 \in S_1$. Ideals that are invariant under the action of any Hamiltonian derivations are called *stable.* In other words, an ideal of a canonical ring $S$ is stable if it is also an ideal for the Lie structure of $S$ determined by the Poisson brackets.

PROPOSITION 5. *Let $q$ be a stable ideal of a canonical ring $S$. Then by the equality $\{\varphi s_1, \varphi s_2\}' = \varphi\{s_1, s_2\}$, where $\varphi\colon S \to S/q$ is the natural projection, a canonical structure on $S/q$ is well defined; and $\varphi$ is then a canonical homomorphism.*

The proof is obvious.

In view of this proposition it is natural to regard the canonical factor ring $S/q$ as a localization of the canonical ring $S$.

The geometrical picture corresponding to a stable ideal $q \subset S$ is that the relevant "submanifold" Spec $S/q$ of the "manifold" Spec $S$ is tangent to all the Hamiltonian vector fields. It would be natural to define the Hamiltonian orbit of a subset $A \subset$ Spec $S$ as the smallest set $A_H \supset A$, $A_H \subset$ Spec $S$, that touches all Hamiltonian vector fields. Intuitively, one can imagine $A_H$ to be the set obtained from $A$ by means of all Hamiltonian transformations, that is, shift operations along the trajectories of Hamiltonian vector fields. Of course, these intuitive geometrical concepts cannot be given, in general, a precise meaning. However, the algebraic version of the set $A_H$ exists and is given by the following definition.

DEFINITION 3. The *algebraic orbit of a point* $p \in$ Spec $S$ is the "submanifold" Spec $S/\Theta(p) \to$ Spec $S$, where $\Theta(p)$ is the maximal stable

---

[1] We mention that in the case of an arbitrary canonical manifold $M$, since the fundamental form is non-degenerate, the set $h^{-1}$(Can $S$) coincides with the set of all closed 1-forms on $M$, and $h^{-1}$(Ham $S$) with the set of all exact 1-forms. Therefore, Can $S$/Ham $S = H^1(M, \mathbf{R})$.

ideal of $S$ contained in the ideal $p$.

Intuitively, we can interpret $S/\Theta(p)$ as the "ring of functions on $p_H$".

PROPOSITION 6. *For every prime ideal let $p \subset S$ the ideal $\Theta(p)$ exists, is uniquely determined, and is prime.*

PROOF. The first two assertions are obvious. Let us only prove that $\Theta(p)$ is prime.

(a) If $\Theta(p) = p$, there is nothing to prove.

(b) We write $q$ for $\Theta(p)$. Suppose that $q$ is a proper subset of $p$. First we show that $q$ is its own radical. For the set $R(q)$ of elements $s \in S$ with $s^n \in q$ for some $n$ is a stable ideal that contains $q$ and is contained in $p$ (because $p$ is prime). Thus, by the maximality of $q$, $R(q) = q$.

(c) We consider an element $s \in S/q$ and the set $q_s$ of elements $s' \in S$ such that $s^n s' \in q$ for some $n$. It is clear that $q_s$ is an ideal and $q_s \supset q$. The identity $H_\sigma(s')\cdot s^{n+1} = H_\sigma(s's^{n+1}) - H_\sigma(s^{n+1})\cdot s' = H_\sigma(s's^{n+1}) - (n+1)s^n s' H_\sigma(s)$ shows that $H_\sigma(s')s^{n+1} \in q$ if $s' \in q_s$, $\sigma \in S$, that is, that $q_s$ is stable. Note also that $q_s \subset p$ if $s \notin p$, because $p$ is prime.

(d) Let $s, s' \in S$ with $s\, s' \in q$. We assume that one of $s$ and $s'$, say $s$, does not lie in $q$. We consider the ideal $q_s$. If there is an element $s'' \in q_s$ with $s'' \notin p$, then $q \subset q_{s''} \subset p$ and so $q_{s''} = q$. Since $s^n s'' \in q$, we see that $s^n \in q_{s''}$ and so $s^n \in q$. As a consequence of (b) we have $s \in q$, which contradicts our assumptions. Thus, $q \subset q_s \subset p$, therefore $q = q_s$. And since $s' \in q_s$ we have $s' \in q$.

COROLLARY 1. *A Hamiltonian structure on $S$ defines a mapping $\Theta$:* Spec $S \to$ Spec $S$ *under which every ideal $p \subset S$ is associated with $\Theta(p)$. This mapping $\Theta$ is a projection of the space* Spec $S$ *onto the subspace* $\mathrm{Spec}_H$ $S$ *consisting of all stable prime ideals of $S$.*

COROLLARY 2. *The set of algebraic orbits in* Spec $S$ *is in one-to-one correspondence with the set of stable prime ideals of $S$, that is,* $\mathrm{Spec}_H$ $S$ *is the "orbit space".*

Even when the geometrical concept of the Hamiltonian orbit $p_H$ of a point $p \in$ Spec $S$, which we have discussed above, can be given a meaning, it can happen that nevertheless $p_H \neq$ Spec $S/\Theta(p)$ (obviously, we always have $p_H \subset$ Spec $S/\Theta(p)$). This has to do with the fact that Spec $S/\Theta(p)$ is only the "algebraic closure" of $p_H$ and therefore can be considerably larger than $p_H$. In that case the algebraic orbit Spec $S/\Theta(p)$ contains some set of geometrical orbits of the form $q_H$, $q \in$ Spec $S$. Simple examples illustrating this behaviour are provided by the Hamiltonian formalism in Lie algebras (see **15.**).

The following definition corresponds intuitively to the operation of excising the geometrical orbits $q_H$ of non-maximal dimension from the algebraic orbit Spec $S/\Theta(p)$.

DEFINITION 4. Let $p \in$ Spec $S$ be a prime ideal of $S$. The *principal part* of the algebraic orbit of $p$ is defined as the set

$$O(p) = \mathcal{O}^{-1}(\mathcal{O}(p)).$$

PROPOSITION 7. 1) *The partition of* Spec $S$ *into the principal parts of algebraic orbits is a partition into disjoint classes.*

2) *The algebraic orbit* Spec $S/q$ *coincides with its principal part if and only if $q$ is a maximal stable ideal.*

The proof is obvious.

The *Poisson centre* $Z(S)$ of a canonical ring is the set of those $s \in S$ for which $Hs = 0$. Obviously, $Z(S)$ is a subring of $S$. The following proposition shows that the Poisson centre of $S/q$ is algebraically irreducible when $q$ is a maximal stable ideal.

PROPOSITION 8. *If $q$ is a maximal stable ideal, then the Poisson centre of $S/q$ (that is, the "ring of functions" on the orbit corresponding to $q$) is a field.*

PROOF. The ring $S/q$ is endowed with a Hamiltonian structure and is characterized by the fact that it has no proper stable ideals. Clearly, $Z(S/q) \neq \varnothing$. Let $z \in Z(S/q)$. Then the principal ideal $(z)$ is stable; consequently, it coincides with the whole ring $S/q$. Thus, $z$ is invertible in $S/q$; evidently, its inverse $z^{-1}$ belongs to $Z(S/q)$.

Thus, with an algebraic orbit of the indicated type there is connected a field $Z(S/q)$, an extension of the base ring $K$. It is natural to call this the field of *infinitesimal characters* of the algebraic orbit Spec $S/\mathcal{O}(q)$.

We end this section by remarking that the concepts introduced here can be taken as the basis of a theory of "Hamiltonian affine schemes", an algebraic-geometric analogue to the concept of a canonical manifold.

**8. Lagrangian manifolds and the Hamilton-Jacobi equation.** It is known that every Lagrangian manifold $L \subset T^*(M)$ lying on the level surface $\mathcal{H} = 0$ of the Hamiltonian function is invariant under the corresponding Hamiltonian vector field. This absorption principle, translated into the algebraic language, leads to the following definition: an ideal $J \subset S$ is called *autostable* if it is invariant under all operators $Hs$, $s \in J$, and a "maximal autostable ideal" is *Lagrangian*. The submanifold $L_J \subset T^*(K)$ corresponding to a Lagrangian ideal is also said to be *Lagrangian*. We mention that in the classical case $K = C^\infty(M, \mathbf{R})$ the algebraic concept introduced above of a Lagrangian ideal corresponds to the geometric concept of a connected Lagrangian manifold in $T^*(M)$.

Thus, the fact that some (connected) Lagrangian manifold lies on the level surface $\mathcal{H} = 0$ can be regarded as "geometrical realization" of the situation where the principal ideal of $S$ generated by $\mathcal{H} \in S$ is contained in the corresponding Lagrangian ideal.

Below we use this interpretation to determine the algebraic meaning of the Hamilton-Jacobi partial differential equation. As a preliminary we have to give a geometrical meaning to the Hamilton-Jacobi equation in the classical context:

$$(4) \qquad H\left(q, \frac{\partial u}{\partial q}\right)=0, \quad q=(q_1, \ldots, q_n), \quad \frac{\partial u}{\partial q}=\left(\frac{\partial u}{\partial q_1}, \ldots, \frac{\partial u}{\partial q_n}\right), \quad u=u(q).$$

Here $q_1, \ldots, q_n$ are local coordinates in $U = M$, and $(q, p)$ are the naturally arising canonical coordinates in $\pi^{-1}(U) \subset T^*(M)$, where $\pi: T^*(M) \to M$ is the natural projection. With this equation we can associate a function $H(q, p)$ on $T^*(M)$. Then the fact that a function $f \in C^\infty(M)$ satisfies (4) means that the manifold $L = L(f) \subset T^*(M)$ defined by the equations $p_i = \frac{\partial f}{\partial q_i}$ $(i = 1, \ldots, n)$ in $\pi^{-1}(U) \subset T^*(M)$ lies on the level surface $H(q, p) = 0$. Submanifolds of the form $L(f)$ are uniquely characterized by the following two properties: 1) $L(f)$ is Lagrangian; 2) $\pi|_{L(f)}$ is a diffeomorphism. Therefore, the differentials of solutions of (4) are in one-to-one correspondence with those Lagrangian manifolds $L \subset \{H = c\}$ for which $\pi|_L$ is a diffeomorphism.

This geometrical interpretation of the concept of a solution of the Hamilton-Jacobi equation is not invariant under canonical transformations of the natural canonical structure on $T^*(M)$. The obstruction is the requirement that $\pi|_L$ be a diffeomorphism. Discarding it we come to regard as a solution of (4) (more accurately, its differential) any Lagrangian manifold $L$ lying entirely on the level surface $H = 0$.

NOTE. There are a number of other weighty reasons, both of a physical and mathematical character for using our general concept of a solution of the Hamilton-Jacobi equation.

This geometrical formulation, in turn, has the following "algebraization" by virtue of the indicated algebraic meaning of the inclusion $L \subset \{H = 0\}$: a solution of the "Hamilton-Jacobi" equation $\mathcal{H} = 0$, where $\mathcal{H} \in S$, is a Lagrangian ideal $J \subset S$ containing the principal ideal $(\mathcal{H})$.

Let us call the problem of finding all Lagrangian ideals $J \subset S$ containing a given principal ideal $(\mathcal{H})$ the "Hamilton-Jacobi problem". Then in the classical case $K = C^\infty(M, \mathbf{R})$ the Hamilton-Jacobi problem is that of finding all (generalized) solutions of the Hamilton-Jacobi equation. In view of this it is natural to regard the Hamilton-Jacobi equation as the "geometrical" way of expressing the Hamilton-Jacobi problem.

We end this section by indicating the algebraic meaning of an "ordinary" solution of the Hamilton-Jacobi equation; we restrict ourselves, for simplicity, to rings $K$ whose additive group is divisible. With this aim we draw attention to the following method of constructing a manifold $L(f)$. Let $X_f$ be the Hamiltonian field on $T^*(M)$ corresponding to the Hamiltonian $\pi^*(f)$, where $\pi: T^*(M) \to M$ is the natural projection. We write $A_t: T^*(M) \to T^*(M)$ for the shift diffeomorphism along the trajectories of $X_f$. Then, in the coordinates considered above, $A_t(q, p) = (q, p - t\frac{\partial f}{\partial q})$. In particular, $A_1(L(f)) = L(0)$, and $L(0)$ is the zero section of the vector bundle $\pi: T^*(M) \to M$. Since $X_f$ has global trajectories, we represent the automorphism $A_t^*: C^\infty(T^*(M), \mathbf{R}) \to C^\infty(T^*(M), \mathbf{R})$ in the form $A_t^* = e^{tX_f}$, where $e^{tX_f} = \sum_{n \geq 0} \frac{t^n X^n f}{n!}$ and $X_f$ is interpreted as the derivation operator $X_f: C^\infty(T^*(M), \mathbf{R}) \to C^\infty(T^*(M), \mathbf{R})$, or in "algebraic" notation, $A_t^* = e^{tH(\pi^* f)}$. Therefore, the operator $A_{-1}^* = e^{-H(\pi^* f)}$ carries the Lagrangian ideal $J_0$ corresponding to $L(0)$ into the Lagrangian ideal $J_f$ corresponding to $f$. Thus, the description of $J_f$ reduces to that of $J_0$. In turn, clearly $J_0 = \sum_{i>0} \mathrm{Smbl}_i\, K$. Thus, in general, we have defined $J_f$ as the inverse image of the ideal $\sum_{i>0} \mathrm{Smbl}_i\, K \subset S$ under the automorphism $e^{-H_{j(f)}}$, where $f \in K$ and $j: K \to \mathrm{Smbl}_*\, K$ is the natural embedding. We remark here that $e^{-H_{j(f)}}(s) = \sum_{i=0}^{m} \frac{(-1)^i H_{j(f)}^i(s)}{i!}$ for any $s \in \mathrm{Smbl}_m\, K$ since $H_{j(f)}^i(s) = 0$ for $i > m$, so that the operator $e^{-H_{j(f)}}$ has a meaning for any ring whose additive group is divisible.

Thus, an element $f \in K$ is an "ordinary" solution of the Hamilton-Jacobi equation if the Lagrangian ideal $J_f$ contains the ideal $(\mathscr{H})$.

NOTE. What we have said in this section demonstrates the purely algebraic nature of the Hamilton-Jacobi problem. However, we have not touched on the vastly more important functorial aspect of this problem, that is, "why" it needs to be solved. To do this we would have to develop the general theory of differential operators on commutative rings in vastly greater volume than we can allow in this paper. A partial functorial meaning of the formulation of the Hamilton-Jacobi problem can be extracted from [2] and [3].

**9. An example. The cotangent spaces of cyclic groups.** We consider the cyclic group $Z_n$ of order $n$, a field $F_p$ of characteristic $p$, $p$ a prime, and the group ring $K = F_p[Z_n]$. The "cotangent space" $T^*(Z_n, F_p)$ of $Z_n$ over $F_p$ is naturally understood to be the cotangent space of $K$. Below we characterize $T^*(Z_n, F_p)$ and also obtain the "fundamental theorem of mechanics" for the rings $K$.

In what follows we use the representation of $K$ as the factor ring $F_p[x]/(x^n - 1)$. Then the $F_p$-algebra of differential operators $\mathrm{Diff}_*\, K$ is isomorphic to $K$ if $n$ is prime to $p$, and is generated by the elements, $X$, $\partial_X$ subject to the relations $X \cdot \partial_X - \partial_X \cdot X = -1$, $\partial_X^n = 0$ if $n$ is a multiple of $p$.

Therefore,

$$S = \mathrm{Smbl}_* K = \begin{cases} K \text{ with the trivial Lie structure if } n \text{ is prime to } p, \\ F_p[x, y]/(x^n - 1, y^n),\ \{x, y\} = -1, \text{ if } n \text{ is a multiple of } p. \end{cases}$$

Thus, $T^*(Z_n, F_p)$ is a "regular $n$-gon" embedded in the affine space Spec $F_p[x]$ if $n$ is prime to $p$, and an "$n$-fold regular $n$-gon" embedded in Spec $F_p[x, y]$ if $n$ is a multiple of $p$.

An easy calculation shows that the "fundamental theorem of mechanics" (that is, Can $S$/Ham $S$) is trivial if $n$ is prime to $p$, and is isomorphic to the Abelian $2m^2$-dimensional Lie algebra over $F_p$ if $n = mp$, $m$ integral.

**10. The Hamiltonian formalism in Hamiltonian rings.** As we have seen, from the formal-structural point of view, we need only the existence of Poisson brackets to construct a "Hamiltonian formalism". On the other hand, the above method of defining Poisson brackets was based only on the fact that in the filtered ring $\mathrm{Diff}_* K \supset \ldots \supset \mathrm{Diff}_s K \ldots \supset K$ the commutator of elements $\Delta_i (i = 1, 2)$ of filtration $\leqslant \nu_i$ is an element of filtration $\leqslant \nu_1 + \nu_2 - 1$. Therefore, we can construct a "Hamiltonian formalism" for any unital associative filtered ring $F \supset \ldots \supset F_s \supset \ldots \supset F_0$ in which the commutation relation is linked up with a filtration in the indicated manner. We call such rings *Hamiltonian.* Then the graded ring $\mathrm{Smbl}_* F = \sum_{i \geqslant 0} F_i/F_{i-1}$ associated with $F$ under the assumptions made is commutative and unital. The Poisson bracket $\{\sigma_1, \sigma_2\} = \mathrm{smbl}_{m_1+m_2-1}[\delta_1, \delta_2]$ of the elements $\sigma_i = \mathrm{smbl}_{m_i} \delta_i$ $(i = 1, 2)$ is well defined if by the symbol $\mathrm{smbl}_m \delta$ where $\delta$ has filtration $\leqslant m$ we mean the image of $\delta$ under the projection $F_m \to F_m/F_{m-1}$. From here onwards the construction of a "Hamiltonian formalism" for $F$ proceeds word-for-word as before.

Thus, we can construct a "Hamiltonian formalism" for any Hamiltonian ring $F$. The invariant Hamiltonian structures so obtained, along with the "differential-topological" invariants of $\mathrm{Smbl}_* F$ (see [2]), are important invariants of $F$.

**11. The Hamiltonian formalism in Lie algebras.**[1] The most important examples of Hamiltonian rings are the universal enveloping algebras of Lie algebras over a unital commutative ring $K$. Let $L$ be a Lie algebra over $K$, $U = U(L)$ its universal enveloping algebra and $U_0 \subset U_1 \subset \ldots \subset U_m \subset \ldots \subset U$ its standard filtration. When $L$ is the free $K$-module with the generators $y_1, y_2, \ldots, y_n$, the Poincaré-Birkhoff-Witt theorem gives a precise description of the corresponding "ring of symbols" $S = \sum_{i \geqslant 0} \mathrm{Smbl}_i U$. Namely, the natural mapping $j\colon L \to S$ is injective and $S$ is the ring of polynomials $K[x_1, x_2, \ldots, x_n]$ where $x_i = j(y_i)$ $(i = 1, 2, \ldots, n)$. In other words, $S$ is isomorphic to a ring of polynomial functions on the space $L^* = \mathrm{Hom}_K(L, K)$.

In this case the module $D(S)$ of ring derivations is the free $S$-module with the generators $\partial_1, \partial_2, \ldots, \partial_n$, where the $\partial_i$ are defined by

---

[1] In this context, see [1].

$\partial_i(x_k) = \delta_{ik}$ $(i, k = 1, 2, \ldots, n)$ $\left(\text{that is, } \partial_i = \frac{\partial}{\partial x_i}\right)$ and $\Lambda(S)$ is also free and is generated by elements of the form $d_i = dx_i$ $(i = 1, 2, \ldots, n)$. Moreover, $\langle d_i, \partial_k \rangle = \delta_{ik}$, that is, the bases described above are conjugate (we recall that $D(S) = \mathrm{Hom}_s(\Lambda(S), S)$). Also, there are natural isomorphisms $D(S) \sim L^* \otimes_K S$ and $\Lambda(S) \sim L \otimes_K S$. The latter allows us to rewrite the fundamental form $\Omega_h$ on $\Lambda(S)$ as follows: $\Omega_h(l \otimes s, l' \otimes s') = ss'\{j(l), j(l')\}$.

We can now clarify what derivations $\partial \in D(S)$ are Hamiltonian. Let

$$[y_i, y_j] = \sum_{k=1}^{n} C_{ij}^k y_k \qquad (i, j = 1, 2, \ldots, n)$$

be the structure equations of the Lie algebra $L$. The Poisson brackets are obviously defined by their values on the generators $x_i$, and

$$\{x_i, x_j\} = \sum_{k=1}^{n} C_{ij}^k x_k \qquad (i, j = 1, 2, \ldots, n).$$

Thus,

$$H_{x_i} = \sum_{j=1}^{n} \Big(\sum_{k=1}^{n} C_{ij}^k x_k\Big) \partial_j \qquad (i = 1, 2, \ldots, n).$$

The derivation $\partial = \sum_{i=1}^{n} \alpha_i \partial_i$ is Hamiltonian if and only if there is an element $s \in S$ such that $\alpha_i = \{s, x_i\}$ $(i = 1, 2, \ldots, n)$, that is,

$$\alpha_i = -\sum_{j=1}^{n} \Big(\sum_{k=1}^{n} C_{ij}^k x_k\Big) \frac{\partial s}{\partial x_j} \qquad (i = 1, 2, \ldots, n).$$

The element $s$ is the "Hamiltonian" of $\partial$.

Now we write out the conditions for the derivation $\partial = \sum_{i=1}^{n} \alpha_i \partial_i$ to be canonical. As we know (see 4.), $\partial$ is canonical if and only if $\partial\{s_1, s_2\} = \{\partial s_1, s_2\} + \{s_1, \partial s_2\}$ for all $s_1, s_2 \in S$. In our situation the verification that $\partial$ is canonical obviously reduces to checking the following equalities:

$$\partial\{x_i, x_j\} = \{\partial x_i, x_j\} + \{x_i, \partial x_j\} \qquad (i, j = 1, 2, \ldots, n)$$

or

$$(*) \qquad \sum_{k=1}^{n} C_{ij}^k \alpha_k = \sum_{m=1}^{n} \Big(\sum_{k=1}^{n} C_{im}^k x_k\Big) \frac{\partial \alpha_j}{\partial x_m} + \sum_{m=1}^{n} \Big(\sum_{k=1}^{n} C_{jm}^k x_k\Big) \frac{\partial \alpha_i}{\partial x_m}$$
$$(i, j = 1, 2, \ldots, n).$$

We consider the matrices $C = \| c_{ij} \|$ and $A = \| a_{ij} \|$, where

$$c_{ij} = \sum_{k=1}^{n} C_{ij}^k x_k, \quad a_{ij} = \frac{\partial \alpha_i}{\partial x_j} \qquad (i, j = 1, 2, \ldots, n).$$

Note that $C$ is skew-symmetric. We write $\partial C = \| \partial c_{ij} \|$. Then the conditions (*) can be rewritten in the compact form $\partial C = C \cdot A^t - A \cdot C$, where $A^t$ is the transpose of $A$.

12. **The "fundamental theorem of mechanics" for Lie algebras.** Let $L$ be a Lie algebra, $U$ its universal enveloping algebra, and $S$ the corresponding ring of symbols. As already remarked, a "fundamental theorem of mechanics" for the ring $U$ is to be understood as a theorem describing the factor module Can $U$/Ham $U$, equipped with the natural structure of a Lie ring. Let us write $\gamma(L) = \text{Can } U/\text{Ham } U$.

We remark that when $L$ is a Lie algebra over a ring $K$, then $\gamma(L)$ can be made into a Lie algebra over $K$.

EXAMPLE. Let $L$ be an Abelian Lie algebra. It is clear that Ham $U(L) = 0$ and Can $U(L) = D(S)$. Hence $\gamma(L) = D(S)$.

EXAMPLE. Let $L$ be the two-dimensional Lie algebra over a ring $K$ with two generators $y_1$ and $y_2$ and the Lie bracket $[y_1, y_2] = y_2$; its ring of symbols is the polynomial ring $K[x_1, x_2]$ with the Poisson bracket $\{x_1, x_2\} = x_2$. The derivation $\partial = \alpha_1 \partial_1 + \alpha_2 \partial_2$ is Hamiltonian with Hamiltonian $s$ if $\alpha_1 = -x_2 \frac{\partial s}{\partial x_1}$, and $\alpha_2 = x_2 \frac{\partial s}{\partial x_2}$, while $\partial$ is canonical if

$$\alpha_1 = c + x_2 \frac{\partial s}{\partial x_2}, \quad \alpha_2 = -x_2 \frac{\partial s}{\partial x_1}, \qquad c \in K.$$

Therefore, $\gamma(L) = K$ with the trivial Lie structure.

We remark that the natural filtration $S = \bigcup_{i \geq 0} S_i$ of the ring of symbols induces a filtration of its module of derivations $D(S)$; namely

$$D(S) = \bigcup_{i \geq 0} D_i(S),$$

where $D_i(S)$ consists of precisely those derivations that map $S_1$ into $S_i$.

It is not difficult to check that Can $U = \bigcup_{i \geq 0} \text{Can}_i\ U$ and Ham $U = \bigcup_{i \geq 0} \text{Ham}_i\ U$ where $\text{Can}_i\ U = \text{Can } U \cap D_i(S)$, and $\text{Ham}_i\ U = \text{Ham } U \cap D_i(S)$ $(i = 0, 1, 2, \ldots, m, \ldots)$. Since $\text{Ham}_i\ U \subset \text{Can}_i\ U$, we have, in turn

$$\gamma(L) = \bigcup_{i \geq 0} \gamma_i(L),$$

where

$$\gamma_i(L) = \text{Can}_i\ U/\text{Ham}_i\ U, \quad \gamma_i(L) \subset \gamma_{i+1}(L) \qquad (i = 0, 1, 2, \ldots, m).$$

It follows immediately from the definition that $\text{Can}_1\ U$ is, modulo elements of $\text{Can}_0\ U$, the Lie algebra of endomorphisms of $L$, and $\text{Ham}_1\ U$ the Lie algebra of inner endomorphisms of $L$. Therefore,

$$\gamma_1(L)/\gamma_0(L) = H^1(L; L),$$

where the space $L$ is endowed with its natural $L$-module structure by means of the adjoint representation ad: $L \to \text{End } L$.

Thus, the Lie algebra $\gamma(L)$ is in a certain sense an "enveloping object" for the first cohomology group $H^1(L; L)$.

**13. The "fundamental theorem of mechanics" for semisimple Lie algebras.**

THEOREM 1. *If $L$ is a finite-dimensional semisimple Lie algebra over a field of characteristic zero, then $\gamma(L) = 0$.*

PROOF. We define the *Poisson commutator* of $S$ as the subspace $\{S, S\}$ of $S$ consisting of the finite linear combinations of all elements of the form $\{s_1, s_2\}$, $s_1, s_2 \in S$ with coefficients in $K$. We consider the linear representation $R$ of $L$ on Can $U$:

$$R: L \to \mathrm{End}(\mathrm{Can}\ U),$$

where $R(l)\psi = [\psi, Hj(l)]$, $\psi \in \mathrm{Can}\ U$, $l \in L$ ; this turns Can $U$ into an $L$-module. Further, we consider the filtration of Can $U$ defined in **12.** and set $C_i = \mathrm{Can}_i\ U$. Then $C_i \subset C_{i+1}$ and $\mathrm{Can}\ U = \bigcup_{i \geq 0} C_i$. It is easy to see that $\psi \in C_i$ if and only if $\psi$ is a canonical derivation and $\psi j(l) \in S_i$, $l \in L$. Let us check that the $C_i$ are invariant subspaces of $R$.

For let $l_1, l_2 \in L$, $\psi j(l_1) \in S_i$. Then

$$(R(l_1)\psi)(j(l_2)) = [\psi, Hj(l_1)](j(l_2)) =$$
$$= \psi\{j(l_1), j(l_2)\} - \{j(l_1), \psi j(l_2)\} = \{\psi j(l_1), j(l_2)\},$$

and since $\psi j(l_1) \in S_i$, we have $(R(l_1)\psi)(j(l_2)) = \{\psi j(l_1), j(l_2)\} \in S_i$. Thus, $R_i = R|_{C_i}: L \to \mathrm{End}\ C_i$ is finite-dimensional and is therefore completely reducible (see [16], Ch. VI, §3). Hence it follows ([17], Ch. V) that $C_i = C_i^\natural \oplus C_i^0$, where $C_i^\natural$ is the subspace of $C_i$ formed by all vectors that are annihilated by the endomorphisms in $R_i(L)$ and $C_i^0$ is generated by the vectors of the form $R_i(l)(c)$, where $l \in L$ and $c \in C_i$. It is clear that $C_i^\natural \subset C_{i+1}^\natural$ and $C_i^0 \subset C_{i+1}^0$.

Let $\mathrm{Can}^\natural\ U$ and $\mathrm{Can}^0\ U$ be the subspaces of $R$ defined similarly. Then

$$\mathrm{Can}^\natural\ U = \bigcup_{i\geq 0} C_i^\natural, \quad \mathrm{Can}^0\ U = \bigcup_{i\geq 0} C_i^0$$

and

$$\mathrm{Can}\ U = \mathrm{Can}^\natural\ U \oplus \mathrm{Can}^0\ U.$$

LEMMA 1. *If* $\psi \in \mathrm{Can}\ U$, *then* $\psi\{S, S\} \subset \{S, S\}$, $\psi Z(S) \subset Z(S)$.

PROOFS. 1) $\psi\{s_1, s_2\} = \{\psi s_1, s_2\} + \{s_1, \psi s_2\} \in \{S, S\}$. 2) The fact that $z \in Z(S)$ means that $\{s, z\} = 0$ for all $s \in S$, and so $0 = \psi\{z, s\} = \{\psi z, s\} + \{z, \psi s\} = \{\psi z, s\}$ , that is, $\psi(z) \in Z(S)$.

LEMMA 2. $S = Z(S) \oplus \{S, S\}$.

We consider the representation $R'$: $L \to \mathrm{End}\ S$, where $R'(l) = Hj(l)$, $l \in L$. An argument quite similar to that by means of which we have established above the decomposition $\mathrm{Can}\ U = \mathrm{Can}^0\ U \oplus \mathrm{Can}^\natural U$, shows that $S = S^\natural \oplus S^0$, where $S^0 = \{s \in S: s = \sum_i H_j(l_i)(s_i)\}$ and

$S^{\natural} = \{s \in S \colon Hj(l)(s) = 0,\ l \in L\}$. Therefore, it only remains to remark that $S^0 = \{S, S\}$ and $S^{\natural} = Z(S)$.

We return to the proof of the theorem. Let $\psi \in \mathrm{Can}^{\natural} L$. This means that $\psi$ is a canonical derivation and that for all $l \in L$ and $s \in S$

$$0 = (R(l)(\psi))(s) = [\psi, Hj(l)](s) = \{\psi j(l), s\}.$$

Thus, $\mathrm{Can}^{\natural} U$ consists of those $\psi \in \mathrm{Can}\, U$ that are carried by $j(L)$ into $Z(S)$. But since $L$ is semisimple, $[L, L] = L$, so that $j(L) \subset \{S, S\}$, and according to the lemmas just proved, $\mathrm{Can}^{\natural}\, U = 0$. So we arrive at the following fundamental equation: $\mathrm{Can}^0\, U = \mathrm{Can}\, U$.

Now let $\psi \in \mathrm{Can}^0\, U$. Then there exist $l_1, l_2, \ldots, l_k \in L$ and $\psi_1, \psi_2, \ldots, \psi_k \in \mathrm{Can}\, U$ such that $\psi(s') = \sum_{i=1}^{k} (R(l_i)\psi_i)(s') = \sum_{i=1}^{k} \{\psi_i(j(l_i)), s'\} = \{\sum_{i=1}^{k} \psi_i(l_i), s'\}$ for all $s' \in S$. Consequently $\psi = Hs \in \mathrm{Ham}\, U$, where $s = \sum_{i=1}^{k} \psi_i j(l_i)$. Therefore,

$\mathrm{Ham}\, U \subset \mathrm{Can}\, U \subset \mathrm{Can}^0\, U \subset \mathrm{Ham}\, U$, from which it follows that $\mathrm{Ham}\, U = \mathrm{Can}\, U$. The theorem is now proved.

**14. Example. The fundamental theorem of mechanics and symmetries of linear vector fields.** Let us consider in the $n$-dimensional Euclidean space $\mathbf{R}^n$ a linear vector field

$$X = \sum_{i=1}^{n} A_i \frac{\partial}{\partial x_i}, \text{ where } A_i = \sum_{k=1}^{n} a_{ik} x_k \qquad (i = 1, 2, \ldots, n),$$

and the matrix $A = \| a_{ij} \|$ is non-zero. Speaking invariantly, we regard the field $X$ as an element of $D(K)$, where $K$ is the ring of smooth functions on $\mathbf{R}^n$, $\mathbf{R}^n \subset \mathrm{Spec}\, K$.

Throughout this section we agree to understand by *symmetries* of $X$ those diffeomorphisms $\mathbf{R}^n \to \mathbf{R}^n$ that carry trajectories of the field in question into trajectories of the same field. The set of diffeomorphisms of $\mathbf{R}^n$ that carry each trajectory of $X$ into itself form in the Lie algebra Sym $X$ an ideal $\widehat{\mathrm{Sym}}\, X$ ("proper" or "inner" symmetries of $X$) whose elements, evidently, are of the form $D = \mu X$, $\mu \in C^\infty(\mathbf{R}^n)$.

In what follows we confine ourselves to the algebraic case, that is, to those symmetries whose coefficients are polynomial functions on $\mathbf{R}^n$ (for such symmetries, $\lambda$ and $\mu$ belong to $\mathbf{R}[x_1, \ldots, x_n]$). Speaking invariantly: we regard $X$ as an element of $D(K)$, where $K$ is a ring of polynomials on $\mathbf{R}^n$, $\mathbf{R}^n \subset \mathrm{Spec}\, K$. From now on the notation Sym $X$ and $\widehat{\mathrm{Sym}}\, X$ refers to the sets of precisely these symmetries. Let

$$S(X) = \mathrm{Sym}\, X / \widehat{\mathrm{Sym}}\, X.$$

We associate with $X$ the $(n + 1)$-dimensional Lie algebra $L(X)$ with the generators $y_1, y_2, \ldots, y_{n+1}$ and the relations $[y_i, y_j] = 0$, $[y_{n+1}, y_i] = \sum_{k=1} a_{ik} y_k$, $1 \leqslant i, j \leqslant n$. The algebra $L(X)$ is obviously solvable and is isomorphic to the subalgebra of $S = \mathrm{Smbl}_* K$ (relative to the Poisson brackets) spanned by the coordinate functions $x_1, x_2, \ldots, x_n$ and the Hamiltonian function $\sum_{i,k\geqslant 1} a_{ik} p_i x_k$ is the natural lifting of $X$ in $T^*(\mathbf{R}^n)$.

THEOREM 2. *The Lie algebra $\gamma(L(X))$ is the semidirect product of $S(X)$ and a commutative ideal $J$ that can be identified with the space* Coker $X$ *(the field is regarded as an operator on the ring* $\mathbf{R}[x_1, \ldots, x_n]$).

PROOF. Let $z_1, z_2, \ldots, z_{n+1}$ be the images of the elements $y_1, y_2, \ldots, y_{n+1}$ under the canonical embedding $j\colon L(X) \to \mathrm{Smbl}_*\, L(X)$. We consider the derivation $\partial = \sum_{i=1}^{n+1} \alpha_i \frac{\partial}{\partial z_i}$ of $\mathrm{Smbl}_*\, L(X)$; $\partial$ is Hamiltonian with Hamiltonian $s \in \mathrm{Smbl}_*\, L(X)$ if and only if

$$\text{(N)}\qquad \begin{cases} \alpha_i = B_i \dfrac{\partial s}{\partial z_{n+1}}, \quad \alpha_{n+1} = -\displaystyle\sum_{k=1}^{n} B_k \frac{\partial s}{\partial z_k}, \\ B_i = \displaystyle\sum_{k=1}^{n} a_{ik} z_k \qquad (i = 1, 2, \ldots, n). \end{cases}$$

In the present case the conditions for $\partial$ to be canonical, according to the formula (*) in **11.** can be written in the form of the following system of equations:

$$\text{(S)}\qquad \begin{cases} B_i \dfrac{\partial \alpha_j}{\partial z_{n+1}} - B_j \dfrac{\partial \alpha_i}{\partial z_{n+1}} = 0, \\ B_i \dfrac{\partial \alpha_{n+1}}{\partial z_{n+1}} + \displaystyle\sum_{k=1}^{n} B_k \frac{\partial \alpha_i}{\partial z_k} = \sum_{k=1}^{n} a_{ik} \alpha_k, \\ 1 \leqslant i, \quad j \leqslant n. \end{cases}$$

Solving (S) we obtain

$$\text{(S')}\qquad \begin{cases} \alpha_i = \beta_i + B_i \dfrac{\partial s}{\partial z_{n+1}}, \\ \alpha_{n+1} = \beta z_{n+1} + \varphi - \displaystyle\sum_{k=1}^{n} B_k \frac{\partial s}{\partial z_k} \qquad (i = 1, 2, \ldots, n), \end{cases}$$

where $\beta_1, \beta_2, \ldots, \beta_n, \beta, \varphi \in \mathbf{R}[z_1, z_2, \ldots, z_n]$ and $s \in \mathbf{R}[z_1, z_2, \ldots, z_{n+1}]$ are arbitrary polynomials satisfying the relations

$$\sum_{k=1}^{n} \left( a_{ik} \beta_k - B_k \frac{\partial \beta_i}{\partial z_k} \right) = \beta B_i \qquad (i = 1, 2, \ldots, n).$$

In this context we note that the condition $[D, X] = \lambda X$ for $D = \sum_{i=1}^{n} \lambda_i \frac{\partial}{\partial x_i}$

to be a symmetry of $X$ has the following form in coordinates:

$$\sum_{k=1}^{n} \left( a_{ik}\lambda_k - A_k \frac{\partial \lambda_i}{\partial x_k} \right) = \lambda A_i \qquad (i = 1, 2, \ldots, n).$$

If we now associate with every $D \in \mathrm{Sym}\ X$ the derivation

$$\rho(D) = \sum_{i=1}^{n} \lambda_i (z_1, \ldots, z_n) \frac{\partial}{\partial z_i} + \lambda (z_1, \ldots, z_n) z_{n+1} \frac{\partial}{\partial z_{n+1}}$$

of $\mathrm{Smbl}_* \ L(X)$, then by the remark just made we get a Lie algebra monomorphism $\rho{:}\mathrm{Sym}\ X \to \mathrm{Can}\ U(L(X\Lambda))$. Further, it is easy to check that $\rho(\mathrm{Sym}\ X) \cap \mathrm{Ham}\ U(L(X)) = \rho(\mathrm{Sym}\ X)$, that is, $\rho'{:}\ S(X) \to \gamma(L(X))$ is an embedding of Lie algebras.

There is also the embedding

$$\tau{:}\ \mathbf{R}[x_1, x_2, \ldots, x_n] \to \mathrm{Can}\ U(L(X)); \quad \tau(\varphi) = \varphi \frac{\partial}{\partial z_{n+1}}.$$

Then $\{\tau(\varphi_1), \tau(\varphi_2)\} = 0$, that is, $\mathrm{Im}\ \varphi$ is a commutative subalgebra of $\mathrm{Can}\ U(L(X))$. Furthermore, it is immediately clear from (N) and (S′) that: 1) $\mathrm{Can}\ U(L(X)) = \mathrm{Ham}\ U(L(X)) + \mathrm{Im}\ \rho + \mathrm{Im}\ \tau$; 2) $\mathrm{Im}\ \tau \cap \mathrm{Im}\ \rho = 0$;

3) $\{\rho(D), \tau(\varphi)\} = \sum_{k=1}^{n} \left( \lambda_k \frac{\partial \varphi}{\partial z_k} - \lambda\varphi \right) \frac{\partial}{\partial z_{n+1}} = \tau\left( \sum_{k=1}^{n} \lambda_k \frac{\partial \varphi}{\partial z_k} - \lambda\varphi \right)$ (that is, $\{\mathrm{Im}\ \rho, \mathrm{Im}\ \tau\} \subset \mathrm{Im}\ \tau$); and 4) $\mathrm{Ham}\ U(L(X)) \cap \mathrm{Im}\ \tau$ consists of the Hamiltonian fields with Hamiltonians of the form $\psi_n \in \mathbf{R}[z_1, \ldots, z_n]$, that is, fields of the form $-X(\psi) \frac{\partial}{\partial z_{n+1}}$, where $X(\psi) = \sum_{i,k=1} a_{ik} z_k \frac{\partial \psi}{\partial z_i}$. If we now write $\tau'$ for the composition of the embedding $\tau$ with the projection $\mathrm{Can}\ U(L(X)) \to \gamma(L(x))$, then from our remarks it follows at once that $\gamma(L(X)) = \mathrm{Im}\ \rho' \oplus \mathrm{Im}\ \tau'$, where $\mathrm{Im}\ \tau' \simeq \mathrm{Coker}\ X$ and $\mathrm{Im}\ \rho'$ are subalgebras of $\gamma(L(X))$, and $\mathrm{Im}\ \tau'$ an Abelian ideal.

NOTE 1. We restrict our attention to the set of elements of filtration 1. Then, if $A$ is invertible, $\gamma_0(L(X)) = 0$ and $S_1(X) = \gamma_1(L(X))$. Thus, the Lie algebra of linear symmetries of $X$, factored by the "proper" linear symmetries, is isomorphic to the first cohomology group $H^1(L(X); L(X))$.

NOTE 2. This theorem together with the theorem of **12.** and the first example of **11.** describes completely the invariant $\gamma(L)$ for all 3-dimensional Lie algebras.

**15.** In this section we examine the connection between the concept of an algebraic Hamiltonian orbit (see 7.) of $S(U)$ and that of an orbit of the ("geometrical") coadjoint operation $\mathrm{Ad}^*$ of the Lie group $G$ on $L^*$, the conjugate space of $L$, $U = U(L)$. These latter orbits are of interest owing to their role in representation theory (the "orbit method", see [9] and [10]). In connection with this we also introduce below a version of the "orbit method", which comes immediately from our view of the Hamiltonian formalism.

First of all we recall that $S = S(U)$ can be identified naturally with a ring of polynomial functions on $L^*$; this gives an embedding of topological spaces $L^* \subset \operatorname{Spec} S$.

The following statement describes the geometrical orbits in $L^*$ in "Hamiltonian" terms.

PROPOSITION 9. *If $p \in L^*$, then the geometrical orbit $G(p)$ of $p$ is obtained by shifting $p$ along the trajectories of all possible Hamiltonian vector fields* (that is, $G(p) = p_H$, see 7).

PROOF. Since shifts along trajectories of Hamiltonian vector fields include, in particular, shifts with respect to trajectories of fields of the form $Hx$, where $x \in \operatorname{Im} j$, $j$: $L \to S(U)$ is the canonical embedding, it is obvious that every manifold of the form $p_H$ contains the geometrical orbit $G(p)$ of $p \in L^*$. On the other hand, any Hamiltonian field $Hs$, $s \in S(U)$, being an $S$-linear combination of fields of the form $Hx$, $x \in \operatorname{Im} j$ (this follows from the relation $H(s_1 s_2) = s_1 Hs_2 + s_2 Hs_1$), touches each of the geometrical orbits in $L^*$. Thus, $G(p) \supset p_H$.

COROLLARY 1. *The geometrical orbits $G(p)$ depend only on L, and not on G.*

COROLLARY 2. *The embedding $G(p) \subset \mathcal{O}(p)$ holds, with equality when $G(p)$ is an algebraic variety.*

If the Lie group $G$ has the structure of an algebraic group, then its orbit in some vector space $V$ is an algebraic variety with an algebraic hypersurface excised. More precisely, each orbit is given by a system $P_0(v) \neq 0$, $P_i(v) = 0$ $(i = 1, 2, \ldots, k)$, where $v \in V$ and $P_0, P_1, \ldots, P_k$ are polynomial functions on $V$.

This proves the next proposition.

PROPOSITION 10. *When G is an algebraic Lie group, each geometrical (=Hamiltonian) orbit $G(p)$, $p \in L^*$, coincides with the principal part of the corresponding algebraic orbit $\mathcal{O}(p)$.*

The study of Lie algebras of the form $L(X)$ shows that $\gamma(L(X))$ can indicate very sensitively whether orbits are algebraic. Without going into details of the calculations, we remark, that when $n = 2$, the Lie algebra $\gamma(L(X))$ is commutative if and only if the geometrical Hamiltonian orbits of $L(X)$ are not algebraic varieties.

We complete our study of Hamiltonian orbits of Lie algebras with the following remarks. Let $\Omega_h$: $\Lambda(U) \otimes \Lambda(U) \to S(U)$ be a canonical form, $U = U(L)$, and $\Lambda(U) = \Lambda(S(U))$. This form, in general, is degenerate and has a "kernel" $\operatorname{Ker} \Omega_h$ consisting of those elements $\lambda \in \Lambda(U)$ for which $\Omega_h(\lambda, \lambda') = 0$ for all $\lambda' \in \Lambda(U)$. However, the Hamiltonian map $h$: $\Lambda(U) \to D(S(U))$ allows us to consider in $D(S(U))$ the submodule $\operatorname{Im} h$, which as is easy to see is formed by the linear combinations of the form $\sum_{i=1}^{h} s_i' Hs_i$, $s_i, s_i' \in S(U)$. Thus, $\Omega_h$ determines on $\operatorname{Im} h$ a non-degenerate closed bilinear skew-symmetric form $\widetilde{\Omega}_h$. These arguments, together with Proposition 9 demonstrate the following result.

PROPOSITION 11. *The orbits $G(p)$ are the maximal submanifolds of $L^*$ on which $\widetilde{\Omega}_h$ is non-degenerate.*

COROLLARY: *The orbit $G(p)$ is a canonical manifold.*

In conclusion we consider briefly a general principle of constructing representations of Lie groups, which is a paraphrase of the "orbit method". The classical problem in representation theory is to search for all irreducible representations on an a priori given class of spaces (Hilbert spaces, for example). It seems to us that this contains an element of arbitrariness and can hardly be called natural. We therefore replace the postulation of a "natural" class of spaces by a natural principle suggested by our view of the Hamiltonian formalism.

Roughly speaking, the classical quantization procedure is a method of reconstructing a differential operator from its symbol (the classical Hamiltonian). In general, there is no natural map $\mathrm{Diff}_* K \to \mathrm{Smbl}_* K$ (or vice versa); to make the reconstruction possible, $K$ must have further structure. In the case $K = C^\infty(M)$ such a structure is the connectivity of the manifold $M$ (see [15], for example).
In particular, a Lie group has a left-invariant connection, which can be used naturally.

Thus, let $L$ be a Lie algebra and $G$ its corresponding simply-connected Lie group. If $x_1, \ldots, x_n$ is a basis in $L$, then we call the coordinates of a point $x \in L$ with respect to this basis the exponential coordinates of $\exp(x) \in G$. Again, let $x \in L$ and let $l_x$ be the operator of left translation by $x$ in $G$, that is, $l_x(g) = \exp(x) \cdot g$, $g \in G$. Then the left-invariant differential operators on $G$ (that is, the elements of $U(L)$) are defined by the condition

$$l^*_{-x} D_0 l^*_x = D_x, \tag{5}$$

where $D_x$ is the value of the differential operator $D$ at the point with exponential coordinates $x$. There is obviously a one-to-one correspondence between the values at 0 of the differential operators on $G$ and the elements of $S(U)$. Thus, (5) establishes[1] the required map $Q$: $S(U) \to U$. We also set $\mathrm{Sm} = Q^{-1}$. The element $\mathrm{Sm}(D)$, $D \in U$, is called the *full symbol* of $D$.

Let $J$ be an ideal in $S(U)$ that is stable under Hamiltonian derivations. It determines a left ideal $QJ \subset U$ in the universal enveloping algebra in the following way: the ideal $QJ$ consists of those elements $u \in U$ for which $\mathrm{Sm}(u'u) \in J$ for all elements $u' \in U$. The left $U$-module $U/QJ$ determines a representation of L. It is natural to expect that the so arising "simplest" (possibly irreducible) representations then come from "really simple", namely maximal stable ideals of $S(U)$.

We call the method described here of obtaining representations of Lie

---

[1] See Gel'fand [5].

lgebras the *principle of localization by symbol.* We illustrate it by one xample.

EXAMPLE. Let $L$ be the three-dimensional nilpotent Lie algebra consist-ng of the matrices of the form $\begin{pmatrix} 0 & x_1 & x_3 \\ 0 & 0 & x_2 \\ 0 & 0 & 0 \end{pmatrix}$ (the Lie algebra of the Heisenberg group).

Then each left-invariant differential operator on the corresponding Lie roup is, when expressed in exponential coordinates, a linear combination f operators of the form $\varepsilon_1^{n_1}\varepsilon_2^{n_2}\varepsilon_3^{n_3}$, , where

$$\varepsilon_1 = \frac{\partial}{\partial x_1} - \frac{x_2}{2}\frac{\partial}{\partial x_3}, \quad \varepsilon_2 = \frac{\partial}{\partial x_2} + \frac{x_1}{2}\frac{\partial}{\partial x_3}, \quad \varepsilon_3 = \frac{\partial}{\partial x_3}.$$

t is easy to see that

$$\mathrm{Sm}\,(\varepsilon_1^{n_1}\varepsilon_2^{n_2}\varepsilon_3^{n_3}) = \sum_{i=0}^{\min(n_1,\, n_2)} \cdot \frac{i!\, C_{n_1}^i C_{n_2}^i}{2^i}\, y_1^{n_1-i} y_2^{n_2-i} y_3^{n_3+i},$$

vhere $y_1, y_2, y_3$ generate $S(U)$ and $\mathrm{Sm}\,\varepsilon_i = y_i$ $(i = 1, 2, 3)$.

There are two types of maximal stable ideals in $S(U)$. Firstly, the rincipal ideals $J_r$ generated by irreducible polynomials $r(y_3)$. The relation $\mathrm{Sm}(u\varepsilon_3) = \mathrm{Sm}(u)y_3$, $u \in U$, shows that $QJ_r$ is the left ideal of $U$ enerated by $r(\varepsilon_3)$. The corresponding representation space $R(J_r)$ is the pace of differential operators on the plane with real or complex coefficients, depending on the form of the polynomial $r$. The image of $\varepsilon_3$ in this epresentation is the operator of multiplication by $\lambda$, $r(\lambda) = 0$. If we denote by $\{e_{n_1,\, n_2}\}$ $(n_1, n_2 = 0, 1, 2, \ldots)$ a basis of $R(J_r)$, then the operators corresponding to $\varepsilon_1$ and $\varepsilon_2$, act in the following way:

$$R(\varepsilon_1)\, e_{n_1,\, n_2} = e_{n_1+1,\, n_2}, \quad R(\varepsilon_2)\, e_{n_1,\, n_2} = e_{n_1,\, n_2+1} - \lambda n_1 e_{n_1,\, n_2}.$$

The maximal stable ideals in $S(U)$ of the second type are $J_{p,q} = (p, q, y_3)$, where $p$ and $q$ are irreducible polynomials depending nly on $y_1$ and $y_2$, respectively. Observing that the commutator of any two lements $u_1, u_2 \in U$ lies in the left ideal of $U$ generated by $\varepsilon_3$, we can verify that $QJ_{p,q}$ is the left ideal of $U$ generated by $p(\varepsilon_1)$, $q(\varepsilon_2)$ and $\varepsilon_3$. The corresponding representation spaces $R(J_{p,q})$ are the tensor products $\mathbf{R}[\varepsilon_1]/(p) \otimes \mathbf{R}[\varepsilon_2]/(q)$ on which $R(\varepsilon_3) = 0$ and the actions of $R(\varepsilon_1)$ and $R(\varepsilon_2)$ are multiplication by the cosets of $\varepsilon_1$ and $\varepsilon_2$ in $\mathbf{R}[\varepsilon_1]/(p)$ and $\mathbf{R}[\varepsilon_2]/(q)$ respectively.

Thus, we can observe better the correspondence between our principle and known results on representations of the Heisenberg group (see [9]).

We shall study this method in greater detail elsewhere.

In conclusion, the authors apologise for using the word "natural" xcessively often.

## References

[1] F.A. Berezin, Some remarks on the enveloping algebra of a Lie algebra, Funktsional. Anal. i Prilozhen. **1**:2 (1967), 1–14. MR **36** # 2750.
= Functional Anal. Appl. **1** (1967), 91–102.

[2] A.M. Vinogradov, The algebra of logic for the theory of linear differential operators, Dokl. Akad. Nauk SSSR **205** (1972), 1025–1028. MR **46** # 3498.
= Soviet Math. Dokl. **13** (1972), 1058–1062.

[3] A.M. Vinogradov, Many-valued solutions and a classification principle of non-linear differential equations, Dokl. Akad. Nauk SSSR **210** (1973), 11–14.
= Soviet Math. Dokl. **14** (1973), 661–665.

[4] P. Gabriel, Propriétés générales des schémas en groupes, Lecture Notes in Math. **151** Exposé VIIA. Springer-Verlag, Berlin-Heidelberg-New York 1970.

[5] I.M. Gel'fand, The centre of an infinitesimal group ring, Mat. Sb. **26** (1950), 103–112 MR **11**–498.

[6] A. Grothendieck, Catégories cofibrées additives et complexe cotangent relatif, Lecture Notes in Math. **79**, Springer-Verlag, Berlin-Heidelberg-New York 1968. MR **39** # 283

[7] A. Grothendieck, Eléments de géometrie algébrique. IV. Etude locale des schémas et des morphismes schémas, Inst. Hautes Études Sci. Publ. Math. **32** (1967). MR **39** # 22

[8] Yu. V. Egorov, On the solubility of differential equations with simple characteristics, Uspekhi Mat. Nauk **26**:2 (1971), 183–199. MR **45** # 5541.
= Russian Math. Surveys **26**:2 (1971), 113–130.

[9] A.A. Kirillov, Unitary representations of nilpotent Lie groups, Uspekhi Mat. Nauk **17**:4 (1962), 57–101. MR **25** # 5396.
= Russian Math. Surveys **17**:4 (1962) 53–104.

[10] A.A. Kirillov, *Lektsii po teorii predstavlenii grupp, IV. Predstavleniya i mekhanika* (Lectures on the theory of group representations, IV. Representations and mechanics) Izdat. Moskov. Gos. Univ., Moscow 1971.

[11] B. Kostant, Quantization and unitary representations, Lectures in modern analysis and applications, II, Springer-Verlag, Berlin-Heidelberg-New York 1970, 87–208.
= Uspekhi Mat. Nauk **28**:1 (1973), 163–225.

[12] Yu. I Manin, *Lektsii po algebraicheskoi geometrii, I. Affinnye skhemy* (Lectures on algebraic geometry, Part I. Affine schemes), Izdat. Moskov. Gos. Univ., Moscow 1970.

[13] V.P. Maslov and B. Yu. Sternin, *Kanonicheskii operator(kompleksnyi sluchai)* (Canonical operators (The complex case)), in the coll. "Problems of modern mathematics", vol. I, All-Union Inst. Sci. Tech. Information (VINITI) 1973, 169–159.

[14] V.P. Maslov and M.V. Fedoryuk, *Kanonicheskii operator* (*veshchestvennyi sluchai*) (Canonical operators (The real case)), in the coll. "Problems of modern mathematics", vol. I, All-Union Inst. Sci. Tech. Information (VINITI) 1973, 85–168.

[15] R. Palais, Seminar on the Atiyah-Singer index theorem, Annals of Mathematics Studies No.57, University Press, Princeton, N.J., 1965. MR **33** # 6649.
Translation: *Seminar po teoreme At'i-Zingera ob indekse,* Mir, Moscow 1970.

[16] J-P. Serre, Lie algebras and Lie groups, Benjamin, New York 1965.
Translation: *Gruppy Li i algebry Li,* Mir, Moscow 1969.

[17] Séminaire "Sophus Lie", Théorie des algèbres de Lie. Topologie des groupes de Lie, École Normale Supérieure, Paris 1955. MR **17**–384.
Translation: *Teoriya algebr Li. Topologiya grupp Li,* Izdat. Inost. Lit., Moscow 1962.

Received by the Editors, 19 July 1974

Translated by Ph. Spain